IMO Problems, Theorems, and Methods

Geometry

Mathematical Olympiad Series

ISSN: 1793-8570

Series Editors: Lee Peng Yee *(Nanyang Technological University, Singapore)*
Xiong Bin *(East China Normal University, China)*

Published

The complete list of the published volumes in the series can be found at
http://www.worldscientific.com/series/mos

Vol. 27 | Mathematical Olympiad Series

IMO Problems, Theorems, and Methods

Geometry

Authors

Tianqi Lin
Bin Xiong
East China Normal University, China

Translator

Xinyuan Yang
East China Normal University, China

Proofreader

Jiu Ding
School of Mathematics and Natural Sciences,
University of Southern Mississippi, USA

Copy Editors

Lingzhi Kong, Liyu Zhang, and Ming Ni
East China Normal University Press, China

East China Normal

World Scientific

Published by

East China Normal University Press
3663 North Zhongshan Road
Shanghai 200062
China

and

World Scientific Publishing Co. Pte. Ltd.
5 Toh Tuck Link, Singapore 596224
USA office: 27 Warren Street, Suite 401-402, Hackensack, NJ 07601
UK office: 57 Shelton Street, Covent Garden, London WC2H 9HE

Library of Congress Control Number: 2025004996

British Library Cataloguing-in-Publication Data
A catalogue record for this book is available from the British Library.

Mathematical Olympiad Series — Vol. 27
IMO PROBLEMS, THEOREMS, AND METHODS
Geometry

Copyright © 2026 by East China Normal University Press and
World Scientific Publishing Co. Pte. Ltd.

ISBN 978-981-98-0330-9 (hardcover)
ISBN 978-981-98-0689-8 (paperback)
ISBN 978-981-98-0331-6 (ebook for institutions)
ISBN 978-981-98-0332-3 (ebook for individuals)

For any available supplementary material, please visit
https://www.worldscientific.com/worldscibooks/10.1142/14099#t=suppl

Desk Editors: Nambirajan Karuppiah/Angeline Husni

Typeset by Stallion Press
Email: enquiries@stallionpress.com

Preface

It is generally believed that formal mathematics competitions began with a contest held in Hungary in 1894, an event that gradually garnered attention worldwide. People aptly liken mathematics competitions to "Mental Gymnastics." In 1934, the Soviet Union straightforwardly termed it the "Mathematical Olympiad," a designation that reflects the Olympic spirit of pursuing excellence in intellect more vividly than the previous term, "mathematics competition."

By 1959, the internationalization of mathematics competitions had matured, leading to the inception of the "International Mathematical Olympiad" (IMO). The first IMO was held in Brasov, Romania in 1959. As of 2023, the IMO has successfully been held 64 times, except for 1980 when it was not conducted.

The IMO is typically held in July each year, and the format has become standardized: the official competition spans two days, with contestants tackling three problems in 4.5 hours each day, each problem worth a maximum of 7 points, totaling 42 points. Each participating team consists of six contestants, accompanied by a Leader and a Deputy Leader. Approximately half of the contestants receive medals, with about 1/12 of the contestants earning gold medals, 2/12 receiving silver medals, and 3/12 obtaining bronze medals.

The IMO is currently one of the most influential secondary school mathematics competitions worldwide. In recent years, over 100 countries and regions have participated in this event, including all major nations globally.

Problems for the IMO are submitted by the participating teams and then reviewed and selected by a problem selection committee organized by the host country. This committee narrows down the submissions to approximately 30 Shortlist problems, covering algebra, geometry, number theory, and combinatorics, with about seven to eight problems on each topic. These are then presented to the Jury Meeting, composed of team leaders, who discuss and vote to decide on the six problems that will constitute the official competition paper. The host country does not provide any problems.

This event has played a significant role in promoting the exchange of mathematical education among nations, enhancing the level of mathematical education, facilitating mutual learning and understanding among young students worldwide, stimulating a broad interest in mathematics among secondary school students, and identifying and nurturing mathematically gifted students.

The development over more than 60 years is the result of the collective efforts of mathematicians, organizers, and contestants, and is worthy of reflection and study. Particularly deserving of study are the evolution of the competition problems, the mathematical ideas, and methods involved. Indeed, several colleagues from the International Mathematical Olympiad Research Center at East China Normal University had envisioned research and publication before the 60th IMO. For this purpose, we initiated several seminars involving over ten people. For special reasons, this work was delayed. Based on the mathematical domains covered by the IMO problems — algebra, geometry, number theory, and combinatorics — we planned to compile the work into four volumes, with the general title *IMO Problems, Theorems, and Methods*, to be included in the "IMO Study Series."

Each volume begins with an introduction that provides an overview of the IMO. Subsequent chapters introduce relevant foundational knowledge and methods, followed by a reclassification and organization of past IMO problems. For some problems, multiple solutions are provided, along with a difficulty analysis. It is worth noting that some problems do not fit neatly into a single topic, as they may involve both algebra and number theory, or algebra and combinatorics. We primarily categorize them based on the topic under which they were placed on the Shortlist.

The four volumes titled *IMO Problems, Theorems, and Methods* were conceived with an overall writing plan proposed by myself, with the authors collectively discussing and refining the plan. The majority of the initial drafts were completed by Jinhua Chen (Algebra), Tianqi Lin (Geometry), Gengyu Zhang (Number Theory), and Guangyu Xu (Combinatorics). The first three volumes were supplemented, consolidated, and finalized by myself, while the combinatorics volume was supplemented, consolidated, and finalized by Zhenhua Qu.

We extend our gratitude to the leaders and contestants of the Chinese IMO teams over the years, as some elegant solutions included in the book were contributed by them. During the compilation of this book, we consulted various domestic and international sources, which are too numerous to acknowledge individually here.

While the authors have diligently studied the IMO problems and provided thoughtful strategies and solutions, errors and inaccuracies may occur due to our limitations. We sincerely invite readers to offer corrections and feedback.

The translation of the algebra volume in this series was done by Jinhua Chen and Bin Xiong; the geometry volume was translated by Xinyuan Yang; the number theory volume was translated by Gengyu Zhang; and the combinatorics volume was translated by Zhenhua Qu and Jinhua Chen. Jiu Ding revised the translations of all four books.

Bin Xiong
June 2024

About the Authors

Tianqi Lin is a mathematics teacher at High School Affiliated to Fudan University, engaged in mathematics education research, and enjoys solving and creating problems in plane geometry. He has participated in the proposition work of the Chinese national training team for the International Mathematical Olympiad in 2019, 2018, and 2016, and has participated in the proposition work of various competitions multiple times. He is passionate about communication, and has published articles in publications such as *High-School Mathematics* to exchange ideas with readers on solving and proposing plane geometry problems.

Bin Xiong is a professor and doctoral supervisor at the School of Mathematical Sciences, East China Normal University. He also serves as the director of the Shanghai Key Laboratory of Core Mathematics and Practice, and the International Mathematical Olympiad Research Center. Professor Xiong is an expert with the State Council Special Allowance, and has been honored with the Shanghai May 1 Labor Medal as well as the Shanghai Model of Teaching and Educating. He has published over 100 scholarly papers in renowned national and international journals and has authored or co-authored more than 150 books. Additionally, Professor Xiong has served as the leader and head coach of the Chinese IMO team more than 10 times, and received the prestigious Paul Erdös Award in 2018 for his contribution to the development of mathematics competitions at the national and international level.

About the Translator

Xinyuan Yang is a Ph.D. from the School of Mathematical Sciences, East China Normal University. His research focuses on mathematics education and mathematics competitions. During the doctoral period, he was sent to the University of Haifa for one year by the officially sponsored study abroad program. Over the years, he has maintained a keen interest in mathematical competitions, especially in geometry and number theory. He served as an external mentor for mathematics competition courses at several high schools and participated multiple times in the grading work for major mathematics competitions in China.

Contents

Introduction to the IMO

The International Mathematical Olympiad (IMO), established in the year 1959, represents one of the foremost intellectual endeavors at the highest tier for youth on a global scale. Prior to 1959, numerous countries around the world had already initiated the organization of mathematics competitions, thereby laying the groundwork for the inception of the IMO.

In 1891, the renowned physicist and President of the Hungarian Academy of Sciences, Loránd Eötvös (also known as Roland Eötvös), founded the Hungarian Mathematical and Physical Society. In 1894, he assumed the position of Minister of Education, and under his enthusiastic support, the Hungarian Mathematical and Physical Society initiated secondary school mathematics competitions. This competition, also known as the Eötvös Competition, offered winners the Eötvös Prize and the opportunity to pursue higher education. Subsequently, the Eötvös Competition was not held during 1919–1921 and 1944–1946 due to the world political events. In 1947, under the leadership of János Surányi, the Eötvös Competition was reinstated and renamed the Kürschák Competition (named after József Kürschák). This competition has played a significant role in Hungary in nurturing numerous mathematicians and scientists, including Győző Zemplén, Lipót Fejér, Theodore von Kármán, Alfréd Haar, Dénes Kőnig, Marcel Riesz, Gábor Szegő, Tibor Radó, Edward Teller, and Tibor Szele. Interestingly, George Pólya also participated in the competition, but did not hand in his paper.

With the aim of identifying and nurturing mathematical talents, prominent mathematicians such as Boris Delaunay (also known as Delone), Grigorii Fikhtengol'ts, Dmitry Faddeev, and others organized the inaugural Leningrad Mathematical Olympiad (LMO) in 1934, under the initiative of

Boris Delaunay. Winners of this competition were granted the privilege of direct admission to the Mathematics Department of Leningrad State University without the need for entrance examinations. Following the example set by the LMO, in 1935, renowned mathematicians Pavel Aleksandrov and Andrey Kolmogorov, alongside the entire faculty of the Mathematics Department at Moscow State University, organized the first Moscow Mathematical Olympiad (MMO).

Subsequently, various regions throughout the Soviet Union started hosting their own Mathematical Olympiads, ultimately laying the foundation for the All-Russian Mathematical Olympiad, which was first conducted in 1961. In 1967, the responsibility for organizing All-Russian Mathematical Olympiad was assumed by the Ministry of Education of the Soviet Union, leading to a renaming of the All-Russian Mathematical Olympiad as the All-Soviet-Union Mathematical Olympiad.

In fact, almost all of the best mathematicians born in the Soviet Union after 1930 had participated in Mathematical Olympiads, usually achieving first prizes. This distinguished group includes Fields Medal awardees such as Sergei Novikov, Grigory Margulis, Vladimir Drinfeld, Maxim Kontsevich, Grigori Perelman, and Stanislav Smirnov. Although Sergei Novikov having claimed that he was never particularly interested in Mathematical Olympiads, Sergei Novikov did secure a second prize in the MMO when he was in eighth grade.

While the United States of America Mathematical Olympiad (USAMO) was first held in 1972, the United States had a longstanding tradition of organizing mathematics competitions prior to that. In 1921, William Lowell Putnam published an article in the Harvard Graduates' Magazine, proposing the idea of conducting a university-level mathematics team competition. Following his passing, the Putnam family established the William Lowell Putnam Intercollegiate Memorial Fund to support the organization of the William Lowell Putnam Mathematical Competition (Putnam Competition), administered by the Mathematical Association of America.

With the assistance of George David Birkhoff, the first Putnam Competition took place in 1938, and it has been held annually since then; the top five ranking contestants are designated as Putnam Fellows. Due to wartime conditions, the competition was not held from 1943 to 1945. In 1946, George Pólya, Tibor Radó, and Irving Kaplansky (Putnam Fellow in 1938) formed the Putnam Competition Committee, thus reestablishing the competition, but the responsibility for administration was undertaken by

Garrett Birkhoff, the son of George David Birkhoff, and his colleagues in the Harvard University Department of Mathematics.

Many contestants in the Putnam Competition have gone on to become prominent mathematicians and scientists. John Milnor, David Mumford, Daniel Quillen, Paul Cohen, John G. Thompson, and Manjul Bhargava have been recipients of Fields Medal. Richard Feynman, Kenneth Geddes Wilson, Steven Weinberg, and Murray Gell-Mann have received Nobel Prize in Physics, while John Nash was awarded the Nobel Prize in Economic Sciences. Additionally, numerous Putnam Fellows have been elected as members of the National Academy of Sciences in the United States.

Building upon the foundation of existing mathematics competitions in many countries, particularly in Eastern European nations, Romania proposed in 1956 the organization of an international mathematics competition involving seven Eastern European countries. This proposal led to the inaugural IMO held in 1959.

The first IMO was held in Braşov, Romania, in 1959. As of 2023, the IMO has been successfully held 64 times, with the exception of the year 1980 when it was not held for specific reasons. Apart from the 61st IMO, which was postponed to September in 2020 due to the impact of the COVID-19 pandemic, IMO typically takes place in July each year.

The IMO has emerged as the most influential secondary school mathematics competition at present. In recent years, the number of countries and regions participating in this event has exceeded 100.

1 Evolution of the IMO

The first IMO, held in 1959, saw the participation of only 52 contestants from seven countries, including Romania, Hungary, Czechoslovakia, Bulgaria, Poland, the Soviet Union, and the German Democratic Republic. Subsequently, new countries and regions gradually joined this prestigious event. By the 20th IMO, also hosted by Romania in 1978, approximately 20 countries and regions participated (with 21 in the 19th, 17 in the 20th, and 23 in the 21st). The number of participating contestants also reached 132. The historical participation trends are depicted in Figure 1.

As the influence of the IMO continued to expand, the number of participating countries and regions, as well as the number of contestants, grew rapidly. The most significant increase in the number of participating countries and regions occurred in the 34th IMO, which was held in Turkey in 1993. In comparison to the 33rd IMO held in Russia in 1992, the number

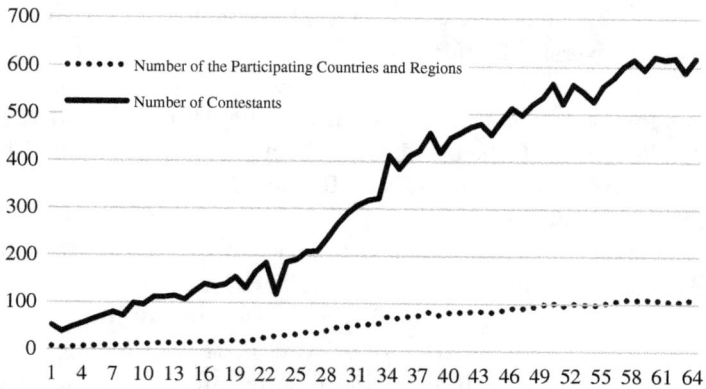

Figure 1 Numbers of Participating Countries and Regions, as Well as the Contestants, in the First 64 IMOs

increased by 17 countries and regions, reaching a total of 73, with 413 contestants. By the 40th IMO, which was still hosted by Romania in 1999, the number of participating countries and regions had reached 81, with 450 contestants.

The first instance of the number of participating countries and regions surpassing one hundred occurred during the 50th IMO, held in Germany in 2009, with a total of 565 contestants. Among the first 64 IMOs, the biggest number of participating countries and regions, as well as the largest number of contestants, was observed in the 60th IMO, hosted by the United Kingdom in 2019, where 621 contestants from 112 countries and regions took part. In the 64th IMO held in Japan in 2023, there were 618 contestants from 112 countries and regions.

As evident from Appendix One, the IMO is primarily hosted by European countries. Moreover, as the number of participating countries and regions in the IMO has increased, it is no longer confined to the seven founding member countries, and many new participating countries and regions have also begun to organize the IMO.

2 Problems in the IMO

The IMO is scheduled to take place annually in July. Each participating country or region officially sends a delegation consisting of six contestants, along with one team leader and one deputy leader. The official competition spans two days, with each day featuring three problems to be solved within a four-and-a-half-hour timeframe. Each problem carries a maximum score of 7 points, resulting in a total maximum score of 42 points, while the total maximum score of the team is 252 points.

In the early IMOs, the number of problems and their individual point values varied from one session to another. For instance, the 2nd and 4th IMOs featured 7 problems, while all other IMOs had 6 problems each. Additionally, in the 13th IMO, although the total score remained at 42 points, the 6 problems were allocated point values of 5, 7, 9, 6, 7, and 8 respectively. It was only from the 22nd IMO, held in the United States in 1981, that the IMO problems have become standardized, with each problem carrying 7 points and a total of 6 problems.

The number of contestants in each delegation has also become stable at six individuals starting from the 24th IMO held in France in 1983.

2.1 The number of problems

The mathematical domains covered by IMO problems encompass four major topics: algebra, combinatorics, geometry, and number theory. These are also the primary focus in various national mathematics competitions.

Across the 1st–64th IMOs, a total of 386 problems have been featured. Among them, geometry problems are the most numerous, with 123 problems, while number theory problems are the least, with 75 problems. Algebra problems account for approximately one-quarter of the total, comprising 101 problems. Furthermore, as indicated in Table 1, the quantity of algebra problems has remained relatively stable over each span of 10 IMOs.

Table 1 Numbers of Problems with Different Topics in the First 64 IMOs

Session	Topic			
	Algebra	Combinatorics	Geometry	Number Theory
1st–10th	20	6	29	7
11th–20th	20	12	18	10
21st–30th	14	16	18	12
31st–40th	13	16	15	16
41st–50th	15	11	20	14
51st–60th	13	19	18	11
61st–64th	6	7	6	5
Total	101	87	123	75

Remarkably, in the first 64 IMOs, there were seven sessions when three or more than three geometry problems were presented, specifically in the 1st, 2nd, 3rd, 4th, 8th, 9th, and 15th IMOs. In 42 IMOs, two geometry problems were featured, while in 15 IMOs, only one geometry problem was included.

2.2   The difficulty level of problems

Typically, the difficulty of the problem is correlated with its problem number in the IMO.

Starting from the 24th IMO, the point value of each problem and the number of contestants per team have become standardized. Therefore, an analysis of the average scores of the 246 problems in the 24th–64th IMOs is presented in Table 2. It can be observed that the first and fourth problems in each IMO are relatively easy, with average scores generally exceeding 3 points. The second and fifth problems are relatively challenging, with average scores mainly ranging from 1 to 4 points. The third and sixth problems are exceptionally difficult, with average scores generally falling below 2 points.

Table 2 Numbers of Problems with Different Average Scores in the 24th–64th IMOs

Problem Number	Problem Mean				
	0–1	1–2	2–3	3–4	4–7
Problem 1	0	0	5	13	23
Problem 2	2	6	18	9	6
Problem 3	22	12	4	3	0
Problem 4	0	0	7	17	17
Problem 5	4	10	14	10	3
Problem 6	24	11	6	0	0
Total	52	39	54	52	49

The 246 problems are categorized into four topics: algebra, combinatorics, geometry, and number theory, as shown in Table 3. Notably, there is a relatively larger representation of combinatorics and geometry problems. Combining this information with Table 1, it is evident that in the first 23 IMOs, algebra and geometry problems were predominant.

Furthermore, in early IMOs, geometry problems predominantly appeared in the 1st/4th and 2nd/5th positions. However, starting from the 41st to 50th IMOs, geometry problems were more commonly found in the 1st/4th and 3rd/6th positions. Similarly, algebra problems were more frequent in the 1st/4th and 2nd/5th positions and combinatorics problems were more prevalent in the 2nd/5th and 3rd/6th positions, while the quantity of number theory problems across different problem numbers does not differ significantly.

From Table 4, it can be observed that among the four topics, the numbers of problems with an average score ranging from 2 to 4 points are quite similar. However, in the combinatorics topic, there is a higher quantity of challenging problems, with 31 problems having an average score between 0 and 2 points. Conversely, the geometry topic has the largest number of relatively easy problems, with 23 problems scoring above 4 points. This discrepancy is largely due to the fact that there are 29 combinatorics problems in the 3rd/6th positions, and 31 geometry problems in the 1st/4th positions.

Furthermore, when considering Table 3 and Table 4, it becomes apparent that among the four topics, the number of problems with an average score ranging from 0 to 2 points closely aligns with the number of problems in the 3rd/6th positions. There are slightly more problems with an average score between 2 to 4 points compared to those in the 2nd/5th positions, and slightly fewer problems with an average score exceeding 4 points compared to those in the 1st/4th positions. This indicates that even the seemingly easier problems in the IMO are not as straightforward as they might appear.

Notably, among these 246 problems, the lowest average score is attributed to IMO 58-3 (Combinatorics, proposed by Austria):

A hunter and an invisible rabbit play a game in the Euclidean plane. The rabbit's starting point, A_0, and the hunter's starting point, B_0, are the same. After $n - 1$ rounds of the game, the rabbit is at point A_{n-1} and the hunter is at point B_{n-1}. In the nth round of the game, three things occur in order.

(1) The rabbit moves invisibly to a point A_n such that the distance between A_{n-1} and A_n is exactly 1.
(2) A tracking device reports a point P_n to the hunter. The only guarantee provided by the tracking device to the hunter is that the distance between P_n and A_n is at most 1.
(3) The hunter moves visibly to a point B_n such that the distance between B_{n-1} and B_n is exactly 1.

Table 3 Numbers of Problems with Different Topics in the 24th–64th IMOs

	Topic											
	Algebra			Combinatorics			Geometry			Number Theory		
Session	1st/4th	2nd/5th	3rd/6th	1st/4th	2nd/5th	3rd/6th	1st/4th	2nd/5th	3rd/6th	1st/4th	2nd/5th	3rd/6th
24th–30th	4	1	5	4	3	5	5	7	0	1	3	4
31st–40th	4	5	4	5	4	7	6	9	0	5	2	9
41st–50th	3	8	4	3	2	6	10	6	5	5	4	5
51st–60th	5	6	2	4	7	8	7	3	6	3	4	4
61st–64th	1	3	2	2	2	3	3	1	2	2	2	1
Total	17	23	17	18	18	29	31	26	13	16	15	23
	57			65			70			54		

Table 4 Numbers of Problems with Different Average Scores by Topics in the 24th–64th IMOs

Topic	Problem Mean					Total
	0–1	1–2	2–3	3–4	4–7	
Algebra	7	14	19	6	11	57
Combinatorics	20	11	12	14	8	65
Geometry	13	5	14	15	23	70
Number Theory	12	9	9	17	7	54
Total	52	39	54	52	49	246

Is it always possible, no matter how the rabbit moves, and no matter what points are reported by the tracking device, for the hunter to choose her moves so that after 10^9 rounds she can ensure that the distance between her and the rabbit is at most 100?

This unconventional problem received an average score of only 0.042 points. Only two contestants, Mikhail Ivanov from Russia and Linus Cooper from Australia, achieved a perfect score of 7 points. Joe Benton from the United Kingdom scored 5 points, Pavel Hudec from Czech Republic earned 4 points, Hadyn Ka Ming Tang from Australia, Yahor Dubovik from Belarus, and Jeonghyun Ahn from South Korea each scored 1 point.

Furthermore, among the 20 lowest scoring problems in the 24th–64th IMOs, nearly all of them appeared in the 3rd/6th positions, as indicated in Table 5. There were three algebra problems, eight combinatorics problems,

Table 5 The 20 Problems with the Lowest Average Scores in the 24th–64th IMOs

Problem	Mean	Topic	Problem	Mean	Topic
IMO 58-3	0.042	Combinatorics	IMO 54-6	0.296	Combinatorics
IMO 48-6	0.152	Algebra	IMO 48-3	0.304	Combinatorics
IMO 50-6	0.168	Combinatorics	IMO 52-6	0.318	Geometry
IMO 47-6	0.187	Geometry	IMO 53-6	0.336	Number Theory
IMO 57-3	0.251	Number Theory	IMO 55-6	0.339	Combinatorics
IMO 49-6	0.260	Geometry	IMO 56-6	0.355	Combinatorics
IMO 64-6	0.275	Geometry	IMO 51-6	0.368	Algebra
IMO 59-3	0.278	Combinatorics	IMO 62-3	0.372	Geometry
IMO 61-6	0.282	Combinatorics	IMO 62-2	0.375	Algebra
IMO 58-6	0.294	Number Theory	IMO 60-6	0.403	Geometry

six geometry problems, and three number theory problems among them. These six geometry problems were:

- **(IMO 47-6, proposed by Serbia-Montenegro).** Assign to each side b of a convex polygon P the maximum area of a triangle that has b as a side and is contained in P. Show that the sum of the areas assigned to the sides of P is at least twice the area of P.

- **(IMO 49-6, proposed by Russia).** Let $ABCD$ be a convex quadrilateral with $|BA| \neq |BC|$. Denote the incircles of triangles ABC and ADC by ω_1 and ω_2 respectively. Suppose that there exists a circle ω tangent to the ray BA beyond A and to the ray BC beyond C, which is also tangent to the lines AD and CD. Prove that the common external tangents of ω_1 and ω_2 intersect on ω.

- **(IMO 64-6, proposed by the United States).** Let ABC be an equilateral triangle. Let A_1, B_1, C_1 be interior points of ABC such that $BA_1 = A_1C, CB_1 = B_1A, AC_1 = C_1B$, and

$$\angle BA_1C + \angle CB_1A + \angle AC_1B = 480°.$$

Let BC_1 and CB_1 meet at A_2, let CA_1 and AC_1 meet at B_2, and let AB_1 and BA_1 meet at C_2. Prove that if triangle $A_1B_1C_1$ is scalene, then the three circumcircles of triangles AA_1A_2, BB_1B_2, and CC_1C_2 all pass through two common points. (*Note*: A scalene triangle is one where no two sides have equal length.)

- **(IMO 52-6, proposed by Japan).** Let ABC be an acute triangle with circumcircle Γ. Let l be a tangent line to Γ, and let l_a, l_b, and l_c be the lines obtained by reflecting l in the lines BC, CA, and AB, respectively. Show that the circumcircle of the triangle determined by the lines l_a, l_b, and l_c is tangent to the circle Γ.

- **(IMO 62-3, proposed by Ukraine).** Let D be an interior point of an acute triangle ABC with $AB > AC$ so that $\angle DAB = \angle CAD$. A point E on the segment AC satisfies $\angle ADE = \angle BCD$, a point F on the segment AB satisfies $\angle FDA = \angle DBC$, and a point X on the line AC satisfies $CX = BX$. Let O_1 and O_2 be the circumcenters of the triangles ADC and EXD, respectively. Prove that the lines BC, EF, and O_1O_2 are concurrent.

- **(IMO 60-6, proposed by India).** Let I be the incentre of an acute triangle ABC with $AB \neq AC$. The incircle ω of ABC is tangent to sides BC, CA, and AB at D, E, and F, respectively. The line through D perpendicular to EF meets ω again at R. Line AR meets ω again at P. The circumcircles of triangles PCE and PBF meet again at Q. Prove that lines DI and PQ meet on the line through A perpendicular to AI.

2.3 The classification of problems

In the 1st–64th IMOs, there were 123 geometry problems, which can be categorized into seven specialized subjects: "similarity and congruence" problems, "basic properties of circles and four points on a circle" problems, "power of a point, radical axis, and radical center" problems, "special points and special lines" problems, "trigonometry, areas, and analytic geometry" problems, "solid geometry" problems, and "geometric inequality" problems.

For the 123 geometry problems categorized and analyzed, as shown in Table 6, it can be observed that the subject of "basic properties of circles and four points on a circle" problems had the biggest number of problems, with 32 problems.

In the first 10 IMOs, geometry problems primarily focused on solid geometry problems, accounting for about one third. Subsequently, the number of solid geometry problems decreased rapidly, while problems related to "basic properties of circles and four points on a circle" problems and "geometric inequality" problems gradually increased and maintained a certain frequency.

In the 51st–60th IMOs, geometry problems were primarily centered around "basic properties of circles and four points on a circle" problems, whereas in the 61st–64th IMOs, geometry problems primarily focused on "basic properties of circles and four points on a circle" problems and "power of a point, radical axis, and radical center" problems.

In the first 64 IMOs, there had been a total of 14 similarity and congruence problems, accounting for approximately 11.4% of all geometry problems. These problems can be primarily categorized into three types: (1) existence problems, totaling four problems; (2) position structure problems, totaling five problems; (3) quantity relation problems, totaling five problems.

As shown in Table 7, in the 24th–64th IMOs, there were a total of 8 similarity and congruence problems. These problems were predominantly present in the 1st/4th positions as well as in the 2nd/5th positions, with

Table 6 Numbers of Geometry Problems in the First 64 IMOs

Session	Subject						
	Similarity Congruence	Basic Properties of Circles, etc.	Power of a Point, etc.	Special Points Special Lines	Trigonometry and Others	Solid Geometry	Geometric Inequalities
1st–10th	4	4	0	2	4	10	5
11th–20th	0	2	0	1	6	7	2
21st–30th	2	4	1	1	4	0	6
31st–40th	3	4	2	0	0	0	6
41st–50th	3	4	3	4	1	0	5
51st–60th	2	11	3	0	1	0	0
61st–64th	0	3	3	0	0	0	0
Total	14	32	12	8	16	17	24

Table 7 Numbers of Similarity and Congruence Problems in the 24th–64th IMOs

Similarity and Congruence Problem	Problem Number			Number of Problems in the First 64 IMOs
	1, 4	2, 5	3, 6	
Existence problems	1	0	0	4
Position structure problems	1	2	1	5
Quantity relation problems	1	2	0	5
Total	3	4	1	14

the main focus being on position structure problems and quantity relation problems. The type of existence problems appeared less frequently in the last 40 IMOs.

In the first 64 IMOs, there had been a total of 32 basic properties of circles and four points on a circle problems, approximately accounting for 26.0% of all geometry problems. These problems can be primarily categorized into three types: (1) existence problems, totaling six problems; (2) position structure problems, totaling 18 problems; (3) quantity relation problems, totaling eight problems.

As shown in Table 8, in the 24th–64th IMOs, there were a total of 24 basic properties of circles and four points on a circle problems. These problems were predominantly present in the 1st/4th positions, with the main focus being on position structure problems, followed by quantity relation problems. Existence problems did not appear in the last 40 IMOs.

Table 8 Numbers of Basic Properties of Circles and Four Points on a Circle Problems in the 24th–64th IMOs

Basic Properties of Circles and Four Points on a Circle Problem	Problem Number			Number of Problem in the First 64 IMOs
	1, 4	2, 5	3, 6	
Existence problems	0	0	0	6
Position structure problems	10	3	4	18
Quantity relation problems	4	3	0	8
Total	14	6	4	32

In the first 64 IMOs, there had been a total of 12 power of a point, radical axis, and radical center problems, approximately accounting for 9.8% of all geometry problems. These problems can be primarily categorized into two types: (1) position structure problems, totaling nine problems; (2) quantity relation problems, totaling three problems.

As shown in Table 9, in the 24th–64th IMOs, there were a total of 12 power of a point, radical axis, and radical center problems. The frequency of these problems in each problem number was relatively balanced. And the primary focus of these problems was position structure problems in the last 40 IMOs.

Table 9   Numbers of Power of a Point, Radical Axis, and Radical Center Problems in the 24th–64th IMOs

Power of a Point, Radical Axis, and Radical Center Problem	Problem Number			Number of Problems in the First 64 IMOs
	1, 4	2, 5	3, 6	
Position structure problems	3	3	3	9
Quantity relation problems	1	2	0	3
Total	4	5	3	12

In the first 64 IMOs, there had been a total of 8 special points and special lines problems, approximately accounting for 6.5% of all geometry problems. These problems can be primarily categorized into two types: (1) position structure problems, totaling five problems; (2) quantity relation problems, totaling three problems.

As shown in Table 10, in the 24th–64th IMOs, there were a total of 4 special points and special lines problems, primarily focusing on position structure problems.

Table 10   Numbers of Special Points and Special Lines Problems in the 24th–64th IMOs

Special Points and Special Lines Problem	Problem Number			Number of Problems in the First 64 IMOs
	1, 4	2, 5	3, 6	
Position structure problems	1	1	1	5
Quantity relation problems	1	0	0	3
Total	2	1	1	8

In the first 64 IMOs, there had been a total of 16 trigonometry, areas, and analytic geometry problems, approximately accounting for 13.0% of all geometry problems. These problems can be primarily categorized into three types: (1) existence problems, totaling three problems; (2) position structure problems, totaling six problems; (3) quantity relation problems, totaling seven problems.

As shown in Table 11, in the 24th–64th IMOs, there were a total of 6 trigonometry, areas, and analytic geometry problems, primarily focusing on quantity relation problems.

Table 11 Numbers of Trigonometry, Areas, and Analytic Geometry Problems in the 24th–64th IMOs

Trigonometry, Areas, and Analytic Geometry Problem	Problem Number			Number of Problems in the First 64 IMOs
	1, 4	2, 5	3, 6	
Existence problems	0	0	0	3
Position structure problems	1	1	0	6
Quantity relation problems	2	1	1	7
Total	3	2	1	16

In the first 64 IMOs, there had been a total of 17 solid geometry problems, approximately accounting for 13.8% of all geometry problems. These problems can be primarily categorized into three types: (1) existence problems, totaling five problems; (2) position structure problems, totaling six problems; (3) quantity relation problems, totaling six problems.

As shown in Table 12, in the 24th to 64th IMOs, there were no solid geometry problems. This indicates that solid geometry problems were primarily featured in the first 20 IMOs.

Table 12 Numbers of Solid Geometry Problems in the 24th–64th IMOs

Solid Geometry Problem	Problem Number			Number of Problems in the First 64 IMOs
	1, 4	2, 5	3, 6	
Existence problems	0	0	0	5
Position structure problems	0	0	0	6
Quantity relation problems	0	0	0	6
Total	0	0	0	17

In the first 64 IMOs, there had been a total of 24 geometric inequality problems, approximately accounting for 19.5% of all geometry problems. These problems can be primarily categorized into three types: (1) existence problems, totaling five problems; (2) equality or inequality proof problems, totaling 16 problems; (3) conditional evaluation problems, totaling three problems.

As shown in Table 13, in the 24th–64th IMOs, there were a total of 15 geometric inequality problems. These problems were predominantly present in the 1st/4th positions as well as in the 2nd/5th positions, with the main focus being on equality or inequality proof problems. For the other two types of geometric inequality problems, there have been particularly few occurrences in the last 40 IMOs.

Table 13   Numbers of Geometric Inequality Problems in the 24th–64th IMOs

Geometric Inequality Problem	Problem Number			Number of Problems in the First 64 IMOs
	1, 4	2, 5	3, 6	
Existence problems	0	1	0	5
Equality or inequality proof problems	5	6	3	16
Conditional evaluation problems	0	0	0	3
Total	5	7	3	24

As shown in Table 14, among the geometry problems in the 24th–64th IMOs, the largest proportion of problems is attributed to "basic properties of circles and four points on a circle" problems. However, the average scores of "basic properties of circles and four points on a circle" problems are primarily concentrated between 2 and 7 points, while most problems between 0 and 2 points were "power of a point, radical axis, and radical center" problems, accounting for half of the problems of this type. This suggests

Table 14   Numbers of Geometry Problems with Different Average Scores in the 24th–64th IMOs

Geometry Problem	Problem Mean					Total
	0–1	1–2	2–3	3–4	4–7	
Similarity and congruence	1	1	3	1	2	8
Basic properties of circles and four points on a circle	4	0	5	4	11	24
Power of a point, radical axis, and radical center	3	3	0	2	4	12
Special points and special lines	0	1	1	1	1	4
Trigonometry, areas, and analytic geometry	1	0	0	1	4	6
Solid geometry	0	0	0	0	0	0
Geometric inequalities	4	0	3	6	2	15
Total	13	5	12	15	24	69
	18		27		24	

that "power of a point, radical axis, and radical center" problems tend to be more challenging, while "basic properties of circles and four points on a circle" problems, although also difficult, include a greater number of relatively simple problems.

Overall, it can be seen that geometry problems in the past 10 years mainly involve circles; geometric inequality problems have not appeared in nearly 10 years; solid geometry problems all appeared in the first 20 IMOs. Of course, most geometry problems are comprehensive, and problems related to circles contain similarity or congruence and may also contain special points and special lines in the triangle. In addition, position structure problems are very closely related to quantity relation problems in some problems, such as proving that the angle bisector (figure structure) can be expressed by proving that two angles are equal (quantity relation).

Therefore, it is difficult to classify geometry problems well, and here is a preliminary classification based on the knowledge mainly involved in these problems.

2.4 *The proposal for problems*

Problems in the IMO are proposed by participating countries and regions, except the host. Usually, team leaders are in charge of submitting problems with a limit of six, and these problems are then subjected to the selection by a selection committee composed of experts organized by the host. Approximately 30 problems are chosen as shortlist problems, with around eight problems in each of the four topics: algebra, geometry, combinatorics, and number theory. Subsequently, these problems are submitted to the Jury, which is comprised of team leaders from each participating country or region. The problems are discussed and voted upon to select the official examination problems. Once the problems are finalized, they are translated into five working languages: English, French, German, Russian, and Spanish. Each leader then translates the problems into their respective national languages, and contestants can choose from two languages in which to answer the problems.

Among the first 64 IMOs, geometry problems were contributed by 37 different countries and regions. Poland had the biggest number of problems proposed, with a total of thirteen. The Soviet Union proposed nine problems, followed by the Netherlands, Romania, and Czechoslovakia with eight problems each. Russia proposed seven problems, while the United Kingdom, Hungary, Ukraine, and Bulgaria contributed five problems each. These countries collectively provided 73 problems. Remarkably, the Soviet

Union (the 41th IMO in 2000 and the 49th IMO in 2008), the United States (the 20th IMO in 1978), and the Netherlands (the 23th IMO in 1982) are the three countries that have proposed two geometry problems in the same IMO session.

Furthermore, as indicated in Appendix B, geometry problems in the first 15 IMOs were primarily proposed by the seven founding countries of the IMO. From the 16th–35th IMOs, geometry problems were mainly contributed by the Soviet Union, the Netherlands, and other countries. However, in the 36th–64th IMOs, geometry problems demonstrated a more diverse range of proposing countries and regions. This, to some extent, correlates with the expansion of the IMO's influence and the growth in the number of participating countries and regions.

3 Awards in the IMO

In addition to selecting problems, the Jury has several other responsibilities, including: establishing grading criteria, resolving discrepancies in grading between leaders and coordinators, and determining the number of gold, silver, and bronze medals, as well as the score thresholds. In each IMO, approximately 1/12 of the contestants receive a gold medal, 2/12 receive silver, and 3/12 receive bronze.

Apart from the gold, silver, and bronze medals, contestants who do not receive medals but attains a score of 7 on at least one problem in the IMO will receive an Honorable Mention. Contestants who deliver exceptionally elegant solutions to specific problems in the IMO will receive a Special Prize.

As depicted in Figure 2, starting from the 24th IMO, the cutoff scores for gold, silver, and bronze medals have gradually stabilized. The gold medal cutoff is approximately 29 points, the silver medal cutoff is around 22 points, and the bronze medal cutoff is roughly 15 points. Furthermore, the average score of all contestants closely aligns with the bronze medal cutoff. This indicates that the problem difficulty is well-balanced.

Interestingly, in the first 64 IMOs, there were three occasions where the gold medal cutoff was a perfect score, meaning only those who scored full marks could earn a gold medal. These three IMOs were: the 11th IMO (1969, Romania) with a perfect score of 40 points and three gold medalists; the 14th IMO (1972, Poland) with a perfect score of 40 points and eight gold medalists; and the 28th IMO (1988, Cuba) with a perfect score of 42 points and 22 gold medalists.

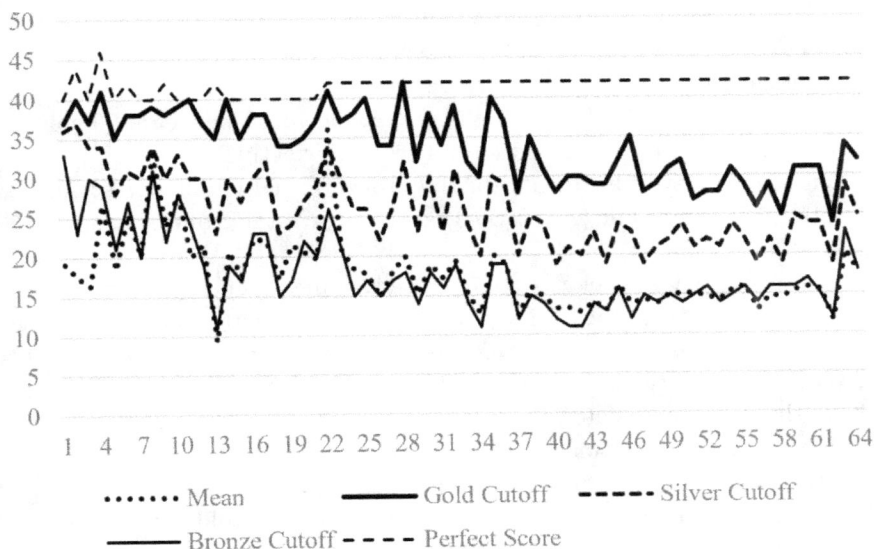

Figure 2   Medal Cutoff Scores in the First 64 IMOs

3.1 *Participation*

In the first 64 IMOs, a total of 269 contestants took part in four or more IMOs. Among them, two contestants attended seven IMOs, four contestants attended six IMOs, 42 contestants attended five IMOs, and 221 contestants attended four IMOs, as shown is Table 15.

Table 15   Contestants with Six or More Participations in the First 64 IMOs

Contestant	Country	Participation Year	Gold	Silver	Bronze	Honorable Mention	Perfect Score
David Kunszenti-Kovács	Norway	1997–2003	1	3	1	1	
Yeoh Zi Song	Malaysia	2014–2020	1	1	4	1	
Zhuo Qun (Alex) Song	Canada	2010–2015	5	0	1	0	1
Teodor von Burg	Serbia	2007–2012	4	1	1	0	
Alexey Entin	Israel	2000–2005	1	3	1	0	
Tan Li Xuan	Malaysia	2016–2021	0	2	2	1	

Coincidentally, in the 43rd IMO held in 2002, the gold medal cutoff was set at 29 points, and David Kunszenti-Kovács achieved a total score of exactly 29 points.

In the 44th IMO held in 2003, the gold medal cutoff was set at 29 points, whereas Alexey Entin attained a total score of exactly 28 points.

In the 51st IMO held in 2010, the bronze medal cutoff was set at 15 points, and Zhuo Qun (Alex) Song achieved a total score of exactly 15 points.

In the 58th IMO held in 2017, the silver medal cutoff was set at 19 points, and Yeoh Zi Song achieved a total score of exactly 19 points. In the 59th IMO held in 2018, the silver medal cutoff was set at 25 points, and Yeoh Zi Song's total score was exactly 24 points. In the 61st IMO held in 2020, the gold medal cutoff was set at 31 points, and Yeoh Zi Song's total score was exactly 31 points.

Additionally, from 2002 to 2005 and in 2007, Sherry Gong participated in the IMO, earning one bronze, two silver, and one gold medals. In the 48th IMO held in 2007, she ranked 7th individually. Notably, from 2002 to 2004, she was a member of the Puerto Rico IMO team, while in 2005 and 2007, she was a member of the United States IMO team.

From 2020 to 2023, Alex Chui participated in the IMO, securing two gold and two silver medals. However, in 2020 and 2021, he was a member of the Chinese Hong Kong IMO team, while in 2022 and 2023, he was a member of the United Kingdom IMO team.

Other than Sherry Gong and Alex Chui, the remaining 267 contestants hailed from 75 different countries and regions. Among them, there were 12 contestants from Cyprus and Moldova each, 11 from Malaysia, eight from Trinidad and Tobago, seven from each of Estonia, Germany, Sri Lanka, and North Macedonia, and six from Japan and Philippines each.

3.2   *Gold medals*

In the first 64 IMOs, a total of 49 contestants achieved three or more gold medals. Among them, one contestant earned five gold medals, six contestants earned four gold medals, and 42 contestants earned three gold medals.

As shown in Table 16, Zhuo Qun (Alex) Song, Reid Barton, and Lisa Sauermann have all achieved perfect scores.

Coincidentally, in the 54th IMO held in 2013, the gold medal cutoff was 31 points, and Nipun Pitimanaaree achieved a total score of exactly 31 points.

Table 16 Contestants with Four or More Gold Medals in the First 64 IMOs

Contestant	Country	Participation Year	Gold Year	Perfect Score Year
Zhuo Qun (Alex) Song	Canada	2010–2015	2011–2015	2015
Reid Barton	The United States of America	1998–2001	1998–2001	2001
Christian Reiher	Germany	1999–2003	2000–2003	
Lisa Sauermann	Germany	2007–2011	2008–2011	2011
Teodor von Burg	Serbia	2007–2012	2009–2012	
Nipun Pitimanaaree	Thailand	2009–2013	2010–2013	
Luke Robitaille	The United States of America	2019–2022	2019–2022	

In the 41st IMO held in 2000, the top four contestants all achieved perfect scores, and Reid Barton ranked fifth with a total score of 39. In the 43rd IMO held in 2002, the top three contestants all achieved perfect scores, and Christian Reiher ranked fourth with a total score of 36. Furthermore, in the 50th IMO held in 2009, Lisa Sauermann achieved a total score of 41, securing the third position.

Moreover, Oleg Golberg participated in the IMO from 2002 to 2004, achieving three gold medals. He consistently ranked within the top 10 in terms of total scores. Notably, in 2002 and 2003, he was a member of the Russia IMO team, and in 2004, he was a member of the United States IMO team.

Apart from Oleg Golberg, the remaining 48 contestants hailed from 22 different countries and regions. Among them, there were four contestants from each of Russia, Bulgaria, Germany, Hungary, Romania, and the United States. Both South Korea and the United Kingdom had three contestants, while Canada, Japan, Singapore, and the Soviet Union were represented by two contestants each.

3.3 *Special prizes*

In the first 64 IMOs, only 44 contestants received special prizes. Among them, one contestant has received the special prize three times, seven contestants have earned twice, and 36 contestants have achieved once. It indicates that achieving a special prize is even more challenging than securing a gold medal.

As shown in Table 17, John Rickard, Imre Ruzsa, and Marc van Leeuwen all achieved two special prizes in one IMO for their elegant solutions. Furthermore, John Rickard, Imre Ruzsa, and László Lovász have all earned a perfect score twice.

Table 17 Contestants with Multiple Special Prizes in the First 64 IMOs

Contestant	Country	Participation Year	Special Prize Year	Gold Year	Perfect Score Year
John Rickard	The United Kingdom	1975–1977	1976, 1977 (2)	1975–1977	1975, 1977
József Pelikán	Hungary	1963–1966	1965, 1966	1964–1966	1966
László Lovász	Hungary	1963–1966	1965, 1966	1964–1966	1965, 1966
László Babai	Hungary	1966–1968	1966, 1968	1968	1968
Simon Phillips Norton	The United Kingdom	1967–1969	1967, 1969	1967–1969	1969
Wolfgang Burmeister	The German Democratic Republic	1967–1971	1970, 1971	1968, 1970, 1971	1970
Imre Ruzsa	Hungary	1969–1971	1971 (2)	1970, 1971	1970, 1971
Marc van Leeuwen	The Netherlands	1977, 1978	1978 (2)		

Coincidentally, in the 11th IMO held in 1969, only three gold medals were awarded, with Imre Ruzsa ranking 4th and receiving a silver medal. Similarly, in the 19th IMO held in 1977, which also resulted in only 13 gold medals, Marc van Leeuwen ranked 14th and earned a silver medal.

Additionally, these 44 contestants hailed from 16 different countries and regions. Among them, there were seven contestants from each of Hungary and the United Kingdom, five from the German Democratic Republic, four from Bulgaria and Poland each, three from each of Czechoslovakia and the Soviet Union, and two from Finland and the United States each.

Special prizes were more frequently granted in the first 20 IMOs, with a total of 27 special prizes earned by 22 contestants from the 11th to 20th IMOs. Subsequently, the frequency of special prize presentations declined. Since Moldovan contestant Iurie Boreico received a special prize for his brilliant solution to IMO 46-3 (Algebra, proposed by South Korea) in 2005, no contestant has achieved this accolade to date.

From 2003 to 2007, Iurie Boreico consistently participated in the IMO, earning three gold and two silver medals. He achieved a perfect score in 2005 and 2006. It's noteworthy that the 44th IMO held in 2003 only yielded 37 gold medals, with Iurie Boreico placing 38th individually and receiving a silver medal.

- **(IMO 46-3, proposed by South Korea).** Let x, y, z be three positive reals such that $xyz \geq 1$. Prove that

$$\frac{x^5 - x^2}{x^5 + y^2 + z^2} + \frac{y^5 - y^2}{y^5 + z^2 + x^2} + \frac{z^5 - z^2}{z^5 + x^2 + y^2} \geq 0.$$

4 Summary

The IMO stands as a distinguished intellectual competition for young minds. According to a study by Agarwal R. and Gaule P., statistical analysis reveals that among contestants in the IMO (including those who did not secure medals), 22% choose to pursue further studies in mathematics, ultimately obtaining doctoral degrees in the field. Additionally, 1% of these contestants become presenters at the International Congress of Mathematicians, and 0.2% attain the Fields Medal. These statistics underscore the vital role of the IMO in identifying and nurturing mathematical talents.

It's essential not to perceive the IMO as a mere selection exam. Rather than focusing solely on the brief two-day competition, the crucial aspect lies in the learning and preparation undertaken before participating. As the mathematician Paul Halmos aptly put it, what mathematics really consists of is problems and solutions. Contestants, through their exploration of Olympiad problems, not only enhance their mathematical abilities but also experience the joy and satisfaction of problem-solving. This experience plants the seeds of a future career in mathematics.

However, it's important to acknowledge that Olympiad problems and research problems in mathematics differ. Research problems often lack readily available answers and may require the investment of countless days and nights. Hence, the IMO is just one pathway in the growth of mathematical talents, and success in the IMO is not the sole qualification for becoming an outstanding mathematician.

Although every contestant aims for a gold medal, their aspirations go far beyond accolades. On this stage, they have the opportunity to showcase their intellectual capabilities, revel in the mathematical exploration, and relish competing with talented young minds from around the world, all without the narrow goal of proving their superiority over others. While the competition results may vary, each contestant stands as a victor in their own right and becomes a companion and witness to one another's life journeys.

In contrast to the Olympics, where athletes' careers are closely intertwined with the Games, the IMO is merely a chapter in the growth of these gifted young individuals. Following the IMO, the door to a new mathematical world has already swung wide open for them.

Chapter 1

Similarity and Congruence

Plane geometry is a subject that studies geometric figures. It is a rich and wonderful category in elementary school mathematics. The significance of plane geometry is to establish axiomatic thinking methods. When we study mathematics, we cannot rely on intuition blindly, but need to set up strict logical reasoning from axioms. In the middle school stage, the plane geometry problem is an excellent subject matter used to train logical thinking and conscious thinking ability. Many mathematics enthusiasts say that since learning plane geometry, they have fallen in love with mathematics.

Plane geometry is a required subject in the IMO. In general, there are 1–2 plane geometry problems out of six problems per year. This section deals with the problem of similarity and congruence in the IMO. Similarity and congruence are the first problems we must face when studying the relationship between the shape and size of two geometric figures. The research basis of similarity and congruence is the similarity and congruence of triangles.

In the first 64th IMOs, there had been a total of 14 problems related to similarity and congruence, approximately accounting for 11.4% of all geometry problems. These problems are mainly categorized into three types: (1) existence problems, totaling four problems; (2) position structure problems, totaling five problems; (3) quantity relation problems, totaling five problems. The statistical distribution of these three types of problems in previous IMOs is presented in Table 1.1.

It can be seen that the proportion of similarity and congruence problems in geometry problems is not high, but there are about two problems every

Table 1.1 Numbers of Similarity and Congruence Problems in the First 64 IMOs

Content	Session							Total
	1–10	11–20	21–30	31–40	41–50	51–60	61–64	
Existence problems	3	0	0	1	0	0	0	4
Position structure problems	0	0	1	0	2	2	0	5
Quantity relation problems	1	0	1	2	1	0	0	5
Geometry problems	29	18	18	15	20	17	6	123
Percentage of similarity and congruence problems among geometry problems	13.8%	0.0%	11.1%	20.0%	15.0%	11.8%	0.0%	11.4%

ten sessions of IMOs. In fact, the position structure problem is very closely related to the quantity relation problem; for example, the structure of the angle bisector can be described by proving the quantity relation that two angles are equal. In particular, some of more difficult geometry problems also contain the structure of similarity and congruence. From this point of view, IMO geometry problems almost all involve the knowledge of similarity and congruence.

Therefore, this chapter will be divided into three parts. The first part introduces properties and determination methods related to similarity and congruence, and help understand the nature and application of these properties and theorems through some examples.

The second part revolves around three types of problems: "existence problems," "position structure problems," and "quantity relation problems." These problems are presented in chronological order, and some problems include various solutions and generalizations.

It is important to note that for each problem, the solutions are followed by information on the scores, including the number of contestants in each score range, the average score, and the scores of the top five teams. However, early IMOs often lacked information on contestant scores, so the number of contestants in each score range only represents the counted number of contestants, and some problems lack scores of the top five teams.

The third part provides a brief summary of this chapter.

1.1 Related Properties, Theorems, and Methods

1.1.1 *Related definitions and properties of congruent triangles*

(1) *The concept of congruent triangles*

If the corresponding angles of two triangles are equal and the corresponding sides are also equal, then we define two triangles to be congruent.

If $\triangle ABC$ and $\triangle DEF$ are two triangles, and

$$\angle A = \angle D, \quad \angle B = \angle E, \quad \angle C = \angle F,$$

$$AB = DE, \quad BC = EF, \quad AC = DF,$$

then $\triangle ABC$ and $\triangle DEF$ are congruent, denoted as $\triangle ABC \cong \triangle DEF$, where (A, D), (B, E), (C, F) are the three pairs of corresponding vertices. When using congruent symbols to represent congruent triangles, it is necessary to pay attention to the correspondence between vertices.

Regarding the congruence relation of general geometric figures, there are the following definitions: In a plane, if a figure can make it completely coincide with another figure through a translation, rotation, and reflection, then the two figures are defined to be "congruent".

(2) *Properties of congruent triangles*

Generally speaking, the corresponding objects (sides, line segments, angles, etc.) of two congruent triangles are the objects for which the coincidence of two congruent triangles will also coincide exactly.

 (i) The corresponding sides of congruent triangles are equal;
 (ii) The corresponding internal angles of congruent triangles are equal;
(iii) The corresponding segments of congruent triangles are equal;
(iv) The corresponding angles of congruent triangles are equal.

Note. (iii) and (iv) contain much important information, such as the corresponding medians, altitudes, and angular bisectors of congruent triangles have equal lengths.

(3) *Determinations of congruent triangles*

Determination method 1: If three pairs of sides of two triangles are equal in length, then the triangles are congruent.

The determination method is referred to as: side-side-side, abbreviated as: "SSS".

Determination method 2: If two pairs of sides of two triangles are equal in length, and the included angles are equal in measurement, then the triangles are congruent.

The determination method is referred to as: side-angle-side, abbreviated as: "SAS".

Determination method 3: If two pairs of angles of two triangles are equal in measurement, and the included sides are equal in length, then the triangles are congruent.

The determination method is referred to as: angle-side-angle, abbreviated as: "ASA".

Determination method 4: If two pairs of angles of two triangles are equal in measurement, and a pair of corresponding non-included sides are equal in length, then the triangles are congruent.

The determination method is referred to as: angle-angle-side, abbreviated as: "AAS".

Determination method 5: If two right-angled triangles have their hypotenuses equal in length, and a pair of other sides are equal in length, then the triangles are congruent.

The determination method is referred to as: hypotenuse-leg, abbreviated as: "HL".

(4) *Congruent triangles related conclusions*

 (i) The distances between any point on the perpendicular bisector of a segment and the two endpoints of the segment are equal. Conversely, points with equal distances to the two endpoints of a segment are on the perpendicular bisector of the segment.

 (ii) The distances between any point on the angle bisector and the two sides of an angle are equal (where the distances between the foot and the vertex are also equal). Conversely, inside an angle, points with equal distances to the two sides of the angle are on the angular bisector.

(iii) Angle bisector theorem: In $\triangle ABC$, the bisector of $\angle BAC$ intersects side BC at point D. Then $\frac{AB}{AC} = \frac{BD}{CD}$. The reverse also holds. Similarly, there exists also the external angle bisector theorem.

(iv) In $\triangle ABC$, if $AB = AC$, then $\angle B = \angle C$ (equal sides correspond to equal angles). Conversely, if $\angle B = \angle C$, then $AB = AC$ (equal angles correspond to equal sides).

 (v) In $\triangle ABC$, if $AB = AC$ (i.e., $\triangle ABC$ is an isosceles triangle), then the altitude, the median, and the angular bisector from point A coincide (three lines of isosceles triangle coincide into one). Conversely, if any two lines of the altitude, the median, and the angular bisector from a vertex coincide, then the triangle is an isosceles triangle.

It follows that the three internal angles of an equilateral triangle are all 60°. Conversely, an isosceles triangle with an internal angle of 60° is an equilateral triangle.

Let the length of an equilateral triangle be a. Then its area is $\frac{\sqrt{3}}{4}a^2$.

Let the interior angles of a right triangle be 30°, 60°, 90°. Then the ratio of their opposite sides is $1 : \sqrt{3} : 2$.

(vi) In $\triangle ABC$, let M, N be the midpoints of sides AB, AC respectively. Then MN is called the midsegment of $\triangle ABC$, then $MN \parallel BC$ and $MN = \frac{1}{2}BC$.

(vii) In a right triangle, the length of the median on the hypotenuse is equal to half the length of the hypotenuse. Conversely, on the hypotenuse of a right triangle, if a point is of the same distance from two vertices, then it must be the midpoint of the hypotenuse.

1.1.2 Related definitions and properties of similar triangles

(1) The concept of similar triangles

If the corresponding internal angles of two triangles are equal and the corresponding sides are proportional, then the two triangles are similar. The ratio of the corresponding sides is called the similarity ratio.

If $\triangle ABC$ and $\triangle DEF$ are similar, then it is denoted as $\triangle ABC \backsim \triangle DEF$. In general, when we write a pair of similar triangles, we should arrange the vertices in the corresponding order, that is, the corresponding points of A, B, C are D, E, F.

Similar triangles with a similarity ratio of 1 are congruent triangles. More generally, two polygons are said to be similar if it is satisfied that all corresponding internal angles are equal and all corresponding sides are proportional.

Regarding the similarity relation of general geometric figures, there are the following definitions: If two figures through a finite number of translations, rotations, symmetries, and contractions, can make them completely coincide, then the two figures are defined to be "similar". This is a general graphically similar definition. The essence of reduction is homothety, which will be explained later.

(2) *Properties of similar triangles*

In the transformation process of two triangles, the angles that can coincide with each other are called corresponding angles, and the segments are called corresponding segments.

The corresponding internal angles of similar triangles are equal, the corresponding sides are proportional (with a similarity ratio), and the area ratio is the square of the similarity ratio.

There are two categories of similar triangles: regular similarity and inverse similarity. For regular similarity, the corresponding included angles of similar triangles maintain a consistent surround order in the plane, while inverse similarity does not maintain a consistent surround order, which is also called mirror similarity.

(i) As shown in Figure 1.1, points E, F are on the sides AB, AC, respectively. If $EF \parallel BC$, then $\triangle AEF$ and $\triangle ABC$ are of regular similarity.

(ii) As shown in Figure 1.2, points E', F' are on the sides AC, AB, respectively. If $\angle AE'F' = \angle ABC$, then $\triangle AEF$ and $\triangle ABC$ are of inverse similarity.

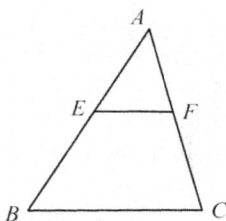
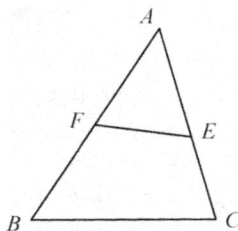

Figure 1.1   Figure 1.2

(3) *Determinations of similar triangles*

Determination method 1: If two pairs of internal angles of two triangles are equal in measurement, then the two triangles are similar.

It follows that:

(i) Two triangles with three pairs of corresponding sides parallel (collinear) are similar; Two triangles with three pairs of corresponding sides perpendicular are similar.

(ii) A right triangle is divided by the altitude of the hypotenuse into two small right triangles. Then the three right triangles are similar to each other.

Determination method 2: If two pairs of sides of two triangles are proportional in length, and the included angles are equal in measurement, then the triangles are similar.

It follows that if the two legs and hypotenuse of two right triangles are proportional, then the two right triangles are similar.

Determination method 3: If three pairs of sides of two triangles are proportional in length, then the triangles are similar.

(4) *Common similar structures and related conclusions*

(i) As shown in Figure 1.3 and Figure 1.4, in $\triangle ABC$, points D, E are on the sides AB, AC (extension line) respectively. If $DE \parallel BC$, then $\triangle ADE \backsim \triangle ABC$.

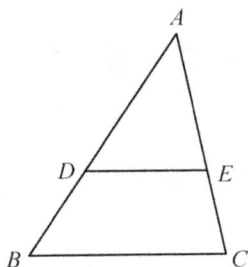 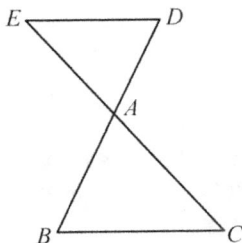

Figure 1.3   Figure 1.4

(ii) As shown in Figures 1.5–1.10, the known points D, E are on sides AC, AB respectively, if $\angle ADE = \angle CBA$, then $\triangle ADE \backsim \triangle ABC$.

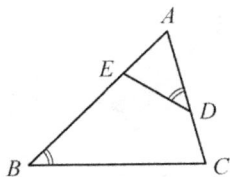 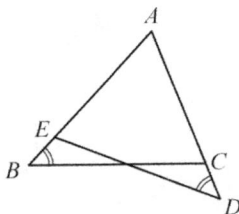 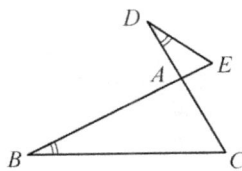

Figure 1.5   Figure 1.6   Figure 1.7

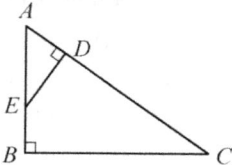

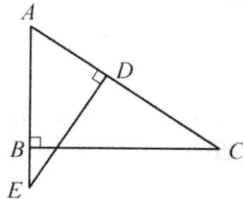

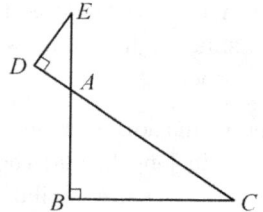

Figure 1.8   Figure 1.9   Figure 1.10

(iii) As shown in Figure 1.11, in $\triangle ABC$, point D is on the side BC, if $\angle CAD = \angle ABC$, then $\triangle ABC \backsim \triangle DAC$. The condition $\angle CAD = \angle ABC$ is also equivalent to $AC^2 = CD \cdot CB$.

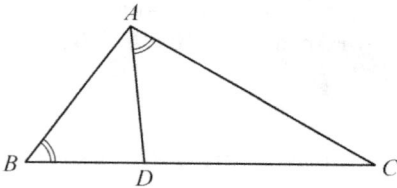

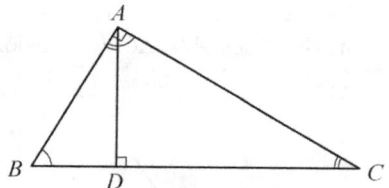

Figure 1.11   Figure 1.12

In particular, as shown in Figure 1.12, if $\angle BAC = \angle ADC = 90°$, then $\triangle ABC \backsim \triangle DBA \backsim \triangle DAC$, and we have:

$$AB^2 = BD \cdot BC, \quad AC^2 = CD \cdot CB,$$
$$AD^2 = BD \cdot CD, \quad AB \cdot AC = BC \cdot AD.$$

(iv) As shown in Figure 1.13–1.18, a point P is on the line AB, if points C, D satisfy $\angle CAB = \angle CPD = \angle ABD$, then $\triangle APC \backsim \triangle BDP$.

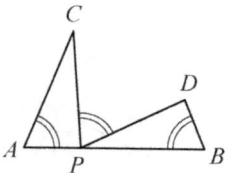

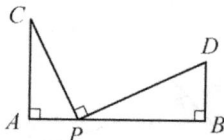

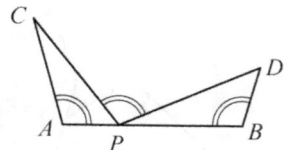

Figure 1.13   Figure 1.14   Figure 1.15

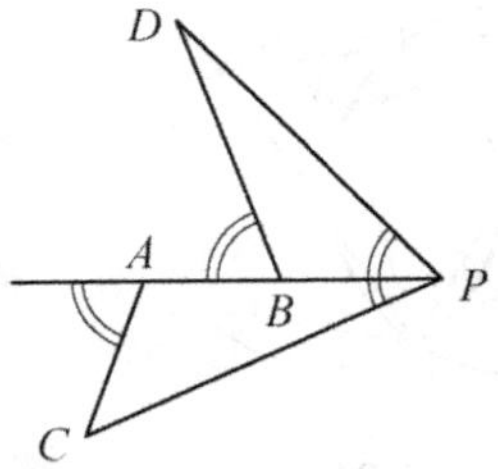

Figure 1.16 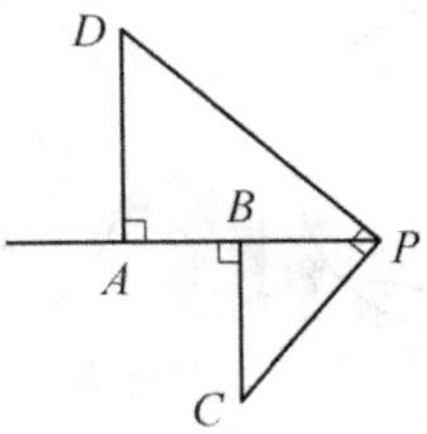   Figure 1.17 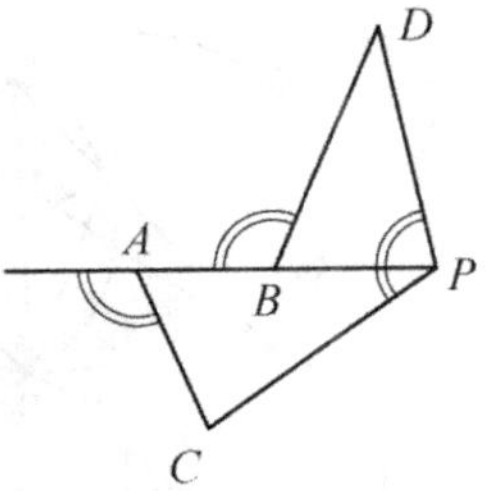   Figure 1.18

(v) As shown in Figure 1.19, if $\triangle OAB \backsim \triangle OA'B'$ (regular similarity) is known, then $\triangle OAA' \backsim \triangle OBB'$. If the points P, P' are corresponding similarity points of $\triangle OAB, \triangle OA'B'$, then $\triangle OPP' \backsim \triangle OAA' \backsim \triangle OBB'$.

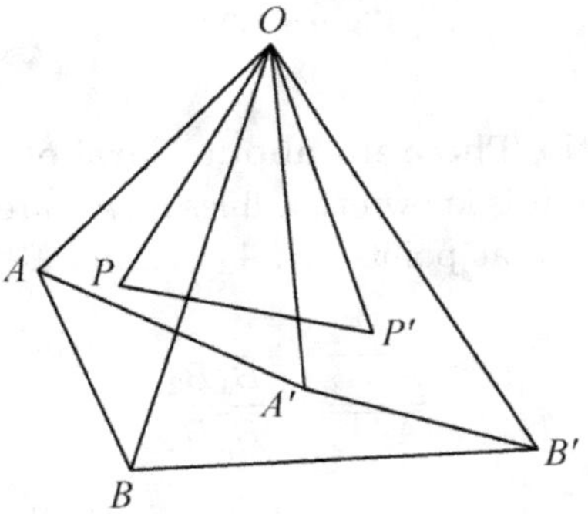

Figure 1.19

(vi) As shown in Figures 1.20–1.22, it is known that two circles ω_1, ω_2 intersect at two points A, B. Through point A, construct a line, which intersects circles ω_1, ω_2 at points $Q, P(\neq A)$, respectively. And through point A, construct another line, which intersects circles ω_1, ω_2 at points $X, Y(\neq A)$, respectively. Then

$$\triangle BPQ \backsim \triangle BYX, \quad \triangle BPY \backsim \triangle BQX.$$

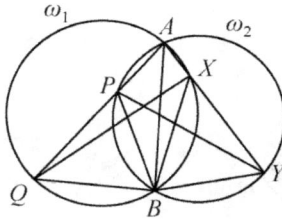

Figure 1.20

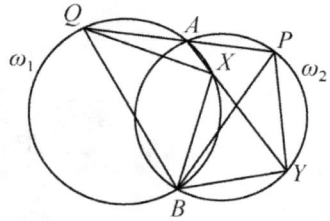

Figure 1.21

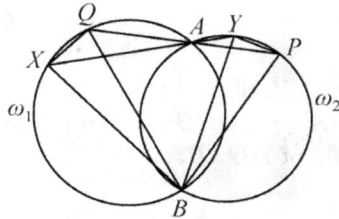

Figure 1.22

Theorem 1.1 (Ratio Theorem about Parallel Lines). As shown in Figures 1.23 and 1.24, it is known that lines m, n, l are parallel in pairs, and lines a, b intersect them at points A_1, A_2, A_3 and B_1, B_2, B_3, respectively. Then

$$\frac{\overline{A_1 A_2}}{\overline{A_2 A_3}} = \frac{\overline{B_1 B_2}}{\overline{B_2 B_3}}.$$

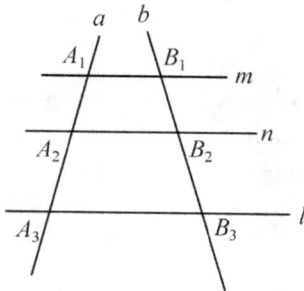

Figure 1.23

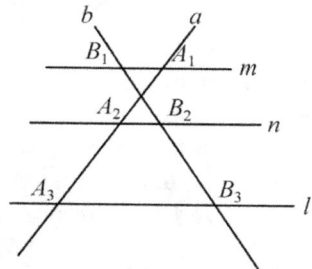

Figure 1.24

It is not difficult to get the following result: a line through point O intersects lines m, n at points $A_1, A_2, B_1, B_2, C_1, C_2$ (see Figures 1.25 and 1.26), respectively. Then

$$\frac{\overline{A_1B_1}}{\overline{B_1C_1}} = \frac{\overline{A_2B_2}}{\overline{B_2C_2}}.$$

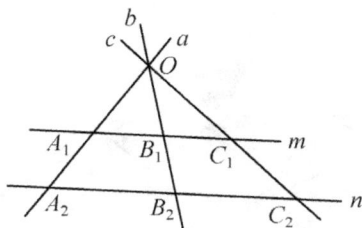

Figure 1.25

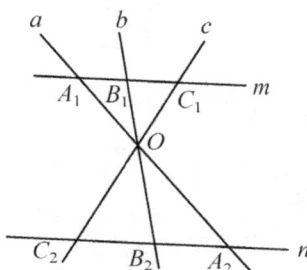

Figure 1.26

Theorem 1.2 (Menelaus's Theorem). As shown in Figures 1.27 and 1.28, in $\triangle ABC$, points P, Q, R are on lines BC, CA, AB, respectively. Then a sufficient and necessary condition that the three points P, Q, R are collinear is

$$\frac{\overline{BP}}{\overline{PC}} \cdot \frac{\overline{CQ}}{\overline{QA}} \cdot \frac{\overline{AR}}{\overline{RB}} = -1.$$

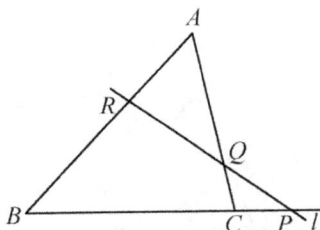

Figure 1.27

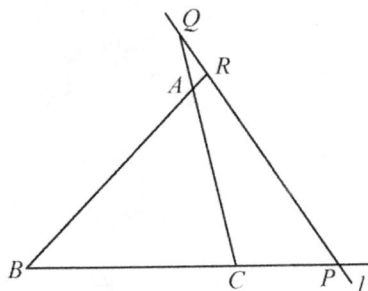

Figure 1.28

Theorem 1.3 (Ceva's Theorem). As shown in Figures 1.29–1.31, in $\triangle ABC$, points P, Q, R are on lines BC, CA, AB, respectively. Then a sufficient and necessary condition that the three lines AP, BQ, CR are concurrent or parallel to each other is

$$\frac{\overline{BP}}{\overline{PC}} \cdot \frac{\overline{CQ}}{\overline{QA}} \cdot \frac{\overline{AR}}{\overline{RB}} = 1.$$

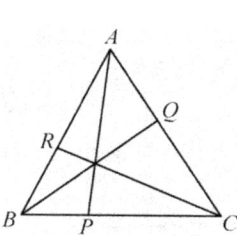
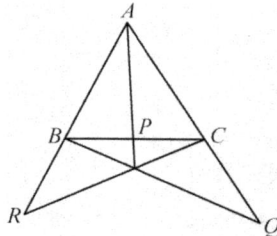
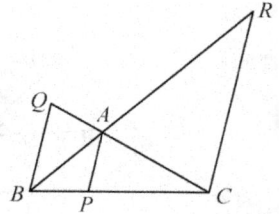

Figure 1.29   Figure 1.30   Figure 1.31

Homothetic. For a given point O in a plane and a non-zero real number λ, a transformation of points in the plane can be constructed: for any point P in the plane, there exists a unique point Q such that $\overrightarrow{OQ} = \lambda \cdot \overrightarrow{OP}$. The correspondence $\varphi : P \mapsto Q$ is a homothetic transformation. Point O is called the homothetic center and λ is called the homothetic ratio.

When $\lambda > 0$, the transformation φ is called externally homothetic; When $\lambda < 0$, it is called internally homothetic.

Two triangles are homothetic if and only if their corresponding sides are parallel to each other (but not different from the translation). The lines of the corresponding vertices of a pair of homothetic triangles are concurrent at the homothetic center. (see Figures 1.32–1.35).

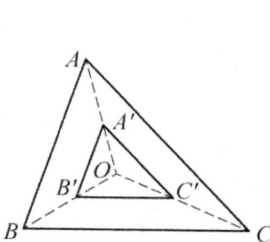
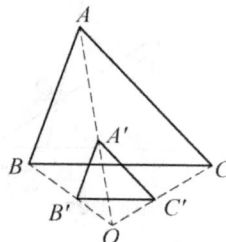

Figure 1.32   Figure 1.33

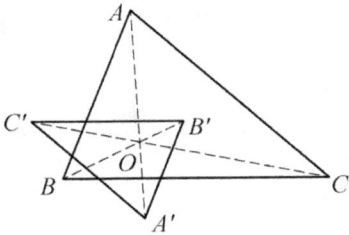

Figure 1.34

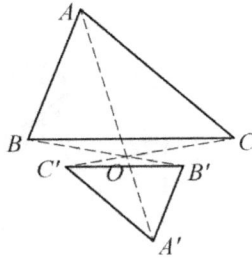

Figure 1.35

Two circles with different radii are externally homothetic. Let the radii of $\odot O_1, \odot O_2$ be r_1, r_2 respectively with $r_1 \neq r_2$. There exists a unique point S on the line O_1O_2 such that $\dfrac{\overline{SO_1}}{\overline{SO_2}} = \dfrac{r_1}{r_2}$, which is the externally homothetic center of the two circles. Take the points P, Q from $\odot O_1, \odot O_2$, and they are externally homothetic corresponding points if and only if $\overrightarrow{O_1P}, \overrightarrow{O_2Q}$ are in the same direction. It is obvious that if $\odot O_1, \odot O_2$ intersect, tangent, or separate, then their externally homothetic center is the intersection of their externally common tangent lines (see Figure 1.36).

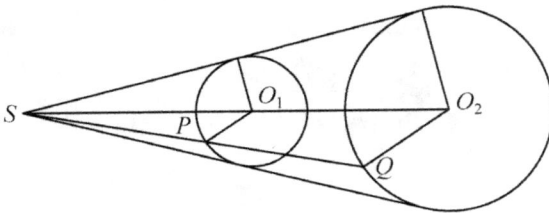

Figure 1.36

Any two circles are internally homothetic. Let the radii of $\odot O_1, \odot O_2$ be r_1, r_2 respectively. There exists a unique point S on the line O_1O_2 such that $\dfrac{\overline{SO_1}}{\overline{SO_2}} = -\dfrac{r_1}{r_2}$, which is the internally homothetic center of the two circles. Take the points P, Q from $\odot O_1, \odot O_2$, and they are internally homothetic corresponding points if and only if $\overrightarrow{O_1P}, \overrightarrow{O_2Q}$ are in the opposite direction. It is obvious that if $\odot O_1, \odot O_2$ separate, then their externally homothetic center is the intersection of their internally common tangent lines (see Figure 1.37).

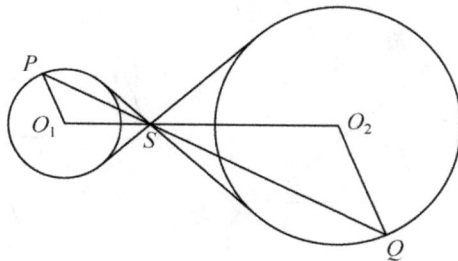

Figure 1.37

Example 1.1. As shown in Figure 1.38, $\triangle ABC$ ($AB > AC$) is inscribed in the circle $\odot O$. The point D is the midpoint of the side BC, and the line AD intersects $\odot O$ at another point E. Construct a parallel line of BC through E, intersecting $\odot O$ at another point F. Construct a perpendicular line to AC through C, intersecting DE at point G. Prove: GC bisects $\angle AGF$.

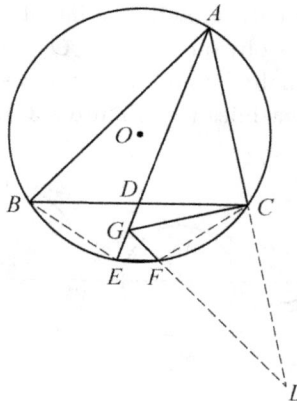

Figure 1.38

Proof.   Extend AC to point L such that $CL = AC$.
   Since $BC \parallel EF$, and A, E, F, C are concyclic,

$$\angle LCF = \angle AEF = \angle ADC,$$

that is, $\angle LCF = \angle ADC$.
   Combining $\triangle BDE \backsim \triangle ADC$ and the known conditions, we get

$$\frac{CL}{AD} = \frac{AC}{AD} = \frac{BE}{BD} = \frac{CF}{CD}.$$

   Thus, $\triangle LCF \backsim \triangle ADC$.

Then $\angle FLC = \angle CAD$. Note that $\triangle AGL$ is an isosceles triangle, $AG = GL$, so

$$\angle FLC = \angle CAD = \angle CAG = \angle CLG.$$

Thus, the three points G, F, L are collinear.

Hence GC bisects angle $\angle AGL$, i.e., GC bisects angle $\angle AGF$.

Example 1.2. As shown in Figure 1.39, For a rhombus $ABCD$, a circle passing the two points A, C intersects with the sides BC, CD at two points E, F, respectively, which are different from point C. Let AE and BD intersect at point G. Prove: $\angle ABF = \angle AFG$.

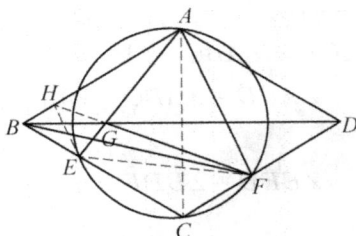

Figure 1.39

Proof. Extend FG so that it intersects AB at point H.

Since $AB \parallel CD$ and $BC \parallel AD$, so $\triangle BGH \backsim \triangle DGF$ and $\triangle BGE \backsim \triangle DGA$. Then

$$\frac{HG}{GF} = \frac{BG}{GD} = \frac{EG}{GA},$$

and thus $EH \parallel AF$.

Therefore,

$$\angle AEH = \angle EAF = 180° - \angle ECF = 180° - \angle BCD = \angle ABE,$$

so $AE^2 = AB \cdot AH$.

While CA bisects $\angle ECF$, we have $AE = AF$. Thus $AF^2 = AB \cdot AH$, so $\triangle AFH \backsim \triangle ABE$. Therefore $\angle AFH = \angle ABF$, i.e., $\angle AFG = \angle ABF$.

Example 1.3. As shown in Figure 1.40, it is known that $\triangle ABC \backsim \triangle ADE$ (regular similarity) and $\angle DBC = 90°$. Prove:

$$\frac{BE^2 - DE^2}{AC^2 - BC^2} = \frac{BD^2}{AB^2}.$$

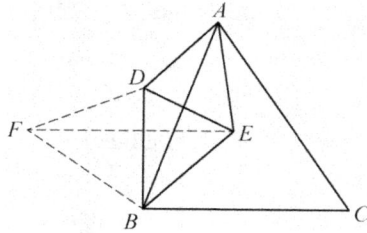

Figure 1.40

Proof.   Take a point F such that $\triangle BDF \backsim \triangle ADE$ (regular similarity) and connect to form EF.

Then $\triangle DAB \backsim \triangle DEF$, so $\angle EFD = \angle ABD$.

Since $\angle BDF = \angle ADE = \angle ABC$, and

$$\angle ABD + \angle ABC = 90°,$$

we get

$$\angle EFD + \angle BDF = 90°.$$

That is, $EF \perp BD$.

Therefore, by the equivalent difference of power of lines theorem,

$$BE^2 - DE^2 = BF^2 - DF^2. \tag{1}$$

Also $\triangle BDF \backsim \triangle ABC$, from which

$$\frac{BF^2 - DF^2}{BD^2} = \frac{AC^2 - BC^2}{AB^2}. \tag{2}$$

By (1) and (2), the conclusion is proved.

Example 1.4. As shown in Figure 1.41, AB, CD are two chords of a circle ω. Points E, F are on the extension lines of AB, CD, respectively, such that

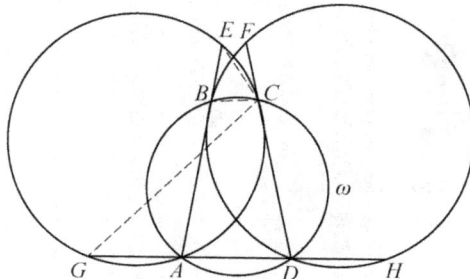

Figure 1.41

$\dfrac{BE}{AB} = \dfrac{CF}{CD}$. The circumcircles of $\triangle ACE, \triangle BDF$ intersect the line AD at the other two points G, H, respectively. Prove: $AG = DH$.

Proof. According to the problem, we can assume that $BE = k \cdot AB$ and $CF = k \cdot CD$.

In $\triangle CDG$ and $\triangle CBE$, we have

$$\angle CDG = \angle CBE,$$

$$\angle CGD = \angle CGA = \angle CEA = \angle CEB.$$

Then $\triangle CDG \backsim \triangle CBE$.

Thus $\dfrac{DG}{BE} = \dfrac{CD}{BC}$, and further

$$DG = \frac{CD \cdot BE}{BC} = k \cdot \frac{AB \cdot CD}{BC}.$$

Similarly, $AH = k \cdot \dfrac{AB \cdot CD}{BC}$. Therefore $DG = AH$, i.e., $AG = DH$.

Example 1.5. In an acute triangle $\triangle ABC$, I is the incenter. N is the midpoint of the arc $\overset{\frown}{BAC}$ on the circumcircle of $\triangle ABC$. A point P makes the quadrilateral $ABPC$ a parallelogram. Let Q be the symmetric point of point A with respect to point N and R be the projection of point A onto QI. Prove: The line AI is tangent to the circumcircle of $\triangle PQR$.

Proof. As shown in Figure 1.42, let M, S be the midpoints of segments BC, AI, respectively. Consider the homothetic transformation with A as

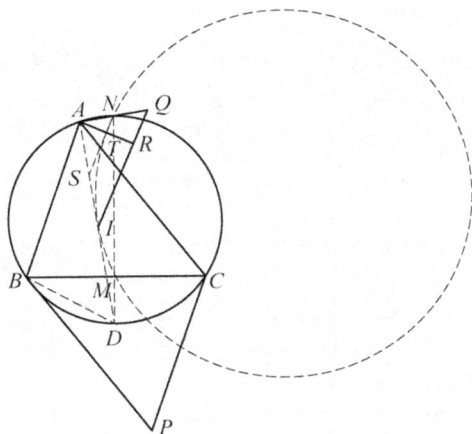

Figure 1.42

the homothetic center and $\frac{1}{2}$ as the homothetic ratio. Then the image of point P is point M, the image of point Q is point N, and the image of point R is point T (T is the projection of point A onto SN). It is only necessary to prove that the circumcircle of $\triangle MNT$ is tangent to AI.

In fact, the circle is tangent to AI at point I. Obviously $\angle NAS = 90°$, and since $\triangle ASN \backsim \triangle TSA$,

$$ST \cdot SN = SA^2 = SI^2.$$

Thus SI is tangent to the circumcircle of $\triangle ITN$. If it can be proved that SI is also tangent to the circumcircle of $\triangle NIM$, then the proof is complete. This is because the circle that passes through points I, N and is tangent to SI is unique.

Let the other intersection of line AI and the circumcircle of $\triangle ABC$ be D, Notice that

$$\angle DBM = \angle DCB = \angle DNB.$$

Therefore,

$$DM \cdot DN = DB^2 = DI^2.$$

Then DI is tangent to the circumcircle of $\triangle NIM$, and we complete the proof.

1.2 Problems and Solutions

1.2.1 *Existence problems*

Problem 1.1 (IMO 2-4, proposed by Hungary). Construct a triangle ABC, given h_a, h_b (the altitudes from A and B), and m_a, the median from vertex A.

Solution 1. First construct a right triangle $\triangle AHM$ such that $AH \perp HM$, $AH = h_a, AM = m_a$.

Then construct a circle $\odot M$ with M as the center and $\frac{1}{2}h_b$ as the radius.

As shown in Figure 1.43, make the tangent line of $\odot M$ through A, and the points of tangency are D, D', respectively.

If the extension AD intersects the line HM, let this intersection point be C, and construct the symmetric point B of C with respect to M. It is easy to know that the distance from B to AC is $2MD = h_b$, so $\triangle ABC$ is the required triangle. As shown in the figure, we can also get $\triangle AB'C'$.

But not all h_a, h_b, and m_a lead to two solutions.

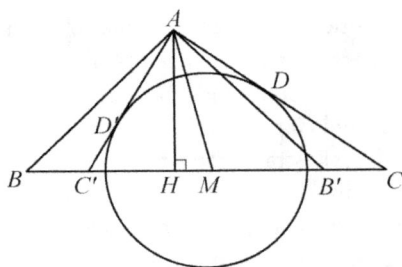

Figure 1.43

First of all, the existence of a solution must satisfy $m_a \geq h_a$. Under this premise, it is easy to know: if $h_a > \frac{1}{2}h_b$, then $m_a > h_a$ has two solutions, while $m_a = h_a$ has one solution; if $h_a \leq \frac{1}{2}h_b \leq m_a$, then $m_a > h_a$ has one solution, while $m_a = h_a$ has no solution; if $\frac{1}{2}h_b > m_a$, then there is no solution.

Solution 2. Let ABC be the required triangle and denote $AH = h_a$, $BK = h_b$, and $AD = m_a$. Crossing point D, we construct a line perpendicular to AC, intersecting AC at L. Then $DL = \frac{1}{2}h_b$, and $\triangle ALD$ is a right triangle.

The construction process of the triangle ABC is as follows (see Figure 1.44): construct the line l and take a point D on l, and construct l' parallel to l at a distance h_a from l. The circle with D as the center and m_a as the radius intersects the line l' at two points A, A' (here $h_a < m_a$ is a necessary condition). Choose point A, and then the right triangle ADL is inscribed in the circle of diameter AD (here $\frac{1}{2}h_b < m_a$ is a necessary condition), and $DL = \frac{1}{2}h_b$. So, we get the points A, D, L. Thus we can determine the triangle ABC.

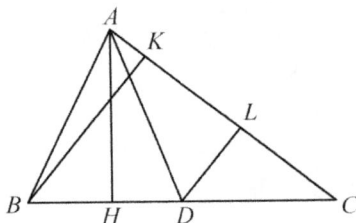

Figure 1.44

【Score Situation】 This particular problem saw the following distribution of scores among contestants: 6 contestants scored 5 points, 1 contestant scored 4 points, no contestant scored 3 points, 1 contestant scored 2 points, no contestant scored 1 point, and 2 contestants scored 0 point. The average score of this problem is 3.600, indicating that it was relatively straightforward.

Among the top five teams in the team scores, the Czechoslovakia team achieved a total score of 257 points, the Hungary team achieved a total score of 248 points, the Romania team achieved a total score of 248 points, the Bulgaria team achieved a total score of 175 points, and the German Democratic Republic team achieved a total score of 38 points.

The gold medal cutoff for this IMO was set at 40 points (with four contestants earning gold medals), the silver medal cutoff was 37 points (with four contestants earning silver medals), and the bronze medal cutoff was 33 points (with four contestants earning bronze medals).

In this IMO, no contestant achieved a perfect score of 44 points.

Problem 1.2 (IMO 9-4, proposed by Italy). Let $A_0B_0C_0$ and $A_1B_1C_1$ be any two acute-angled triangles. Consider all triangles ABC that are similar to $\triangle A_1B_1C_1$ (so that vertices A_1, A_2, and A_3 correspond to vertices A, B, C respectively) and circumscribed about triangle $A_0B_0C_0$ (where A_0 lies on BC, B_0 on CA, and C_0 on AB). Of all such possible triangles, determine the one with the maximum area, and construct it.

Solution 1. Two arcs are constructed respectively, with the sides A_0C_0, A_0B_0 of $\triangle A_0B_0C_0$ as chords, and $\angle B_1$ and $\angle C_1$ as inscribed angles, as shown in Figure 1.45.

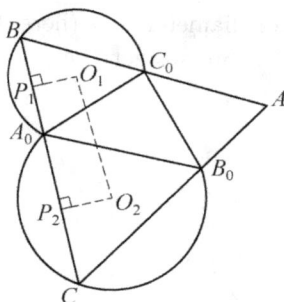

Figure 1.45

Take points B, C on the two arcs respectively such that B, A_0, C are on the same line, and extend BC_0 and CB_0 to intersect at A It is easy to know that $\angle A = \angle A_1$, so $\triangle ABC \sim \triangle A_1 B_1 C_1$. And B is any point on $\overparen{A_0 C_0}$, and the position of C depends on B.

And as shown in the figure, let the centers of the two arcs be O_1, O_2 respectively and construct $O_1 P_1$ and $O_2 P_2$ perpendicular to BC, respectively, where P_1, P_2 are the perpendicular feet and P_1, P_2 are also the midpoint of $BA_0, A_0 C$. Then

$$BC = 2P_1 P_2 \leq 2O_1 O_2,$$

and the equality holds only if $BC \parallel O_1 O_2$, from which it is not difficult to determine the exact position of the largest triangle that satisfies the condition.

Solution 2. Let triangle $A_0 B_0 C_0$ be fixed in the plane and triangle $A_1 B_1 C_1$ change position in the plane. For any position of triangle $A_1 B_1 C_1$, consider constructing three lines l_{AB}, l_{BC}, l_{CA} passing through points C_0, A_0, B_0 and satisfying conditions $l_{AB} \parallel A_1 B_1$, $l_{BC} \parallel B_1 C_1$, $l_{CA} \parallel C_1 A_1$, respectively. Then they form the triangle ABC.

Let $A'B'C'$ be a triangle circumscribed to the triangle $A_0 B_0 C_0$ using the above construction, move the triangle $A_1 B_1 C_1$, and construct the triangle ABC in the same way. We have

$$\frac{S_{ABC}}{S_{A'B'C'}} = \left(\frac{AC}{A'C'} \right)^2 .$$

Since $S_{\triangle A'B'C'}$ is constant, $S_{\triangle ABC}$ reaches its maximum when side AC is the longest. That is, point A lies on an arc of length $\overparen{B_0 A C_0}$ passing through points B_0, C_0. Let O_1 be the center of the circle containing this arc, so we construct the longest secant line $AB_0 C$. And let O_3 be the center of a circle that passes through B_0 and contains an arc of length $\overparen{B_0 C A_0}$, and point C lies on this arc. Then let P_1, P_3 be the projections of the centers O_1, O_3 on AC, respectively, as shown in Figure 1.46, and we have

$$AC = AB_0 + B_0 C = 2B_0 P_1 + 2B_0 P_3 = 2P_1 P_3.$$

The quadrilateral $O_1 O_3 P_3 P_1$ is a right trapezoid where $P_1 P_3$ reaches its maximum value when $P_1 P_3 = O_1 O_3$, in which case AC is parallel to $O_1 O_3$. Point B is the intersection of lines AC_0 and CA_0, where C_0, A_0 are the given points on the two circles.

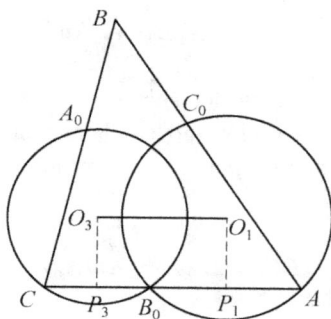

Figure 1.46

【Score Situation】 This particular problem saw the following distribution of scores among contestants: 16 contestants scored 6 points, 6 contestants scored 5 points, 2 contestants scored 4 points, 5 contestants scored 3 points, 4 contestants scored 2 points, no contestant scored 1 point, and 4 contestants scored 0 point. The average score of this problem is 4.243, indicating that it was simple.

Among the top five teams in the team scores, the Soviet Union team achieved a total score of 275 points, the German Democratic Republic team achieved a total score of 257 points, the Hungary team achieved a total score of 251 points, the United Kingdom team achieved a total score of 231 points, and the Romania team achieved a total score of 214 points.

The gold medal cutoff for this IMO was set at 38 points (with 11 contestants earning gold medals), the silver medal cutoff was 30 points (with 14 contestants earning silver medals), and the bronze medal cutoff was 22 points (with 26 contestants earning bronze medals).

In this IMO, a total of 5 contestants achieved a perfect score of 42 points.

Problem 1.3 (IMO 10-1, proposed by Romania). Prove that there is one and only one triangle whose side lengths are consecutive integers, and one of whose angles is twice as large as another.

Proof 1. As shown in Figure 1.47, in the triangle $\triangle ABC$, let $\angle C = 2\angle B$ and the three sides of the triangle be $n-1$, n, $n+1$.

Extend AC to D such that $BC = DC$, and connect BD. It is easy to know that $\angle D = \frac{1}{2}\angle C = \angle ABC$, from which $\triangle ABC \backsim \triangle ADB$. Then

$$AB^2 = AC \cdot AD = AC \cdot (AC + BC). \tag{1}$$

Since $AB > AC$, so $AB = n$ or $n + 1$.

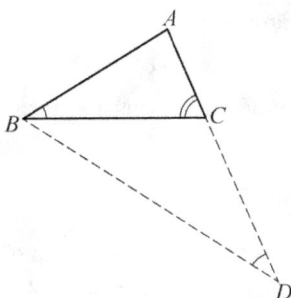

Figure 1.47

When $AB = n$, there can only be $AC = n - 1, BC = n + 1$, and then (1) becomes $n^2 = (n - 1) \cdot 2n$, that is, $n = 2$, in which case the three sides cannot form a triangle.

When $AB = n + 1$, there are two cases. If $AC = n$, $BC = n - 1$, we have $(n+1)^2 = n(2n-1)$, in which case the equation has no positive integer solution. Thus $AC = n - 1$, $BC = n$, that is, $(n + 1)^2 = (n - 1)(2n - 1)$, and we get $n = 5$, so the only solution is the triangle with sides $4, 5, 6$.

Proof 2. Let $\angle B = 2\angle A$ and BD be the internal bisector of $\angle B$ (see Figure 1.48). By the angle bisector theorem,

$$AD = BD = \frac{bc}{a+c}, \quad DC = \frac{ab}{a+c}.$$

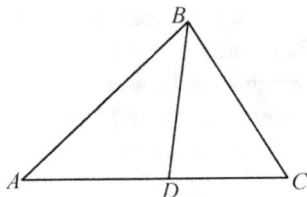

Figure 1.48

On the other hand, $\triangle ABC$ and $\triangle BDC$ are similar, so

$$\frac{BC}{DC} = \frac{AC}{BC} = \frac{AB}{BD},$$

that is

$$\frac{a+c}{b} = \frac{b}{a}.$$

In this way, we get a key equation:

$$a(a + c) = b^2. \qquad (*)$$

In the following we divide the discussion into three cases:

(i) $\angle A < \angle B < \angle C$, and thus $b = a + 1$ and $c = a + 2$.
By the equation $(*)$, $a = 1$, $b = 2$, and $c = 3$. In this case, the triangle ABC is degenerate.

(ii) $\angle A < \angle C < \angle B$, so $c = a + 1$ and $b = a + 2$.
By the equation $(*)$, $a(a - 3) = 4$. In this case, the equation has a positive integer solution $a = 4$. Then $c = 5$ and $b = 6$.

(iii) $\angle C < \angle A < \angle B$, from which $c = a - 1$ and $b = a + 1$.
By the equation $(*)$, $a(a - 3) = 1$. In this case, the equation has no integer solution.

Hence, there exists a unique triangle that satisfies that $a = 4, b = 6$, $c = 5$, and $\cos A = \frac{3}{4}$.

【Score Situation】 This particular problem saw the following distribution of scores among contestants: 36 contestants scored 6 points, 26 contestants scored 5 points, 7 contestants scored 4 points, 6 contestants scored 3 points, 4 contestants scored 2 points, 3 contestants scored 1 point, and 9 contestants scored 0 point. The average score of this problem is 4.429, indicating that it was simple.

Among the top five teams in the team scores, the German Democratic Republic team achieved a total score of 304 points, the Soviet Union team achieved a total score of 298 points, the Hungary team achieved a total score of 291 points, the United Kingdom team achieved a total score of 263 points, and the Poland team achieved a total score of 262 points.

The gold medal cutoff for this IMO was set at 39 points (with 22 contestants earning gold medals), the silver medal cutoff was 33 points (with 22 contestants earning silver medals), and the bronze medal cutoff was 26 points (with 20 contestants earning bronze medals).

In this IMO, a total of 16 contestants achieved a perfect score of 40 points.

Problem 1.4 (IMO 33-4, proposed by France). In the plane let C be a circle, L a line tangent to the circle C, and M a point on L. Find the locus of all points P with the following property: there exist two points Q, R on L such that M is the midpoint of PQ and C is the inscribed circle of triangle PQR.

Solution. As shown in Figure 1.49, let $\odot C$ be tangent to L at point T, and TS be one diameter of $\odot C$.

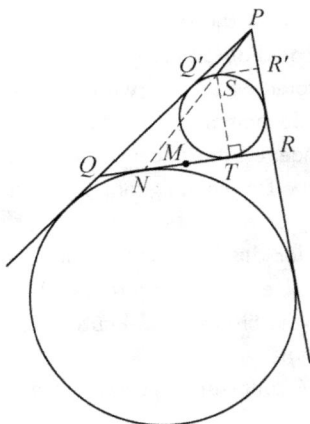

Figure 1.49

Construct the line crossing point S and parallel to QR, intersecting PR at point R' and PQ at point Q' respectively.

It is easy to know that point P is the homothetic center of $\triangle PQ'R'$ and $\triangle PQR$, and then $\odot C$ becomes the excircle of $\triangle PQ'R'$. The intersection point of the lines PS and QR is the point of tangency of excircle of $\triangle PQR$ (inside $\angle P$) on QR.

It is easy to know $QN = TR = \frac{1}{2}(PR + RQ - QP)$.

In this way, M is still the midpoint of NT.

So, we can take a point N such that $MN = MT$ and join NS, and extend it. Then P will be on this extended ray (excluding point S).

On the contrary, find a point P in the above ray and construct the tangent line of the circle $\odot C$. It will intersect the line MN. Let the intersection points be Q, R respectively, and it is easy to prove that M is the midpoint of QR.

In summary, the desired locus is an extension line of NS (the ray with S as its endpoint, but excluding point S).

【Score Situation】 This particular problem saw the following distribution of scores among contestants: 86 contestants scored 7 points, 34 contestants scored 6 points, 36 contestants scored 5 points, 10 contestants scored 4 points, 22 contestants scored 3 points, 39 contestants scored 2 points, 55 contestants scored 1 point, and 68 contestants scored 0 point. The average score of this problem is 3.500, indicating that it had a certain level of difficulty.

Among the top five teams in the team scores, the scores of this problem are as follows: the China team scored 38 points (with a total team score of 240 points), the United States team scored 36 points (with a total team score of 181 points), the Romania team scored 38 points (with a total team score of 177 points), the Commonwealth of Independent States team scored 26 points (with a total team score of 176 points), and the United Kingdom team scored 26 points (with a total team score of 168 points).

The gold medal cutoff for this IMO was set at 32 points (with 26 contestants earning gold medals), the silver medal cutoff was 24 points (with 55 contestants earning silver medals), and the bronze medal cutoff was 14 points (with 74 contestants earning bronze medals).

In this IMO, a total of 4 contestants achieved a perfect score of 42 points.

1.2.2 *Position structure problems*

Problem 1.5 (IMO 23-2, proposed by the Netherlands). A non-isosceles triangle $A_1A_2A_3$ is given with sides a_1, a_2, and a_3 (a_i is the side opposite A_i). For all $i = 1, 2, 3$, suppose M_i is the midpoint of side a_i, and T_i is the point where the incircle touches the side a_i. Denote by S_i the reflection of T_i in the interior bisector of angle A_i. Prove that the lines M_1S_1, M_2S_2, and M_3S_3 are concurrent.

Proof 1.   As shown in Figure 1.50, let A_1B_1 , A_2B_2, A_3B_3 be the three angular bisectors, B_i is on a_i, and I is the incenter. Thus

$$\angle T_2B_2I = \frac{1}{2}\angle A_2 + \angle A_3,$$

$$\angle T_2IB_2 = \angle S_2IB_2 = 90° - \frac{1}{2}\angle A_2 - \angle A_3,$$

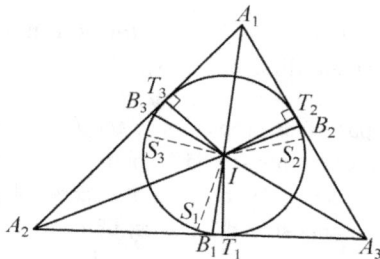

Figure 1.50

and then

$$\angle S_2 I T_1 = \angle T_2 I T_1 - 2 \angle T_2 I B_2$$
$$= 180° - \angle A_3 - (180° - \angle A_2 - 2\angle A_3)$$
$$= \angle A_2 + \angle A_3.$$

Similarly, $\angle S_3 I T_1 = \angle A_2 + \angle A_3$.

Since $IT_1 \perp A_2 A_3$ and $IS_2 = IS_3$ (easy to prove: S_1, S_2, S_3 are on the incircle), so $S_2 S_3 \parallel A_2 A_3$. Similarly, we can get $S_3 S_1 \parallel A_3 A_1$ and $S_1 S_2 \parallel A_1 A_2$.

Thus, the corresponding sides of $\triangle S_1 S_2 S_3$ and $\triangle M_1 M_2 M_3$ (not drawn) are parallel. Since triangle $\triangle A_1 A_2 A_3$ is a non-isosceles triangle, so $\triangle S_1 S_2 S_3$ and $\triangle M_1 M_2 M_3$ have no overlapping vertices, and then $M_1 S_1$, $M_2 S_2$, $M_3 S_3$ have a common point.

Proof 2. Let I be the incenter of $\triangle A_1 A_2 A_3$, and the angular bisectors of $\angle A_1$, $\angle A_2$, $\angle A_3$ intersect their opposite sides at points B_1, B_2, B_3, respectively. Then $\angle T_i I B_i = \angle B_i I S_i$ for $i = 1, 2, 3$ (as shown in Figure 1.51, where points M_1, M_2, M_3 are omitted).

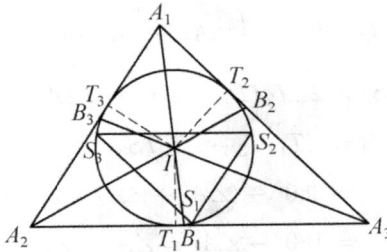

Figure 1.51

We first prove that the three sides of $\triangle M_1 M_2 M_3$ and $\triangle S_1 S_2 S_3$ are parallel and have no common sides. That is, $\triangle M_1 M_2 M_3$ and $\triangle S_1 S_2 S_3$ are similar or congruent, not in any of the cases in Figures 1.52–1.54.

Further, we prove that they are not congruent. In fact, they have a center, which is the intersection of segments $M_1 S_1, M_2 S_2$, and $M_3 S_3$.

Sides $M_1 M_2, M_2 M_3, M_3 M_1$ are parallel to sides $A_1 A_2, A_2 A_3, A_3 A_1$, respectively, and we further prove that sides $S_1 S_2, S_2 S_3, S_3 S_1$ are

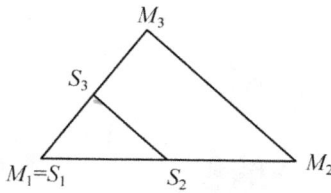

Figure 1.52

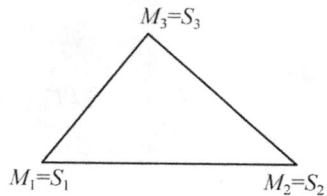

Figure 1.53

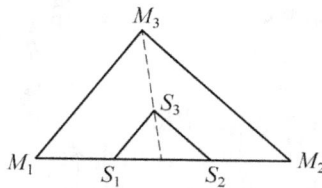

Figure 1.54

perpendicular to the line segments IT_3, IT_1, IT_2 respectively.

$$\angle T_1 I T_3 = 180° - \angle A_2,$$

$$\angle T_3 B_3 A_3 = \angle A_2 + \frac{1}{2}\angle A_3,$$

$$\angle T_3 I B_3 = 90° - \left(\angle A_1 + \frac{1}{2}\angle A_3\right),$$

$$\angle T_3 I S_3 = 2\angle T_3 I B_3 = 180° - 2\angle A_1 - \angle A_3,$$

$$\angle T_1 I S_3 = \angle T_1 I T_3 - \angle T_3 I S_3$$

$$= 360° - (\angle A_1 + \angle A_2 + \angle A_3) - \angle A_1$$

$$= 180° - \angle A_1 = \angle A_2 + \angle A_3.$$

Similarly, we have $\angle T_1 I S_2 = \angle A_2 + \angle A_3$.

Since $\angle T_1 I S_3 = \angle T_1 I S_2$, so IT_1 is perpendicular to $A_2 A_3$, and $IS_2 = IS_3$. Then $S_2 S_3 \parallel A_2 A_3$.

Similarly, we can get $S_1 S_3 \parallel A_1 A_3, S_1 S_2 \parallel A_1 A_2$.

Since $\triangle S_1 S_2 S_3$ is inscribed to the incircle of $\triangle A_1 A_2 A_3$, and $\triangle M_1 M_2 M_3$ is inscribed to the nine-point circle, $\triangle S_1 S_2 S_3$ and $\triangle M_1 M_2 M_3$ are not congruent. And since triangle $\triangle A_1 A_2 A_3$ is not an isosceles triangle, $\triangle S_1 S_2 S_3$ and $\triangle M_1 M_2 M_3$ are similar. Hence $M_1 S_1, M_2 S_2, M_3 S_3$ intersect at one point.

【Score Situation】 This particular problem saw the following distribution of scores among contestants: 8 contestants scored 7 points, 8 contestants scored 6 points, 3 contestants scored 5 points, 1 contestant scored 4 points, 2 contestants scored 3 points, 8 contestants scored 2 points, 9 contestants scored 1 point, and 80 contestants scored 0 point. The average score of this problem is 1.294, indicating that it was relatively challenging.

Among the top five teams in the team scores, the scores of this problem are as follows: the Germany team scored 16 points (with a total team score of 145 points), the Soviet Union team scored 16 points (with a total team score of 137 points), the German Democratic Republic team scored 23 points (with a total team score of 136 points), the United States team scored 9 points (with a total team score of 136 points), and the Vietnam team scored 17 points (with a total team score of 133 points).

The gold medal cutoff for this IMO was set at 37 points (with 10 contestants earning gold medals), the silver medal cutoff was 30 points (with 20 contestants earning silver medals), and the bronze medal cutoff was 21 points (with 31 contestants earning bronze medals).

In this IMO, only three contestants achieved a perfect score of 42 points, namely Bruno Haible from Germany, Grigori Perelman from the Soviet Union and Lê Tự Quốc Thắng from Vietnam.

Problem 1.6 (IMO 48-2, proposed by Luxembourg). Consider five points A, B, C, D, and E such that $ABCD$ is a parallelogram and $BCED$ is a cyclic quadrilateral. Let l be a line passing through A. Suppose that l intersects the interior of the segment DC at F and intersects line BC at G. Suppose also that $EF = EG = EC$. Prove that l is the bisector of angle DAB.

Proof. Draw the altitude lines of two isosceles triangles EGC and ECF as in Figure 1.55.

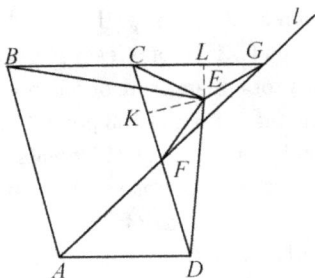

Figure 1.55

In view of the given condition, it is easy to see that $\triangle ADF \backsim \triangle GCF$. Hence

$$\frac{AD}{GC} = \frac{DF}{CF} \Rightarrow \frac{BC}{CG} = \frac{DF}{CF} \Rightarrow \frac{BC}{CL} = \frac{DF}{CK}$$

$$\Rightarrow \frac{BC + CL}{CL} = \frac{DF + FK}{CK}$$

$$\Rightarrow \frac{BL}{CL} = \frac{DK}{CK}.$$

So

$$\frac{BL}{DK} = \frac{CL}{CK}. \tag{1}$$

Since $BCED$ is a cyclic quadrilateral, $\angle LBE = \angle EDK$, and then this yields $\text{Rt}\triangle BLE \backsim \text{Rt}\triangle DKE$, so

$$\frac{BL}{DK} = \frac{EL}{EK}. \tag{2}$$

In view of (1) and (2), $\dfrac{CL}{CK} = \dfrac{EL}{EK}$, which means $\triangle CLE \backsim \triangle CKE$. Thus

$$\frac{CL}{CK} = \frac{CE}{CE} = 1,$$

i.e., $CL = CK$, so $CG = CF$.

It is intuitively obvious that $\angle BAG = \angle GAD$.

【Score Situation】This particular problem saw the following distribution of scores among contestants: 137 contestants scored 7 points, 8 contestants scored 6 points, 1 contestant scored 5 points, 5 contestants scored 4 points, 15 contestants scored 3 points, 26 contestants scored 2 points, 181 contestants scored 1 point, and 147 contestants scored 0 point. The average score of this problem is 2.519, indicating that it had a certain level of difficulty.

Among the top five teams in the team scores, the scores of this problem are as follows: the Russia team scored 42 points (with a total team score of 184 points), the China team scored 42 points (with a total team score of 181 points), the Vietnam team scored 42 points (with a total team score of 168 points), the South Korea team scored 41 points (with a total team score of 168 points), and the United States team scored 36 points (with a total team score of 155 points).

The gold medal cutoff for this IMO was set at 29 points (with 39 contestants earning gold medals), the silver medal cutoff was 21 points (with 83 contestants earning silver medals), and the bronze medal cutoff was 14 points (with 131 contestants earning bronze medals).

In this IMO, no contestant achieved a perfect score of 42 points.

Problem 1.7 (IMO 49-6, proposed by Russia). Let $ABCD$ be a convex quadrilateral with $|BA| \neq |BC|$. Denote the incircles of triangles ABC and ADC by ω_1 and ω_2 respectively. Suppose that there exists a circle ω tangent to the ray BA beyond A and to the ray BC beyond C, which is also tangent to the lines AD and CD. Prove that the common external tangents of ω_1 and ω_2 intersect on ω.

Proof (By Xiaosheng Mu).

Lemma 1. *Let $ABCD$ be a convex quadrilateral, and a circle ω be tangent to ray BA beyond A and to the ray BC beyond C, which is also tangent to the lines AD and CD. Then $AB + AD = CB + CD$.*

Proof of Lemma 1. Let ω meet AB, BC, CD, DA at P, Q, R, S respectively. As show in Figure 1.56,

$$AB + AD = CB + CD \Leftrightarrow AB + (AD + DS) = CB + (CD + DR)$$
$$\Leftrightarrow AB + AS = CB + CR \Leftrightarrow AB + AP = CB + CQ$$
$$\Leftrightarrow BP = BQ.$$

This ends the proof of Lemma 1.

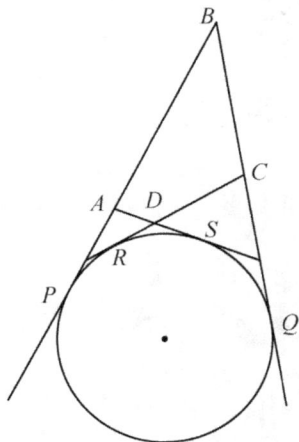

Figure 1.56

Lemma 2. *If the radii of three circles $\odot O_1, \odot O_2, \odot O_3$ differ to each other, then their homothetic centers are collinear.*

Proof of Lemma 2. Let X_3 be the homothetic center of $\odot O_1$ and $\odot O_2$, X_2 be the homothetic center of $\odot O_1$ and $\odot O_3$, X_1 be the homothetic center of $\odot O_2$ and $\odot O_3$, and r_i be the radius $\odot O_i$ $(i = 1, 2, 3)$. By the property of homothety

$$\frac{\overline{O_1 X_3}}{\overline{X_3 O_2}} = -\frac{r_1}{r_2}.$$

Here $\overline{O_1 X_3}$ denotes the direct segment $O_1 X_3$, as show in Figure 1.57.

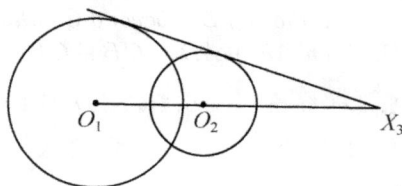

Figure 1.57

Similarly

$$\frac{\overline{O_2 X_1}}{\overline{X_1 O_3}} = -\frac{r_2}{r_3}, \quad \frac{\overline{O_3 X_2}}{\overline{X_2 O_1}} = -\frac{r_3}{r_1},$$

so

$$\frac{\overline{O_1 X_3}}{\overline{X_3 O_2}} \cdot \frac{\overline{O_2 X_1}}{\overline{X_1 O_3}} \cdot \frac{\overline{O_3 X_2}}{\overline{X_2 O_1}} = \left(-\frac{r_1}{r_2}\right)\left(-\frac{r_2}{r_3}\right)\left(-\frac{r_3}{r_1}\right) = -1,$$

By Menelaus's theorem, X_1, X_2, X_3 are collinear.

Let ω_1, ω_2 meet AC at U, V respectively. As show in Figure 1.58,

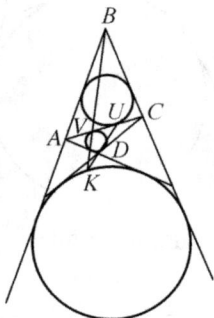

Figure 1.58

$$AV = \frac{AD + AC - CD}{2} = \frac{AC}{2} + \frac{AD - CD}{2}$$

$$= \frac{AC}{2} + \frac{CB - AB}{2} = \frac{AC + CB - AB}{2} \quad \text{(by Lemma 1)}$$

$$= CU.$$

Hence the excircle ω_3 of $\triangle ABC$ on the side AC meets AC at V. Therefore ω_2, ω_3 meet at point V, i.e., V is the homothetic center of ω_2, ω_3. Denote the homothetic center of ω_1, ω_2 by K (i.e., K is the intersection of the external common tangent to ω_2, ω_3), by Lemma 2, K, V, B are collinear.

Similarly, K, D, U are collinear.

Since $BA \neq BC$, we have $U \neq V$ (otherwise, $U = V$ is the midpoint of side AC by $AV = CU$. It is contradictory to $BA \neq BC$). So BV does not coincide with DU, i.e., $K = BV \cap DU$.

Now we prove K is on the circle ω.

Construct the tangent l of ω which is parallel to AC (and on the same side of ω as AC). Let that l meet ω at T. We show B, V, T are collinear.

As shown in Figure 1.59, l intersects BA, BC at A_1, C_1 respectively. Then ω is the excircle of $\triangle BA_1C_1$ on the side A_1C_1 and meets A_1C_1 at T; meanwhile ω_3 is the excircle of $\triangle ABC$ on the side AC and meets AC at V. Since $AC \parallel A_1C_1$, the point B is the homothetic center of $\triangle BAC$ and $\triangle BA_1C_1$, and V and T are corresponding points, so B, V, T are collinear.

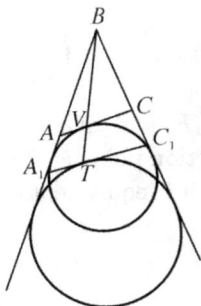

Figure 1.59

Similarly, D, U, T are collinear. This means $K = T$.

【Score Situation】This particular problem saw the following distribution of scores among contestants: 12 contestants scored 7 points, 1 contestant scored 6 points, no contestant scored 5 points, no contestant scored 4 points, no contestant scored

3 points, 9 contestants scored 2 points, 31 contestants scored 1 point, and 482 contestants scored 0 point. The average score of this problem is 0.260, indicating that it was extremely difficult.

Among the top five teams in the team scores, the scores of this problem are as follows: the China team scored 14 points (with a total team score of 217 points), the Russia team scored 5 points (with a total team score of 199 points), the United States team scored 15 points (with a total team score of 190 points), the South Korea team scored 2 points (with a total team score of 188 points), and the Iran team scored 9 points (with a total team score of 181 points).

The gold medal cutoff for this IMO was set at 31 points (with 47 contestants earning gold medals), the silver medal cutoff was 22 points (with 100 contestants earning silver medals), and the bronze medal cutoff was 15 points (with 120 contestants earning bronze medals).

In this IMO, only three contestants achieved a perfect score of 42 points, namely Xiaosheng Mu and Dongyi Wei from China, and Alex Zhai from the United States.

Problem 1.8 (IMO 51-2, proposed by Chinese Hong Kong). Let I be the incentre of triangle ABC and let Γ be its circumcircle. Let the line AI intersect Γ again at D. Let E be a point on the arc $\widehat{BDC}$ and F a point on the side BC such that

$$\angle BAF = \angle CAE < \frac{1}{2}\angle BAC.$$

Finally, let G be the midpoint of the segment IF. Prove that the lines DG and EI intersect on Γ.

Proof.   As show in Figure 1.60, let I_a be the center of the exscribed circle of the triangle ABC with respect to the side BC. Then D is the midpoint of the segment $I\,I_a$, $DG \parallel I_aF$, and $\angle GDA = \angle FI_aA$.

To prove that the intersection point P of the lines DG and EI lies on Γ, i.e., A, E, D, P are concyclic, it is equivalent to prove that $\angle GDA = \angle IEA$, or similarly $\angle FI_aA = \angle IEA$.

By the assumption $\angle BAF = \angle CAE$, we have $\triangle ABF \backsim \triangle AEC$, and hence

$$\frac{AF}{AB} = \frac{AC}{AE}. \tag{1}$$

Since

$$\angle AIB = \angle C + \frac{1}{2}(\angle A + \angle B),$$

$$\angle ACI_a = \angle C + \frac{1}{2}(\angle A + \angle B).$$

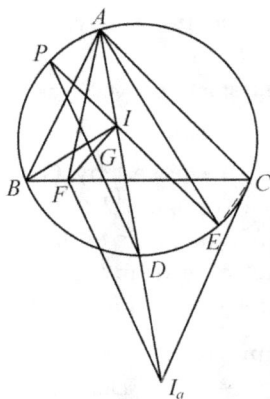

Figure 1.60

Also $\triangle ABI \backsim \triangle AI_a C$, and thus

$$\frac{AI}{AB} = \frac{AC}{AI_a}. \tag{2}$$

From (1) and (2),

$$\frac{AF}{AI} = \frac{AI_a}{AE}. \tag{3}$$

By the assumption $\angle FAI_a = \angle EAI$, we have $\triangle AFI_a \backsim \triangle AIE$, and therefore $\angle FI_a A = \angle IEA$, which completes the proof.

【Score Situation】 This particular problem saw the following distribution of scores among contestants: 161 contestants scored 7 points, 2 contestants scored 6 points, 4 contestants scored 5 points, 2 contestants scored 4 points, 8 contestants scored 3 points, 23 contestants scored 2 points, 93 contestants scored 1 point, and 223 contestants scored 0 point. The average score of this problem is 2.578, indicating that it had a certain level of difficulty.

Among the top five teams in the team scores, the scores of this problem are as follows: the China team scored 42 points (with a total team score of 197 points), the Russia team scored 35 points (with a total team score of 169 points), the United States team scored 42 points (with a total team score of 168 points), the South Korea team scored 42 points (with a total team score of 156 points), the Thailand team scored 42 points (with a total team score of 148 points), and the Kazakhstan team scored 35 points (with a total team score of 148 points).

The gold medal cutoff for this IMO was set at 27 points (with 47 contestants earning gold medals), the silver medal cutoff was 21 points (with 103 contestants

earning silver medals), and the bronze medal cutoff was 15 points (with 115 contestants earning bronze medals).

In this IMO, only one contestant achieved a perfect score of 42 points, namely Zipei Nie from China.

Problem 1.9 (IMO 55-4, proposed by Georgia). Points P and Q lie on side BC of acute-angled triangle ABC so that $\angle PAB = \angle BCA$ and $\angle CAQ = \angle ABC$. Points M and N lie on lines AP and AQ, respectively, such that P is the midpoint of AM and Q is the midpoint of AN. Prove that lines BM and CN intersect on the circumcircle of triangle ABC.

Solution. As show in Figure 1.61, let S be the intersection point of lines BM and CN; see the figure below. Denote $\beta = \angle QAC = \angle CBA$ and $\gamma = \angle PAB = \angle ACB$. Then $\triangle ABP \backsim \triangle CAQ$, and thus

$$\frac{BP}{PM} = \frac{BP}{PA} = \frac{AQ}{QC} = \frac{NQ}{QC}.$$

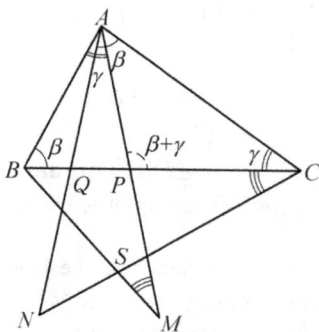

Figure 1.61

By $\angle BPM = \beta + \gamma = \angle CQN$, we have $\triangle BPM \backsim \triangle NQC$, from which $\angle BMP = \angle NCQ$.

Consequently, $\triangle BPM \backsim \triangle BSC$.

Therefore, $\angle CSB = \angle BPM = \beta + \gamma = 180° - \angle BAC$.

【Score Situation】This particular problem saw the following distribution of scores among contestants: 378 contestants scored 7 points, 3 contestants scored 6 points, 3 contestants scored 5 points, 5 contestants scored 4 points, 16 contestants scored 3 points, 28 contestants scored 2 points, 103 contestants scored 1 point, and 24 contestants scored 0 point. The average score of this problem is 5.189, indicating that it was simple.

Among the top five teams in the team scores, the scores of this problem are as follows: the China team scored 42 points (with a total team score of 201 points), the United States team scored 42 points (with a total team score of 193 points), the Chinese Taiwan team scored 42 points (with a total team score of 192 points), the Russia team scored 42 points (with a total team score of 191 points), and the Japan team scored 37 points (with a total team score of 177 points).

The gold medal cutoff for this IMO was set at 29 points (with 49 contestants earning gold medals), the silver medal cutoff was 22 points (with 113 contestants earning silver medals), and the bronze medal cutoff was 16 points (with 133 contestants earning bronze medals).

In this IMO, only three contestants achieved a perfect score of 42 points, namely Jiyang Gao from China, Po-Sheng Wu from Chinese Taiwan and Alexander Gunning from Australia.

1.2.3 *Quantity relation problems*

Problem 1.10 (IMO 6-3, proposed by Yugoslavia). A circle is inscribed in triangle ABC with sides a, b, c. Tangents to the circle parallel to the sides of the triangle are constructed. Each of these tangents cuts off a triangle from $\triangle ABC$. In each of these triangles, a circle is inscribed. Find the sum of the areas of all four inscribed circles (in terms of a, b, c).

Solution 1. As shown in Figure 1.62, let the radii of the four incircles be r, r_a, r_b, r_c respectively. Consider $\triangle APQ$. Since $\triangle APQ \backsim \triangle ABC$,

$$\frac{h_a - 2r}{h_a} = \frac{r_a}{r},$$

where h_a is the altitude on the side of BC in $\triangle ABC$.

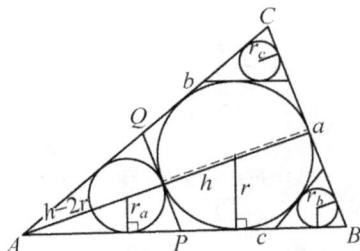

Figure 1.62

Therefore, the sum of the required areas is

$$\pi r^2 + \pi r^2 \sum \left(\frac{h_a - 2r}{h_a}\right)^2 = \pi r^2 \left[1 + \sum \left(\frac{\frac{2S_\triangle}{a} - \frac{2S_\triangle}{p}}{\frac{2S_\triangle}{a}}\right)^2\right]$$

$$= \pi r^2 \left[1 + \sum \left(\frac{p-a}{p}\right)^2\right] = \frac{\pi r^2}{p^2}(a^2 + b^2 + c^2)$$

$$= \frac{\pi}{p^3}(p-a)(p-b)(p-c)(a^2 + b^2 + c^2),$$

where $S_\triangle$ is the area of $\triangle ABC$, $p = \frac{1}{2}(a + b + c)$.

Solution 2. Let KL be the tangent line of the incircle of the triangle ABC and parallel to the side BC, and then, let r be the radius of the incircle of the triangle ABC and r_A be the radius of the incircle of the triangle AKL. The definitions of r_B, r_C are similar to r_A (see Figure 1.63).

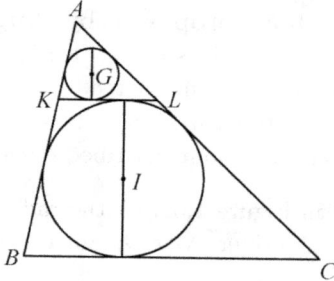

Figure 1.63

Triangles ABC and AKL are similar, and the ratio of the altitudes of these two triangles is the similarity ratio, which is $\frac{r}{r_A}$.

Let h be the altitude of the point A in the triangle ABC. Then

$$\frac{h}{h - 2r} = \frac{r}{r_A}.$$

Thus,

$$r_A = \frac{r(h - 2r)}{h} = r - \frac{2r^2}{h}.$$

By the formula $2S = ah = 2pr$, we get $r_A = \frac{r}{p}(p - a)$. By the same method, $r_B = \frac{r}{p}(p - b)$, $r_C = \frac{r}{p}(p - c)$.

Therefore, the sum of the required areas is

$$T = \pi(r^2 + r_A^2 + r_B^2 + r_C^2) = \pi r^2 \left[1 + \frac{(p-a)^2 + (p-b)^2 + (p-c)^2}{p^2} \right]$$

$$= \pi r^2 \left[4 + \frac{a^2 + b^2 + c^2 - 2p(a+b+c)}{p^2} \right] = \frac{\pi r^2}{p^2}(a^2 + b^2 + c^2)$$

$$= \frac{\pi(a^2 + b^2 + c^2)(p-a)(p-b)(p-c)}{p^3}$$

$$= \frac{\pi(a^2 + b^2 + c^2)(b+c-a)(a+b-c)(a+c-b)}{(a+b+c)^3}.$$

【Score Situation】 This particular problem saw the following distribution of scores among contestants: 11 contestants scored 6 points, 4 contestants scored 5 points, no contestant scored 4 points, no contestant scored 3 points, no contestant scored 2 points, no contestant scored 1 point, and 2 contestants scored 0 point. The average score of this problem is 5.059, indicating that it was simple.

Among the top five teams in the team scores, the scores of this problem are as follows: the Soviet Union team scored 48 points (with a total team score of 269 points), the Hungary team scored 48 points (with a total team score of 253 points), the Romania team scored 40 points (with a total team score of 213 points), the Poland team scored 42 points (with a total team score of 209 points), and the Bulgaria team scored 41 points (with a total team score of 198 points).

The gold medal cutoff for this IMO was set at 38 points (with seven contestants earning gold medals), the silver medal cutoff was 31 points (with nine contestants earning silver medals), and the bronze medal cutoff was 27 points (with 19 contestants earning bronze medals).

In this IMO, only one contestant achieved a perfect score of 42 points, namely David Bernstein from the Soviet Union.

Problem 1.11 (IMO 24-2, proposed by the Soviet Union). Let A be one of the two distinct points of the intersection of two unequal coplanar circles C_1 and C_2 with centres O_1 and O_2 respectively. One of the common tangents to the circles touches C_1 at P_1 and C_2 at P_2, while the other touches C_1 at Q_1 and C_2 at Q_2. Let M_1 be the mid-point of P_1Q_1 and M_2 be the mid-point of P_2Q_2. Prove that $\angle O_1AO_2 = \angle M_1AM_2$.

Proof 1. As shown in Figure 1.64, let O be the intersection of lines P_1P_2 and Q_1Q_2. The two circles are homothetic and O is their homothetic center. Under this homothety, three points M_1, O_1, A correspond to M_2, O_2, C, respectively. Then $AM_1 \parallel CM_2$, $AO_1 \parallel CO_2$. Thus $\angle O_1AM_1 = \angle O_2CM_2$.

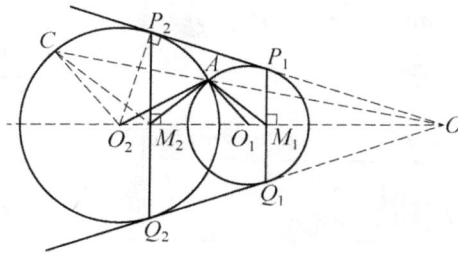

Figure 1.64

And connect O_2P_2. By the geometric mean theorem and tangent-secant theorem,

$$OM_2 \cdot OO_2 = OP_2^2 = OA \cdot OC.$$

Hence, A, C, O_2, M_2 are concyclic. Then

$$\angle O_2AM_2 = \angle O_2CM_2 = \angle O_1AM_1 (= \alpha),$$

so

$$\angle O_1AO_2 = \alpha + \angle M_2AO_1 = \angle M_1AM_2.$$

Proof 2. Let O be the intersection of P_1P_2, Q_1Q_2, O_1O_2. The two circles are homothetic and O is their homothetic center. Let B be the second intersection of the two circles C_1, C_2 (see Figure 1.65).

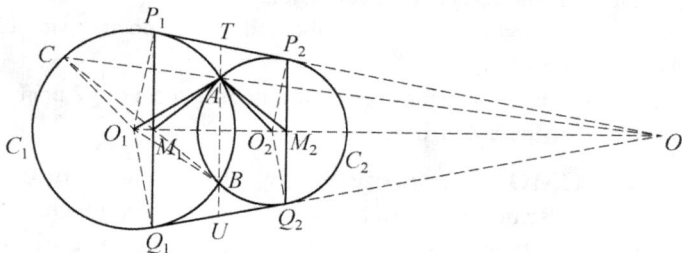

Figure 1.65

The line AB intersects P_1P_2, Q_1Q_2 at points T, U, respectively. By the power of a point, we get

$$TA \cdot TB = TP_1^2 = TP_2^2.$$

Thus, $TP_1 = TP_2$.

Since TU is perpendicular to O_1O and is the midsegment of the trapezoidal $P_1Q_1Q_2P_2$, the segment AB is the perpendicular bisector of the segment M_1M_2. Thus

$$\angle AM_1M_2 = \angle M_1M_2A = x.$$

Let C be the second intersection point of OA and the circle C_1. Since $\angle BM_1M_2 = x$, and C is the image of the homothetic transformation of A, the three points B, M_1, C are collinear. From the reflectivity of line OO_1,

$$\angle O_1AM_1 = \angle O_1BM_1 = \angle O_1CM_1 = y,$$

and $\angle O_1CM_1 = y = \angle O_2AM_2$. Hence $\angle O_1AO_2 = \angle M_1AM_2$.

【Score Situation】 This particular problem saw the following distribution of scores among contestants: 11 contestants scored 7 points, 3 contestants scored 6 points, no contestant scored 5 points, no contestant scored 4 points, 1 contestant scored 3 points, 7 contestants scored 2 points, 2 contestants scored 1 point, and 16 contestants scored 0 point. The average score of this problem is 2.850, indicating that it had a certain level of difficulty.

Among the top five teams in the team scores, the Germany team achieved a total score of 212 points, the United States team achieved a total score of 171 points, the Hungary team achieved a total score of 170 points, the Soviet Union team achieved a total score of 169 points, and the Romania team achieved a total score of 161 points.

The gold medal cutoff for this IMO was set at 38 points (with nine contestants earning gold medals), the silver medal cutoff was 26 points (with 27 contestants earning silver medals), and the bronze medal cutoff was 15 points (with 57 contestants earning bronze medals).

In this IMO, a total of four contestants achieved a perfect score of 42 points.

Problem 1.12 (IMO 31-1, proposed by India). Chords AB and CD of a circle intersect at a point E inside the circle. Let M be an interior point of the segment EB. The tangent line at E to the circle through D, E, and M intersects the lines BC and AC at F and G, respectively. If $\frac{AM}{AB} = t$, find $\frac{EG}{EF}$ in terms of t.

Solution. As shown in Figure 1.66, connect to form AD, MD, and BD.
It is easy to know that

$$\angle CEF = \angle DEG = \angle EMD,$$

$$\angle ECF = \angle MAD.$$

Thus $\triangle CEF \backsim \triangle AMD$, and then $CE \cdot MD = AM \cdot EF$.

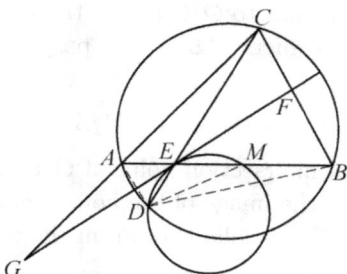

Figure 1.66

In addition,

$$\angle ECG = \angle MBD,$$

$$\angle CEG = \angle 180° - \angle GED$$

$$= 180° - \angle EMD$$

$$= \angle BMD,$$

so $\triangle CGE \backsim \triangle BDM$, from which $GE \cdot MB = CE \cdot MD$.
In this way, we get $GE \cdot MB = AM \cdot EF$. Therefore,

$$\frac{EG}{EF} = \frac{AM}{MB} = \frac{t \cdot AB}{(1-t) \cdot AB} = \frac{t}{1-t}.$$

【Score Situation】 This particular problem saw the following distribution of scores among contestants: 71 contestants scored 7 points, 8 contestants scored 6 points, 14 contestants scored 5 points, 26 contestants scored 4 points, 19 contestants scored 3 points, 28 contestants scored 2 points, 54 contestants scored 1 point, and 88 contestants scored 0 point. The average score of this problem is 2.877, indicating that it had a certain level of difficulty.

Among the top five teams in the team scores, the scores of this problem are as follows: the China team scored 42 points (with a total team score of 230 points), the Soviet Union team scored 26 points (with a total team score of 193 points), the United States team scored 23 points (with a total team score of 174 points), the Romania team scored 37 points (with a total team score of 171 points), and France team scored 30 points (with a total team score of 168 points).

The gold medal cutoff for this IMO was set at 34 points (with 23 contestants earning gold medals), the silver medal cutoff was 23 points (with 56 contestants earning silver medals), and the bronze medal cutoff was 16 points (with 76 contestants earning bronze medals).

In this IMO, a total of four contestants achieved a perfect score of 42 points.

Problem 1.13 (IMO 34-2, proposed by the United Kingdom). Let D be a point inside an acute-angled triangle ABC such that

$$\angle ADB = \angle ACB + 90° \quad \text{and} \quad AC \cdot BD = AD \cdot BC.$$

(a) Calculate the value of the ratio $\frac{AB \cdot CD}{AC \cdot BD}$.
(b) Prove that the tangents at C to the circumcircles of the triangles ACD and BCD are perpendicular.

Solution. (a) As shown in Figure 1.67, construct $\triangle CEB \backsim \triangle CDA$, where $\angle CAD = \angle CBE$, $\angle ACD = \angle BCE$.

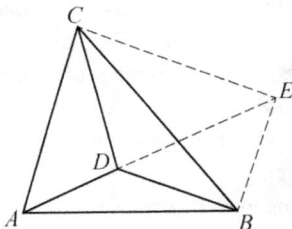

Figure 1.67

By the condition and similarity,

$$\frac{BE}{BC} = \frac{AD}{AC} = \frac{BD}{BC}.$$

Thus, $BE = BD$.
Also

$$\angle ADB = \angle CAD + \angle CBD + \angle ACB$$

$$= \angle DBE + \angle ACB$$

$$= 90° + \angle ACB,$$

and then $\angle DBE = 90°$, $\triangle BDE$ is an isosceles right triangle.
It is easy to know $\angle ACB = \angle DCE$, $\frac{AC}{CD} = \frac{BC}{CE}$, so $\triangle ACB \backsim \triangle DCE$.
Thus

$$\frac{AB \cdot CD}{AC \cdot BD} = \frac{DE}{BD} = \sqrt{2}.$$

(b) Suppose that the tangent lines of the circumcircles of $\triangle ACD, \triangle BCD$ at point C are CG and CF, respectively (not drawn in the figure), such that $\angle FCD = \angle DBC, \angle GCD = \angle CAD$. Thus

$$\angle FCG = \angle FCD + \angle GCD$$
$$= \angle DBC + \angle CAD$$
$$= \angle ADB - \angle ACB$$
$$= 90°,$$

that is, the two tangent lines are perpendicular.

【Score Situation】 This particular problem saw the following distribution of scores among contestants: 57 contestants scored 7 points, 6 contestants scored 6 points, 6 contestants scored 5 points, 12 contestants scored 4 points, 19 contestants scored 3 points, 98 contestants scored 2 points, 32 contestants scored 1 point, and 183 contestants scored 0 point. The average score of this problem is 1.932, indicating that it was relatively challenging.

Among the top five teams in the team scores, the scores of this problem are as follows: the China team scored 42 points (with a total team score of 215 points), the Germany team scored 13 points (with a total team score of 189 points), the Bulgaria team scored 29 points (with a total team score of 178 points), the Russia team scored 23 points (with a total team score of 177 points), and the Chinese Taiwan team scored 23 points (with a total team score of 162 points).

The gold medal cutoff for this IMO was set at 30 points (with 35 contestants earning gold medals), the silver medal cutoff was 20 points (with 66 contestants earning silver medals), and the bronze medal cutoff was 11 points (with 97 contestants earning bronze medals).

In this IMO, only two contestants achieved a perfect score of 42 points, namely Hong Zhou from China and Hung-Wu Wu from Chinese Taiwan.

Problem 1.14 (IMO 45-5, proposed by Poland). In a convex quadrilateral $ABCD$ the diagonal BD does not bisect the angles ABC and CDA. The point P lies inside $ABCD$ and satisfies $\angle PBC = \angle DBA$ and $\angle PDC = \angle BDA$. Prove that $ABCD$ is a cyclic quadrilateral if and only if $AP = CP$.

Proof 1.   When four points A, B, C, D are concyclic, let point P be inside the triangle BCD. As shown in Figure 1.68, take a point E on DC or its extension such that $\angle PEC = \angle PBC$. Then, draw lines PE and BE.

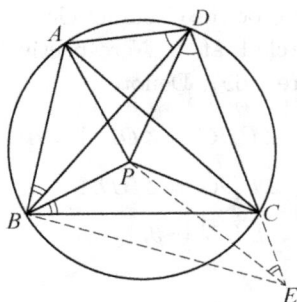

Figure 1.68

It is easy to know that the four points P, B, E, C are concyclic. By the law of sines and the fact that the four points A, B, C, D are concyclic,

$$\frac{BE}{\sin \angle BAD} = \frac{BE}{\sin \angle BCE} = \frac{PC}{\sin \angle PBC} = \frac{PC}{\sin \angle ABD}.$$

Hence

$$\frac{PC}{BE} = \frac{\sin \angle ABD}{\sin \angle BAD} = \frac{AD}{BD}.$$

Note that the coincidence of points C and E does not affect this conclusion.

Since triangles ABD and PED are similar, triangles DAP and DBE are similar, too. So

$$\frac{AP}{BE} = \frac{AD}{BD}.$$

Therefore $AP = CP$.

Conversely, according to the above derivation, if $AP = CP$, then $\sin \angle BAD = \sin \angle BCD$, so if $\angle BAD + \angle BCD = 180°$, then A, B, C, D are concyclic. If $\angle BAD = \angle BCD$, then since point P is not on the line BD,

$$180° > \angle BPD = \angle PBC + \angle PDC + \angle BCD$$

$$= \angle DBA + \angle BDA + \angle BAD = 180°.$$

This is a contradiction.

Note. If A, B, C, D are concyclic, then extend BP, DP to intersect $\overset{\frown}{CD}, \overset{\frown}{BC}$ on Q, R respectively, deducing that the quadrilateral $DBRQ$ is an isosceles trapezoidal and $DQ \parallel AC$. In this way, $AP = CP$ can also be obtained, but the reverse is not possible.

Proof 2. Let the circle Γ be the circumcircle of $\triangle BCD$. Extend BP, DP beyond P to meet the circle Γ at M, N respectively. Draw lines MC, MD, MN, NB, NC (see Figure 1.69). Denote

$$\angle PBC = \angle DBA = \alpha,$$
$$\angle PDC = \angle BDA = \beta,$$
$$\angle BDN = \theta.$$

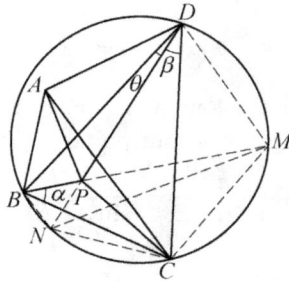

Figure 1.69

Then $\angle CNM = \alpha$, $\angle NMC = \beta$. Thus $\triangle ABD \backsim \triangle CNM$,

$$\frac{AD}{CM} = \frac{BD}{NM}. \tag{1}$$

Since $\triangle BDP \backsim \triangle NMP$,

$$\frac{PD}{PM} = \frac{BD}{NM}. \tag{2}$$

Combining (1) and (2) gives

$$\frac{PD}{PM} = \frac{AD}{CM}.$$

Then $\angle ADP = \beta + \theta$ and $\angle CMP = \beta + \theta$, so $\triangle ADP \backsim \triangle CMP$, and hence

$$AP = CP \Leftrightarrow \triangle ADP \cong \triangle CMP$$
$$\Leftrightarrow PD = PM \Leftrightarrow \angle BCD = \angle PMD = \angle MDP$$
$$\Leftrightarrow \angle BCD + \angle BAD = 180°$$
$$\Leftrightarrow A, B, C, D \quad \text{are concyclic.}$$

【Score Situation】This particular problem saw the following distribution of scores among contestants: 64 contestants scored 7 points, 13 contestants scored 6 points, 18 contestants scored 5 points, 35 contestants scored 4 points, 124 contestants scored 3 points, 18 contestants scored 2 points, 57 contestants scored 1 point, and 157 contestants scored 0 point. The average score of this problem is 2.512, indicating that it had a certain level of difficulty.

Among the top five teams in the team scores, the scores of this problem are as follows: the China team scored 37 points (with a total team score of 220 points), the United States team scored 35 points (with a total team score of 212 points), the Russia team scored 35 points (with a total team score of 205 points), the Vietnam team scored 38 points (with a total team score of 196 points), and the Bulgaria team scored 36 points (with a total team score of 194 points).

The gold medal cutoff for this IMO was set at 32 points (with 45 contestants earning gold medals), the silver medal cutoff was 24 points (with 78 contestants earning silver medals), and the bronze medal cutoff was 16 points (with 120 contestants earning bronze medals).

In this IMO, a total of four contestants achieved a perfect score of 42 points.

1.3 Summary

Similarity and congruence are important contents in the study of geometric structures and quantitative relations.

In the first 64 IMOs, there were 14 problems on similarity and congruence, as depicted in Figure 1.70, which can be broadly categorized into three types.

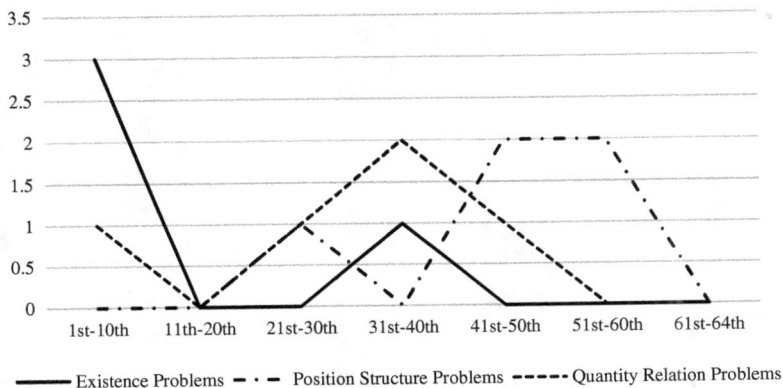

Figure 1.70 Numbers of Similarity and Congruence Problems in the First 64 IMOs

Table 1.2 Score Details of Similarity and Congruence Problems in the First 64 IMOs

Problem	1.1	1.2	1.3	1.4	1.5	1.6	1.7
Full points	5.000	6.000	6.000	7.000	7.000	7.000	7.000
Average score	3.600	4.243	4.429	3.500	1.294	2.519	0.260
Top five mean				5.467	4.050	6.767	1.500
6th–15th mean				4.98	0.975	5.57	0.83
16th–25th mean				4.58	0.8	3.52	0.5
Problem number in the IMO	2-4	9-4	10-1	33-4	23-2	48-2	49-6
Proposing country	Hungary	Italy	Romania	France	Netherlands	Luxembourg	Russia

Problem	1.8	1.9	1.10	1.11	1.12	1.13	1.14
Full points	7.000	7.000	6.000	7.000	7.000	7.000	7.000
Average score	2.578	5.189	5.059	2.850	2.877	1.932	2.512
Top five mean	6.611	6.833	5.475		5.267	4.333	6.033
6th–15th mean	4.96	6.75			3.727	3.71	4.5
16th–25th mean	4.12	6.7			3.093	2.43	3
Problem number in the IMO	51-2	55-4	6-3	24-2	31-1	34-2	45-5
Proposing country or region	Chinese Hong Kong	Georgia	Yugoslavia	Soviet Union	India	United Kingdom	Poland

Note. Top five Mean = Total score of the top five teams / Total number of contestants from the top five teams,

6th–15th Mean = Total score of the 6th–15th teams / Total number of contestants from the 6th–15th teams,

16th–25th Mean = Total score of the 16th–25th teams / Total number of contestants from the 16th–25th teams.

Problems 1.1–1.4 focus on "existence problems;" among these four problems, the one with the lowest average score is Problem 1.4 (IMO 33-4), proposed by France. Problems 1.5–1.9 deal with "position structure problems;" among these 5 problems, the one with the lowest average score is Problem 1.7 (IMO 49-6), proposed by Russia. Problems 1.10–1.14 are about "quantity relation problems;" among these five problems, the one with the lowest average score is Problem 1.13 (IMO 34-2), proposed by the United Kingdom.

These problems were proposed by 14 countries and regions. From Table 1.2, it can be observed that in the first 64 IMOs, there was one problem with an average score of 0–1 point; two problems with an average score of 1–2 points; five problems with an average score of 2–3 points; two problems with an average score of 3–4 points; four problems with an average score above 4 points. Problems with an average score of more than 3 points are mostly concentrated in the first 30 IMOs, and overall, the similarity and congruence problems had a certain level of difficulty. Of these, Problem 1.7 (IMO 49-6, proposed by Russia) had the lowest average score.

In the 24th to 64th IMOs, there were a total of 8 similarity and congruence problems. Among these, one had an average score of 0–1 point; one had an average score of 1–2 points; three had an average score of 2–3 points; one had an average score of 3–4 points; two had an average score above 4 points. Further analysis of the problem numbers of these 8 similarity and congruence problems, as shown in Table 1.3, reveals that these problems frequently appeared as the 1st/4th or 2nd/5th problem. The majority of these problems, totaling 7, were of the type position structure problems and quantity relation problems. The type of existence problems was less frequent, with only one problem appearing.

Table 1.3 Numbers of Similarity and Congruence Problems in the 24th–64th IMOs

Similarity and Congruence Problem	Problem Number			Number of Problems in the First 64 IMOs
	1,4	2,5	3,6	
Existence problems	1	0	0	4
Position structure problems	1	2	1	5
Quantity relation problems	1	2	0	5
Total	3	4	1	14

From Table 1.2, it can be observed that, in most cases, the average score of the top five teams is typically about 2 to 4 points higher than the average score of the problem. The average scores of the 6th–15th teams and 16th–25th teams also tend to be higher than the average score of the problem. However, in the difficult problems, the disparity between the average scores of the 16th–25th teams and the average score of the problem is not significant, as seen in problems such as Problem 1.5 (IMO 23-2), Problem 1.7 (IMO 49-6), Problem 1.12 (IMO 31-1), and Problem 1.13 (IMO 34-2).

In addition, it should be noted that the number of teams participating in early IMOs was small. It was not until the 22nd IMO in 1981 that the number of participating teams exceeded 25. And it was not until the 30th IMO in 1989 that the number of participating teams exceeded 50.

Chapter 2

Basic Properties of Circles and Four Points on a Circle

A circle is the most perfect figure in plane geometry. It is symmetric about every line passing through the center of the circle, and the distance from the center to every point on the circle is the same. Circles also constitute an important content of plane geometry. The basic knowledge of circles includes the definition of a circle, invariance of rotations, symmetry of axes, relations between chords, arcs, and diameters of circles, theorems of angles related to circles (the inscribed angle, central angle, and chord-tangent angle), properties of tangent lines of circles, tangent length formulas, the intersecting chords theorem, tangent-secant theorem, secants theorem, position relations between lines and circles, position relations between two circles, calculation formulas of the area and arc length of circles, etc.

If in a plane, four points are on a circle, then the four points are said to be concyclic, and the four points on a circle have some nice properties that are quite useful in solving problems involving circles. The four points on a circle are not only an important topic in plane geometry, but also a medium for the mutual transformation of a measurement relation or positional relation between straight lines and circles.

In the first 64th IMOs, there had been a total of 32 problems related to basic properties of circles and four points on a circle, accounting for 26.0% of all geometry problems. These problems are mainly categorized into three types: (1) existence problems, totaling six problems; (2) position structure problems, totaling 18 problems; (3) quantity relation problems, totaling eight problems. The statistical distribution of these three types of problems in the previous IMOs is presented in Table 2.1.

Table 2.1 Numbers of Basic Properties of Circles and Four Points on a Circle Problems in the First 64 IMOs

Content	Session							Total
	1–10	11–20	21–30	31–40	41–50	51–60	61–64	
Existence problems	4	1	1	0	0	0	0	6
Position structure problems	0	1	2	1	2	10	2	18
Quantity relation problems	0	0	1	3	2	1	1	8
Geometry problems	29	18	18	15	20	17	6	123
Percentage of basic properties of circles and four points on a circle problems among the geometry problems	13.8%	11.1%	22.2%	26.7%	20.0%	64.7%	50%	26.0%

It can be seen that basic properties of circles and four points on a circle are the focus of geometry, and the number of relevant problems exceeds a quarter of all IMO geometry problems. In the 1st to 20th IMOs, the frequency of problems on basic properties of circles and four points on a circle is about 12.0%. In the 21st to 50th IMOs, the frequency is about 23.0%. The highest frequency of occurrence is 66.7% in the 51st to 64th IMOs.

Among them, in early IMOs, existence problems were the main part, and then they gradually disappeared, while position structure problems and the quantity relation problems were relatively stable, and for both types of problems, there were two problems in every 10 sessions of IMOs, respectively. In particular, in the 51st to 64th IMOs, there were 12 position structure problems. This also shows that the geometric content pays a great attention to position relationships of points and lines, such as collinearity of three points, the common point of three lines, four points on a circle, tangents, etc. In fact, position structure problems also include the quantity relation of geometry, because the position relationship of points and lines can also be described by length, angle, area and other quantities.

Therefore, this chapter will be divided into three parts. The first part introduces basic properties of circles, methods of determining four points on a circle, and helps understand the nature and application of these properties and theorems through some examples.

The second part revolves around three types of problems: "existence problems," "position structure problems," and "quantity relation problems." These problems are presented in chronological order, and some problems include various solutions and generalizations.

It is important to note that for each problem, solutions are followed by information on the scores, including the number of contestants in each score range, the average score, and scores of top five teams. However, early IMOs often lacked information on contestant scores, so the number of contestants in each score range only represents the counted number of contestants, and some problems lack scores of the top five teams.

The third part provides a brief summary of this chapter.

2.1 Related Properties, Theorems, and Methods

2.1.1 *Basic properties of circles*

(1) *Chord in a circle theorem*

Given a circle and one of its chords, a line that passes through the center of the circle and is perpendicular to the chord bisects the chord.

When the chord is non-diameter, the inverse statement also holds.

(2) *Chord length formula*

Let the radius of a circle be r and the distance from the center of the circle to one of its chords be d. Then the length of the chord $l = 2\sqrt{r^2 - d^2}$.

If two circles have a common chord, then it is bisected perpendicularly by the line of centers.

(3) *Position relations about circles*

(i) Position relationships between points and circles: Let P be a point in the plane of $\odot O$, and the radius of $\odot O$ be r. Then

$$\text{Point } P \text{ lies inside } \odot O \Leftrightarrow OP < r;$$

$$\text{Point } P \text{ lies on } \odot O \Leftrightarrow OP = r;$$

$$\text{Point } P \text{ lies outside } \odot O \Leftrightarrow OP > r.$$

(ii) Position relationships between lines and circles: The number of inter-section points of a line and a circle may be $0, 1, 2$. These three position relationships are called: separate, tangent, and intersecting, respectively.

Let the distance from the center of $\odot O$ to the line l be d, and the radius of $\odot O$ be r. Then the position relations of $\odot O$ and l correspond to the following three quantitative relations of d, r:

$$\text{The line } l \text{ and } \odot O \text{ are separate} \Leftrightarrow d > r;$$

$$\text{The line } l \text{ and } \odot O \text{ are tangent} \Leftrightarrow d = r;$$

$$\text{The line } l \text{ and } \odot O \text{ are intersecting} \Leftrightarrow d < r.$$

When a line and a circle are tangent, their only common point is called the tangent point. Obviously, the tangent line of the circle is perpendicular to the line connecting the center of the circle and the tangent point. Pass a point outside a circle to make two tangent lines of the circle, and then the lengths of the two tangent line segments are equal.

(iii) Position relationships between two circles: External separation, exter-nally tangent, intersection, internally tangent, and internal inclusion (see Figure 2.1).

Among them, two circles with external separation or internal inclu-sion have no common point, and the difference is that two circles with external separation are outside each other, and for two circles with internal inclusion, the smaller circle is inside the larger circle.

Two circles that are externally tangent or internally tangent have an only common point, and the difference is that two circles that are externally tangent are outside each other (except the tangent point), and for two circles that are internally tangent, the smaller circle is inside the larger circle (except the tangent point). Two circles with intersection have two different common points.

Let the center distance of $\odot O_1, \odot O_2$ be $O_1 O_2 = d$ and their radii be r_1, r_2 respectively. Then the position relation of $\odot O_1, \odot O_2$ corresponds

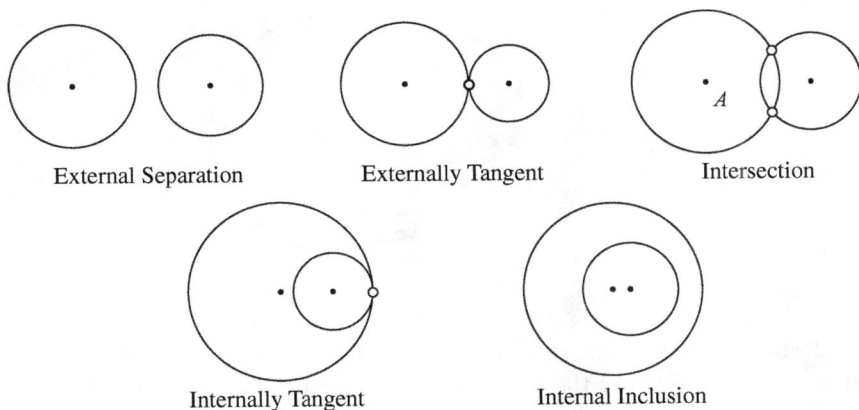

Figure 2.1

to the following five quantitative relations of d, r_1, r_2:

$\odot O_1$ and $\odot O_2$ are externally separate $\Leftrightarrow d > r_1 + r_2$;

$\odot O_1$ and $\odot O_2$ are externally tangent $\Leftrightarrow d = r_1 + r_2$;

$\odot O_1$ and $\odot O_2$ are intersecting $\Leftrightarrow |r_1 - r_2| < d < r_1 + r_2$;

$\odot O_1$ and $\odot O_2$ are internally tangent $\Leftrightarrow d = |r_1 - r_2| > 0$;

$\odot O_1$ and $\odot O_2$ are internally inclusive $\Leftrightarrow d < |r_1 - r_2|$.

When $r_1 = r_2$, the two circles do not have the two positional relations of internally tangent and internal inclusion.

Common Tangent. If a line is tangent to both circles at the same time, it is called the common tangent of the two circles.

If two circles are tangent to each other, then they must have a common tangent that passes through the tangent point and is perpendicular to the line connecting the centers of the two circles.

Two circles with external separation have four common tangent lines, two of which make the circles on the same side of the line (the external common tangent lines), and the other two make the circles on either side of the line (the internal common tangent lines). The length of either of the external common tangent segments (that is, the line segment with two tangent points as the endpoints, the same below) is $\sqrt{d^2 - (r_1 - r_2)^2}$ (*), and the length of either of the internal common tangent segments is $\sqrt{d^2 - (r_1 + r_2)^2}$ (see Figure 2.2).

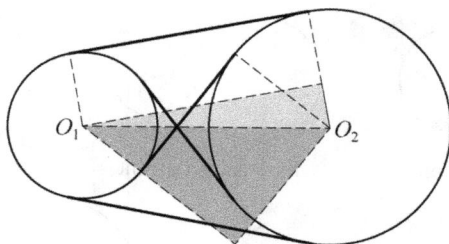

Figure 2.2

Two circles that are externally tangent have three common tangent lines, and the length of the two external common tangent lines is $2\sqrt{r_1 r_2}$.

Two circles that are intersecting have two common tangent lines, both of which are external common tangent lines, and the length formula is the same as (*).

Two circles that are internally tangent have only one common tangent line.

Two circles with internal inclusion have no common tangent line.

(4) *Angles in a circle*

  (i) Inscribed angle theorem: the inscribed angle opposite the same arc is half of the central angle (see Figure 2.3).

   According to the above theorem, inscribed angles opposite the same arc are equal. In particular, an inscribed angle opposite a diameter is a right angle, and vice versa.

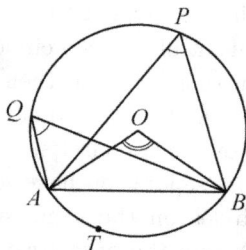

Figure 2.3

 (ii) If the arcs of equal length are called equal arcs, then it is not difficult to get that central angles, inscribed angles, and chords lengths opposite equal arcs are equal, respectively.

(iii) Equal arcs between parallel chords: Let A, C, D, B be four points arranged in order on a circle. Then

$$AB \parallel CD \Leftrightarrow \overparen{AC} = \overparen{BD} \Leftrightarrow AC = BD \Leftrightarrow AD = BC.$$

(iv) Central angles: Suppose two chords AB, CD of $\odot O$ intersect at point P, and the extension lines of AC, DB intersect at point Q (see Figure 2.4) Then

$$\angle BPC = \angle APD = \frac{1}{2}(\overparen{AD} + \overparen{BC}),$$

and

$$\angle Q = \frac{1}{2}(\overparen{AD} - \overparen{BC}),$$

where $\overparen{AC}$ denotes the central angle opposite $\overparen{AC}$, and so on for the other central angles.

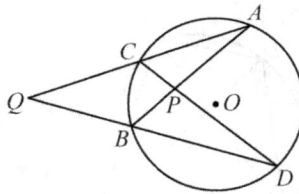

Figure 2.4

(v) Chord-tangent angle theorem: Given that PA and $\odot O$ are tangent at point A, and AB is any chord. Point C is on the circle (on the other side of the chord AB with respect to P). Then $\angle PAB = \angle ACB$ (see Figures 2.5 and 2.6).

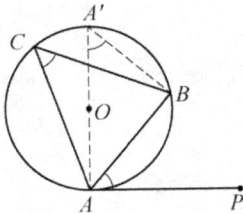

Figure 2.5

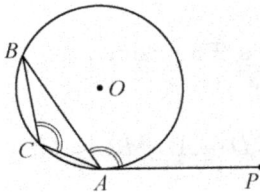

Figure 2.6

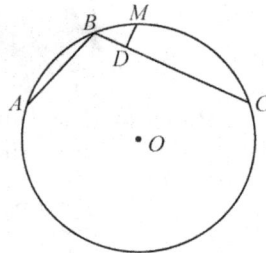

Figure 2.7

Theorem 2.1 (Archimedes's Midpoint Theorem). As shown in Figure 2.7, three points A, B, C are on $\odot O$, and point B is on $\overparen{AC}$, where

$\widehat{AB} < \widehat{BC}$. Given that M is the midpoint of $\widehat{ABC}$, make $MD \perp BC$ at point D. Then $AB + BD = CD$.

2.1.2 Properties and determination of four points on a circle

As shown in Figure 2.8, let $ABCD$ be a convex quadrilateral in a plane. Then the four points A, B, C, D are concyclic if and only if any of the following conditions is satisfied:

 (i) $\angle ADB = \angle ACB$;
 (ii) $\angle BAD + \angle BCD = \angle ABC + \angle ADC = 180°$;
(iii) $\angle CBP = \angle ADC$, where P is any point on the extension line AB (the outer angle equals the inner diagonal angle).

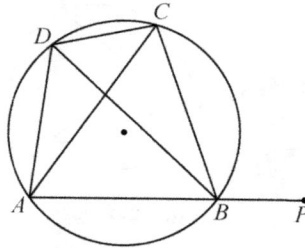

Figure 2.8

If directed angles are used, then any four points A, B, C, D in a plane are concyclic if and only if

$$\measuredangle ACB = \measuredangle ADB.$$

Theorem 2.2 (Ptolemy's Theorem). Let $ABCD$ be a convex quadrilateral inscribed on a circle. Then

$$AB \cdot CD + AD \cdot BC = AC \cdot BD,$$

and the equality holds if and only if the quadrilateral is a cyclic quadrilateral.

Theorem 2.3 (Casey's Theorem). As shown in Figure 2.9, let $\odot O_1, \odot O_2, \odot O_3, \odot O_4$ be internally tangent to $\odot O$ (tangent points are

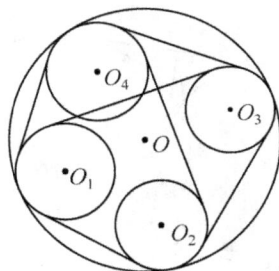

Figure 2.9

arranged in order). Denote t_{ij} as the length of the external common tangent segment of $\odot O_i, \odot O_j (1 \leq i < j \leq 4)$. Then

$$t_{12} \cdot t_{34} + t_{23} \cdot t_{14} = t_{13} \cdot t_{24}.$$

Casey's theorem has the following generalizations:

(i) The condition can be generalized to: $\odot O_1, \odot O_2, \odot O_3, \odot O_4$ are all tangent to $\odot O$. If $\odot O_i, \odot O_j$ are both internally or externally tangent to $\odot O$, then t_{ij} is regarded as the length of the external common tangent segment of $\odot O_i, \odot O_j$. If $\odot O_i, \odot O_j$ are tangent to $\odot O$, one of which is internally tangent and the other is externally tangent, then t_{ij} is regarded as the length of the internal common tangent segment of $\odot O_i, \odot O_j$.

(ii) If $\odot O_i$ can be regarded as a point, then the length of the external common tangent segment should be understood accordingly as the length of the tangent line from the point to the circle, or the distance between two points.

Example 2.1. Let AB, CD be the two external tangent lines of intersecting circles Γ_1, Γ_2 (A, B, C, D are points of tangency, and A, C are on the circle Γ_1), and let M be the midpoint of AB. Through point M, construct the tangent lines of the two circles (different from AB) intersecting CD at points X, Y, respectively, and let I be the incenter of $\triangle MXY$. Prove: $IC = ID$.

Proof. As shown in Figure 2.10, let MX be tangent to the circle Γ_1 at point E and MY be tangent to the circle Γ_2 at point F. Then

$$ME = MA = MB = MF.$$

Since MI bisects $\angle EMF$, so MI is the perpendicular bisector of segment EF, and then $IE = IF$.

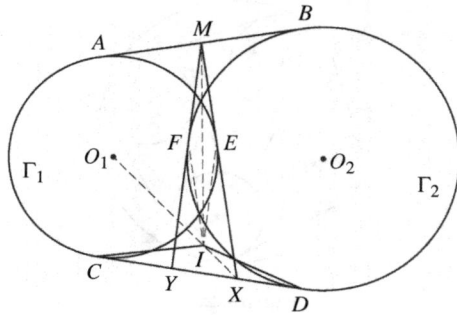

Figure 2.10

Let O_1 be the center of the circle Γ_1. Then O_1X bisects $\angle CXE$, so the incenter I of $\triangle MXY$ is on O_1X. While $XC = XE$, one has $IC = IE$.

Similarly, $ID = IF$, and noticing that $IE = IF$, we obtain $IC = ID$.

Example 2.2. The side BC of an inscribed convex quadrilateral $ABCD$ has a point P that satisfies $\angle PAB = \angle PDC = 90°$. The two medians from point A of $\triangle ABP$ and point D of $\triangle DPC$ intersect at point K, and the two angular bisectors from them intersect at point L. Prove: $KL \perp BC$.

Proof.   As shown in Figure 2.11, let the extension lines of AB, DP intersect at point X, and the extension lines of DC, AP intersect at point Y.

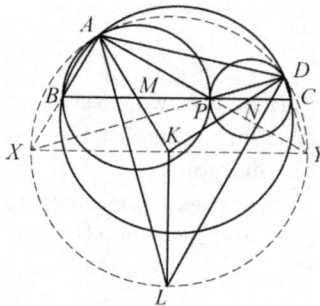

Figure 2.11

By the known condition,

$$\angle XAY = \angle XDY = 90°.$$

Therefore, the four points A, X, Y, D are concyclic, and the diameter is XY.

Since the four points A, B, C, D are concyclic and the four points A, X, Y, D are concyclic,

$$\angle ABC = 180° - \angle ADC = 180° - \angle ADY = \angle AXY,$$

and then $BC \parallel XY$.

Since $\triangle ABP \backsim \triangle AXY$ and $\triangle DPC \backsim \triangle DXY$, the extension lines of AM, ND all pass through the midpoint of XY, so K is the midpoint of XY, which is the center of the circle formed by points A, X, Y, D.

Obviously, the angular bisectors of $\angle BAP, \angle CDP$ pass through the midpoint of the semicircle $\overset{\frown}{XY}$ (excluding point A), and then the point is L.

Therefore, $KL \perp XY$, that is $KL \perp BC$.

Example 2.3. As shown in Figure 2.12, $\triangle ABC$ is inscribed in the circle $\odot O$, where D is the midpoint of $\overset{\frown}{BAC}$ and E is the midpoint of $\overset{\frown}{BC}$. Construct $CF \perp AB$ at point F, and through point F construct the perpendicular line of EF, which intersects the extension line of DA at point G. Prove: $CG = CD$.

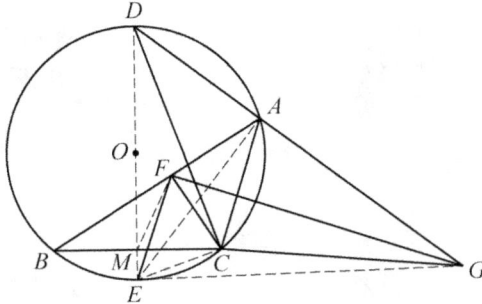

Figure 2.12

Proof. Let BC, DE intersect at point M. Since DE is the diameter of the circle $\odot O$, so

$$\angle EAG = 90° = \angle EFG.$$

Then, A, G, E, F are concyclic. Therefore

$$\angle EGF = \angle EAF = \angle EAB = \angle ECB = \angle ECM.$$

And $\angle EFG = \angle EMC = 90°$, so we get $\text{Rt}\triangle EFG \backsim \text{Rt}\triangle EMC$, and then

$$\frac{EF}{EM} = \frac{EG}{EC}, \quad \angle GEF = \angle CEM.$$

Thus $\angle FEM = \angle CEM - \angle CEF = \angle GEF - \angle CEF = \angle CEG.$
From which $\triangle ECG \sim \triangle EMF$, and consequently

$$\angle CGE = \angle MFE. \tag{1}$$

Since A, G, E, F are concyclic,

$$\angle AGE = \angle BFE. \tag{2}$$

By (2)–(1), we get

$$\angle AGC = \angle BFM. \tag{3}$$

In the right triangle BCF, we know that M is the midpoint of the hypotenuse BC. Thus $BM = MF$, and then

$$\angle BFM = \angle FBM = \angle ABC = \angle ADC. \tag{4}$$

Combined with (3), (4) we get $\angle AGC = \angle ADC$. Hence $CD = CG$.

Example 2.4. As shown in Figure 2.13, $\triangle ABC$ is inscribed in a circle $\odot O$, and the angle bisector of $\angle BAC$ intersects $\overset{\frown}{BC}$ at point D. Construct a parallel line of AD through point O, which intersects side AB at point E. The point H is the orthocenter of $\triangle ABC$ and F is the midpoint of DH. Prove: $\angle CFE = 90°$.

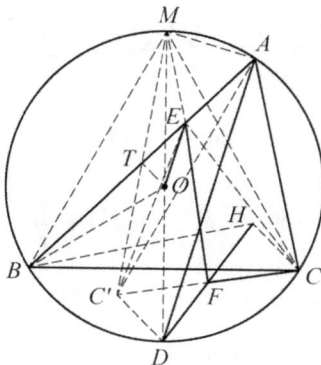

Figure 2.13

Proof. Extend CF to point C' such that $CF = FC'$. Then we only need to prove that $EC = EC'$. Let the midpoint of side AB be T and the midpoint of arc $\overset{\frown}{BAC}$ be M.

Obviously, MD is the diameter of $\odot O$, so $AD \perp AM$. By the known condition, $OE \parallel AD$, so $OE \perp AM$. And $OA = OM$. Thus OE is the perpendicular bisector of segment AM, from which $EM = EA$.

Since F is the midpoint of both CC', HD then $\overrightarrow{CH} = \overrightarrow{DC'}$. By the properties of the circumcenter and the orthocenter, $\overrightarrow{CH} = 2\overrightarrow{OT}$, and so $\overrightarrow{DC'} = 2\overrightarrow{OT}$. Note that O is the midpoint of MD, and thus OT is the midsegment of $\triangle MDC'$. Then T is the midpoint of MC'.

Since T is the midpoint of AB, MC', we see that $AC' = BM$ and $AC' \parallel BM$. While $BM = CM$, so $MC = AC'$.

Combining with $EA = EM$ and $MB = MC$, and the properties of the inscribed angle, we get
$$\angle AME = \angle EAM = \angle BAM = \angle BCM = \angle MBC,$$
so $\angle AME = \angle MBC$ and $\angle AMC = \angle ABC$. Subtracting the two equalities, we obtain $\angle CME = \angle ABM$.

Since $BM \parallel AC'$, so $\angle ABM = \angle C'AE$. Then
$$\angle CME = \angle C'AE.$$

In summary, $\triangle EMC \cong \triangle EAC'$. Thus $EC = EC'$. And F is the midpoint of CC', so $\angle CFE = 90°$.

Example 2.5. As shown in Figure 2.14, AB, CD are two chords in $\odot O$ and $AB \parallel CD$. P is any point of $\overset{\frown}{AB}$. Prove: $\frac{PA+PB}{PC+PD}$ is a definite value.

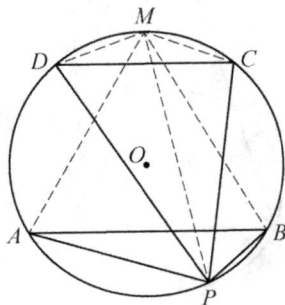

Figure 2.14

Proof. Take the midpoint M of $\overset{\frown}{CD}$. Then M is also the midpoint of $\overset{\frown}{ACB}$.

In the inscribed quadrilateral $APBM$, by Ptolemy's theorem,

$$AP \cdot BM + BP \cdot AM = AB \cdot PM.$$

Combining $AM = BM$, we have

$$AP + BP = \frac{AB \cdot PM}{AM}. \tag{1}$$

In the inscribed quadrilateral $CPDM$, by Ptolemy's theorem,

$$CP \cdot DM + DP \cdot CM = CD \cdot PM,$$

Combining $CM = DM$, we have

$$CP + DP = \frac{CD \cdot PM}{CM}. \tag{2}$$

Combining the two equalities (1) and (2), we know that $\frac{AP+BP}{CP+DP} = \frac{AB \cdot CM}{AM \cdot CD}$ is definite.

Example 2.6. As shown in Figure 2.15, $AB = AC$ in $\triangle ABC$, point D is on the side AC, and point K is on the minor arc $\overset{\frown}{CD}$ of the circumcircle of $\triangle BCD$. Construct a parallel line of BC through point A, which intersects side CK at point T. And M is the midpoint of the segment DT. Prove: $\angle AKT = \angle CAM$.

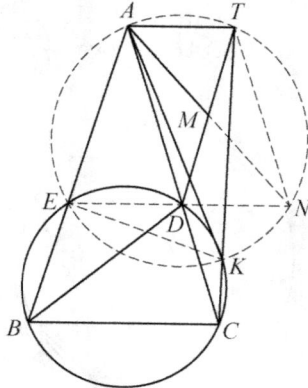

Figure 2.15

Proof. Extend AM to point N such that $MN = AM$. Then the quadrilateral $ADNT$ is a parallelogram.

Let the circle passing through B, C, D intersect AB at another point E. Then $DE \parallel BC$, $AT \parallel DN$. By the known condition, $AT \parallel BC$, so the three points D, E, N are collinear.

Since $AD = AE = TN$, the quadrilateral $AENT$ is an isosceles trapezoid, and then A, E, N, T are concyclic. And

$$\angle TKE = \angle EBC = \angle AEN = \angle TNE,$$

so, the four points E, K, N, T are concyclic. Thus, the five points A, E, K, N, T are concyclic.

Since $AC \parallel TN$ and A, T, N, K are concyclic,

$$\angle AKT = \angle ANT = \angle CAM.$$

2.2 Problems and Solutions

2.2.1 *Existence problems*

Problem 2.1 (IMO 1-6, proposed by Czechoslovakia). Two planes, P and Q, intersect along a line p. A point A is given in the plane P, and a point C in the plane Q; neither of these points lies on the straight line p. Construct an isosceles trapezoid $ABCD$ (with AB parallel to CD) in which a circle can be inscribed, with vertices B and D lying in the planes P and Q respectively.

Solution. First, we prove that AB, CD are parallel to p. This is because if let M be a point on p, then there is only one line l through M, which is parallel to AB, CD. Since l is in both the plane P and the plane Q, we have $l = p$.

In Figure 2.16, crossing points A and C, respectively, make parallel lines AB and CD of p.

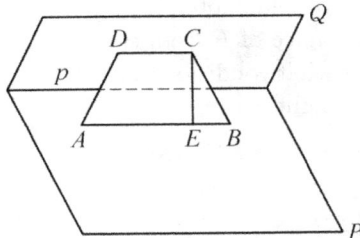

Figure 2.16

In the plane formed by AB and CD, let $CE \perp AB$, and the foot of the altitude is E. It is easy to know

$$AE = \frac{1}{2}(AB + CD) = \frac{1}{2}(AD + BC) = AD.$$

Thus, it is possible to make a circle with A as the center and AE as the radius intersecting CD at point D, and then further determine point B.

When $AE > CE$, there are two solutions; when $AE = CE$, there is only one solution (then the trapezoid $ABCD$ becomes a square); when $AE < CE$, there is no solution.

【Score Situation】 This particular problem saw the following distribution of scores among contestants: 1 contestant scored 7 points, no contestant scored 6 points, no contestant scored 5 points, 2 contestants scored 4 points, 1 contestant scored 3 points, 1 contestant scored 2 points, 3 contestants scored 1 point, and 1 contestant scored 0 point. The average score of this problem is 2.556, indicating that it had a certain level of difficulty.

Among the top five teams in the team scores, the Romania team achieved a total score of 249 points, the Hungary team achieved a total score of 233 points, the Czechoslovakia team achieved a total score of 192 points, the Bulgaria team achieved a total score of 131 points, and the Poland team achieved a total score of 122 points.

The gold medal cutoff for this IMO was set at 37 points (with three contestants earning gold medals), the silver medal cutoff was 36 points (with three contestants earning silver medals), and the bronze medal cutoff was 33 points (with five contestants earning bronze medals).

In this IMO, only one contestant achieved a perfect score of 40 points, namely Bohuslav Diviš from Czechoslovakia.

Problem 2.2 (IMO 2-7, proposed by the German Democratic Republic). An isosceles trapezoid with bases a, b, and altitude h is given.

(a) On the axis of symmetry of this trapezoid, find all points P such that both legs of the trapezoid subtend right angles at P.

(b) Calculate the distance of P from either base.

(c) Determine under what conditions such points P actually exist. (Discuss various cases that might arise.)

Solution. As shown in Figure 2.17, let the upper side AD and lower side BC of the isosceles trapezoid be a and b respectively, the altitude be h, and MN be the symmetry axis of the trapezoid $ABCD$.

If $\angle DPC = 90°$, then the point P must lie on the circle with CD as its diameter.

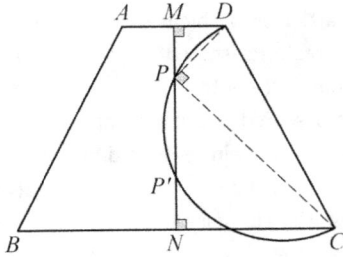

Figure 2.17

(a) Now make a circle with CD as its diameter, intersecting the symmetry axis MN at two points (P and P'), or at a point P, or with no intersection.

(b) It is easy to know that $\text{Rt}\triangle DMP \backsim \text{Rt}\triangle PNC$, so

$$\begin{cases} PM \cdot PN = MD \cdot NC = \frac{ab}{4}, \\ PM + PN = h. \end{cases}$$

According to Vieta's theorem, PM, PN are the two roots of the equation

$$x^2 - hx + \frac{ab}{4} = 0.$$

And according to Vieta's theorem, as long as the above quadratic equation has roots, it must have two positive roots. Thus, point P exists only if $\Delta \geq 0$, i.e.,

$$h^2 - ab \geq 0,$$

$$x = \frac{h \pm \sqrt{h^2 - ab}}{2},$$

and, one of the roots is PM and the other is PN.

(c) As known by (b), we have:

(i) If $h^2 < ab$, then the intersection point P does not exist, that is, the circle with the diameter CD does not intersect the segment MN.

(ii) If $h^2 = ab$, then there is only one point P at the midpoint of MN, and the circle is tangent to MN.

(iii) If $h^2 > ab$, then there are two points P and P' that satisfy the condition, that is, the circle intersects the segment MN. As shown in Figure 2.17, we can assume that

$$PM = P'N = \frac{h}{2} - \sqrt{h^2 - ab},$$

$$PN = P'M = \frac{h}{2} + \sqrt{h^2 - ab}.$$

【Score Situation】This particular problem saw the following distribution of scores among contestants: 4 contestants scored 5 points, 1 contestant scored 4 points, no contestant scored 3 points, 2 contestants scored 2 points, no contestant scored 1 point, and 3 contestants scored 0 point. The average score of this problem is 2.800, indicating that it had a certain level of difficulty.

Among the top five teams in the team scores, the Czechoslovakia team achieved a total score of 257 points, the Hungary team achieved a total score of 248 points, the Romania team achieved a total score of 248 points, the Bulgaria team achieved a total score of 175 points, and the German Democratic Republic team achieved a total score of 38 points.

The gold medal cutoff for this IMO was set at 40 points (with four contestants earning gold medals), the silver medal cutoff was 37 points (with four contestants earning silver medals), and the bronze medal cutoff was 33 points (with four contestants earning bronze medals).

In this IMO, no contestant achieved a perfect score of 44 points.

Problem 2.3 (IMO 3-5, proposed by Czechoslovakia). Construct triangle ABC such that $AC = b$, $AB = c$, and $\angle AMB = \omega$, where M is the midpoint of segment BC and $\omega < 90°$. Prove that a solution exists if and only if $b\tan\frac{\omega}{2} \le c < b$. In what case does the equality hold?

Solution 1. As shown in Figure 2.18, let D is the midpoint of AC. Then $MD = \frac{AB}{2} = \frac{c}{2}$ and $\angle AMC = 180° - \omega$. Construct arc $\overparen{AC}$ with AC as a chord, $180° - \omega$ as the inscribed angle, and then construct a circle with D as the center and DM as the radius. The intersection point of this circle and arc $\overparen{AC}$ is M(generally there may be two intersection points). Connect to form CM and extend it twice to B, i.e., $\triangle ABC$ is requested.

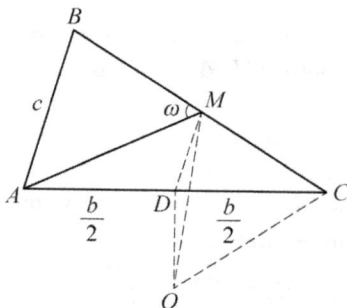

Figure 2.18

According to the above construction, if $\triangle ABC$ is to exist, only if the point M, the intersection of the arc and the arc, exists. Now find the circumcenter O of $\triangle AMC$, which is outside $\triangle AMC$. It is easy to know $\angle DOC = \omega < 90°$.

The point M exists only if $0 \leq \angle DOM < \omega$ or $\cos \omega < \cos \angle DOM \leq 1$. By the law of cosines

$$\cos \omega < \frac{DO^2 + MO^2 - DM^2}{2DO \cdot MO} \leq 1.$$

It is also easy to know $MO = CO = \frac{b}{2 \sin \omega}$, $DO = \frac{b}{2} \cot \omega$, and $DM = \frac{c}{2}$. Substituting them into the inequality, we get

$$\cos \omega < \frac{\frac{b^2}{\sin^2 \omega} + b^2 \cot^2 \omega - c^2}{2b^2 \cdot \frac{\cos \omega}{\sin^2 \omega}} \leq 1,$$

and simplifying,

$$\cos \omega < \frac{1 + \cos^2 \omega - \frac{c^2}{b^2} \sin^2 \omega}{2 \cos \omega} \leq 1.$$

This is

$$\frac{c^2}{b^2} \sin^2 \omega < 1 - \cos^2 \omega = \sin^2 \omega,$$

and

$$1 - \cos \omega \leq \frac{c}{b} \sin \omega.$$

Hence

$$b \tan \frac{\omega}{2} = b \frac{1 - \cos \omega}{\sin \omega} \leq c.$$

So, the proposition is proved.

Solution 2. As shown in Figure 2.18, let ABC be the required triangle and AM be the midline of the crossing point A. Let D be the midpoint of side AC and l be the vertical bisector of segment AC. Then $DM = \frac{c}{2}$.

So M is the intersection point of two circles, one with D as the center and $\frac{c}{2}$ as the radius, and the other with O ($O \in l$) as the center. Passing through two points A, C, construct arc $\overgroup{AMC} = \pi - \omega$. When both point M and segment AC are given, the required triangle is easily obtained.

Such a triangle exists if and only if the two circles intersect. The condition for the two circles to intersect is: $OM \leq OD + DM$.

The above formula is equivalent to $\frac{b}{2} \tan \frac{\omega}{2} \leq \frac{c}{2}$.

By $\angle BMA < \angle CMA$, we get $b = AC > AB = c$, so $b \tan \frac{\omega}{2} \leq c < b$.

【Score Situation】This particular problem saw the following distribution of scores among contestants: 1 contestant scored 7 points, no contestant scored 6 points, 1 contestant scored 5 points, 3 contestants scored 4 points, 1 contestant scored 3 points, no contestant scored 2 points, no contestant scored 1 point, and 3 contestants scored 0 point. The average score of this problem is 3.000, indicating that it had a certain level of difficulty.

Among the top five teams in the team scores, the Hungary team achieved a total score of 270 points, the Poland team achieved a total score of 203 points, the Romania team achieved a total score of 197 points, the Czechoslovakia team achieved a total score of 159 points, and the German Democratic Republic team achieved a total score of 146 points.

The gold medal cutoff for this IMO was set at 37 points (with three contestants earning gold medals), the silver medal cutoff was 34 points (with four contestants earning silver medals), and the bronze medal cutoff was 30 points (with four contestants earning bronze medals).

In this IMO, only one contestant achieved a perfect score of 40 points, namely Béla Bollobás from Hungary.

Problem 2.4 (IMO 4-5, proposed by Bulgaria). On a circle K there are given three distinct points A, B, C. Construct (using only a straightedge and compass) a fourth point D on K such that a circle can be inscribed in the quadrilateral thus obtained.

Solution 1. As shown in Figure 2.19, assume that point D has been made and the incenter I of the inscribed circle lies on the bisector of $\angle A$, $\angle B$, and $\angle C$.

It is easy to know

$$\angle AIC = \angle IAB + \angle ABC + \angle ICB$$
$$= \frac{1}{2}(\angle A + \angle C) + \angle ABC$$
$$= 90° + \angle ABC$$

(When $\angle ABC > 90°$, we see that $\angle AIC$ is the reflex angle).

Then, with AC as a chord and $\angle ABC + 90°$ as a known inscribed angle, arc $\overgroup{AC}$ can be constructed by using the straightedge and compass, and the

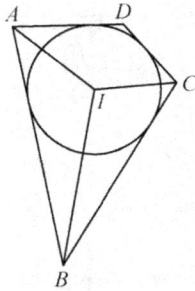

Figure 2.19

angular bisector of $\angle ABC$ intersects the arc $\overset{\frown}{AC}$ at point I. Join AI, CI. We can construct $\angle DAI = \angle BAI$ and $\angle DCI = \angle BCI$. Both of one side of the two angles, AD, CD, intersect on the circle K, so the intersection point D is the required point.

The proof is as follows: since

$$\angle AIC = 90° + \angle ABC = \frac{1}{2}(\angle DAB + \angle DCB) + \angle ABC,$$

so, $\angle DAB + \angle DCB = 180°$.

Thus, A, B, C, D are concyclic. The point D on the circle is the requested point.

Discussion. Let P be any point of arc $\overset{\frown}{ACK}$. When $P = A$, there holds

$$AB + PC > AP + BC,$$

when $P = C$,

$$AB + PC < AP + BC.$$

By the continuity, there will be a point P' on the arc $\overset{\frown}{AC}$ that satisfies

$$AB + CP' = AP' + BC.$$

So, this is the solution.

Solution 2. A sufficient and necessary condition for the convex quadrilateral $ABCD$ to have an inscribed circle is $AB + CD = AD + BC$.

As shown in Figure 2.20, without loss of generality, let $BC \geq AB$, so

$$CD - AD = BC - AB = \text{constant},$$

and the point D is on one of the branches of the hyperbola.

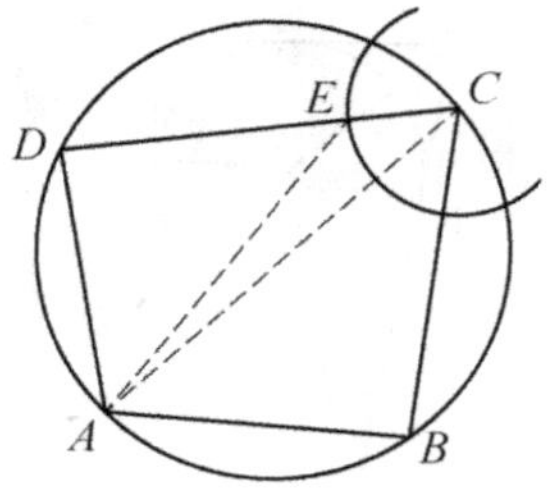

Figure 2.20

Geometric construction:

(1) If $BC = AB$, then $CD = AD$, and point D is the intersection of the perpendicular bisector of AC and arc $\overset{\frown}{AC}$.

(2) If $BC > AB$, then take a point E on CD such that $DE = AD$, and $EC = CD - DE = DC - AD = BC - AB$.

$$\angle AEC = 180° - \angle AED = 180° - (90° - \frac{1}{2}\angle ADC)$$

$$= 90° + \frac{1}{2}\angle ADC = 90° + \frac{1}{2}(180° - \angle B)$$

$$= 180° - \frac{1}{2}\angle B.$$

The construction of point D depends on point E. Construct circle C with C as the center and $CE = BC - AB$ as the radius, and then construct arc $\overset{\frown}{AEC}$ with AC as the chord and $180° - \frac{1}{2}\angle B$ as the inscribed angle. The intersection point of this arc $\overset{\frown}{AEC}$ and $\odot C$ is E. Then we can construct point D, and this proof is omitted.

【Score Situation】 This particular problem saw the following distribution of scores among contestants: 6 contestants scored 7 points, 2 contestants scored 6 points, 1 contestant scored 5 points, 3 contestants scored 4 points, no contestant scored 3 points, no contestant scored 2 points, no contestant scored 1 point, and 5 contestants scored 0 point. The average score of this problem is 4.176, indicating that it was simple.

Among the top five teams in the team scores, the Hungary team achieved a total score of 289 points, the Soviet Union team achieved a total score of 263 points, the Romania team achieved a total score of 257 points, the Czechoslovakia team achieved a total score of 212 points, and the Poland team achieved a total score of 212 points.

The gold medal cutoff for this IMO was set at 41 points (with four contestants earning gold medals), the silver medal cutoff was 34 points (with 12 contestants

earning silver medals), and the bronze medal cutoff was 29 points (with 15 contestants earning bronze medals).

In this IMO, only one contestant achieved a perfect score of 46 points, namely Iosif Bernstein from the Soviet Union.

Problem 2.5 (IMO 14-2, proposed by the Netherlands). Prove that if $n \geq 4$, then every quadrilateral that can be inscribed in a circle can be dissected into n quadrilaterals each of which is inscribable in a circle.

Proof. If the quadrilateral is an isosceles trapezoid, then it can be divided into n isosceles trapezoids by lines parallel to the base, because any isosceles trapezoid has a circumcircle, and thus the proposition is proved.

In the following let an inscribed quadrilateral $ABCD$ of a circle be not trapezoidal. Since the opposite angles are supplementary, without loss of generality, let

$$\angle A \geq \angle C, \angle D \geq \angle B, \text{ so } \angle A \geq \angle B, \angle D \geq \angle C.$$

Let $\angle BAP = \angle B$, $\angle CDQ = \angle C$, and $PQ \parallel AD$. Then, let $PR \parallel AB$ intersect BC at point R, and let $QS \parallel DC$ intersect BC at point S, as shown in Figure 2.21. Since

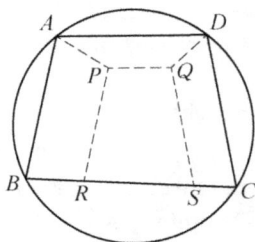

Figure 2.21

$$\angle DAP = \angle BAD - \angle ABC$$

$$= (180° - \angle BCD) - (180° - \angle CDA)$$

$$= \angle CDA - \angle BCD = \angle ADQ,$$

so, quadrilateral $APQD$ is an isosceles trapezoid.

Both the quadrilaterals $ABRP$ and $CDQS$ are isosceles trapezoids.

And since $PQ \parallel AD$, $PR \parallel AB$, so $\angle QPR = \angle DAB$. From $QS \parallel DC$, we know that $\angle RSQ = \angle BCD$. Then

$$\angle QPR + \angle RSQ = \angle DAB + \angle BCD = 180°.$$

Thus, the quadrilateral $PQSR$ has a circumscribed circle.

From the above, we can see that the quadrilateral is divided into four quadrilaterals with circumcircles, and three of them are isosceles trapezoids. For $n > 4$, we can further divide the isosceles trapezoids such that there are exactly n quadrilaterals that all have circumcircles.

【Score Situation】 This particular problem saw the following distribution of scores among contestants: 13 contestants scored 6 points, 5 contestants scored 5 points, 1 contestant scored 4 points, 3 contestants scored 3 points, 4 contestants scored 2 points, no contestant scored 1 point, and 7 contestants scored 0 point. The average score of this problem is 3.758, indicating that it was relatively straightforward.

Among the top five teams in the team scores, the scores of this problem are as follows: the Soviet Union team scored 42 points (with a total team score of 270 points), the Hungary team scored 41 points (with a total team score of 263 points), the German Democratic Republic team scored 29 points (with a total team score of 239 points), the Romania team scored 31 points (with a total team score of 208 points), and the United Kingdom team scored 32 points (with a total team score of 179 points).

The gold medal cutoff for this IMO was set at 40 points (with eight contestants earning gold medals), the silver medal cutoff was 30 points (with 16 contestants earning silver medals), and the bronze medal cutoff was 19 points (with 30 contestants earning bronze medals).

In this IMO, a total of eight contestants achieved a perfect score of 40 points.

Problem 2.6 (IMO 21-3, proposed by the Soviet Union). Two circles in a plane intersect. Let A be one of the points of intersection. Starting simultaneously from A, two points move with constant speeds, each point travelling along its own circle in the same sense. The two points return to A simultaneously after one revolution. Prove that there is a fixed point P in the plane such that, at any time, the distances from P to the moving points are equal.

Proof. By the conditions, two moving points have the same angular velocity.

As shown in Figure 2.22, there are two moving points Q_1 and Q_2 respectively on $\odot O_1$ and $\odot O_2$. Then $\angle AO_1Q_1 = \angle AO_2Q_2$ (which is θ).

Let the other intersection point of the two circles be B. It is easy to know

$$\angle ABQ_1 = \frac{\theta}{2}, \angle ABQ_2 = 180° - \frac{\theta}{2},$$

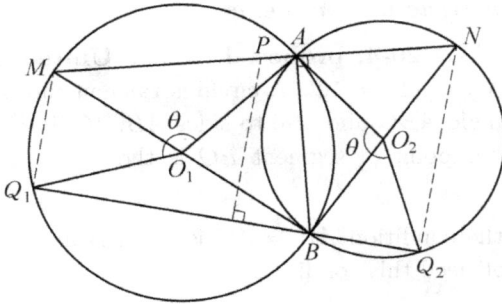

Figure 2.22

so $\angle ABQ_1 + \angle ABQ_2 = 180°$.

Thus, the points Q_1, B, Q_2 are collinear.

Through the point A make $MAN \perp AB$, where the points M, N are on $\odot O_1$ and $\odot O_2$, respectively. So $\angle MQ_1Q_2 = \angle NQ_2Q_1 = 90°$, that is, the quadrilateral MQ_1Q_2N is a right trapezoid.

So, the perpendicular bisector of Q_1Q_2 passes through the midpoint of MN. Let this midpoint be P, and then $PQ_1 = PQ_2$. Obviously, P is the required fixed point.

【Score Situation】 This particular problem saw the following distribution of scores among contestants: 69 contestants scored 7 points, 4 contestants scored 6 points, 5 contestants scored 5 points, 12 contestants scored 4 points, 6 contestants scored 3 points, 9 contestants scored 2 points, 16 contestants scored 1 point, and 45 contestants scored 0 point. The average score of this problem is 3.807, indicating that it was relatively straightforward.

Among the top five teams in the team scores, the scores of this problem are as follows: the Soviet Union team scored 49 points (with a total team score of 267 points), the Romania team scored 44 points (with a total team score of 240 points), the Germany team scored 50 points (with a total team score of 235 points), the United Kingdom team scored 48 points (with a total team score of 218 points), and the United States team scored 32 points (with a total team score of 199 points).

The gold medal cutoff for this IMO was set at 37 points (with eight contestants earning gold medals), the silver medal cutoff was 29 points (with 32 contestants earning silver medals), and the bronze medal cutoff was 20 points (with 42 contestants earning bronze medals).

In this IMO, a total of four contestants achieved a perfect score of 40 points.

2.2.2 *Position structure problems*

Problem 2.7 (IMO 20-4, proposed by the United States). In triangle ABC, we have $AB = AC$. A circle is tangent internally to the circumcircle of triangle ABC and also to sides AB, AC at P, Q, respectively. Prove that the midpoint of segment PQ is the centre of the incircle of triangle ABC.

Note. In fact, the condition $AB = AC$ is superfluous, and the following two proofs do not need this condition.

Proof 1.   As shown in Figure 2.23, let the midpoint of PQ be I, so we know that AI bisects $\angle BAC$. Then just prove that CI bisects $\angle ACB$.

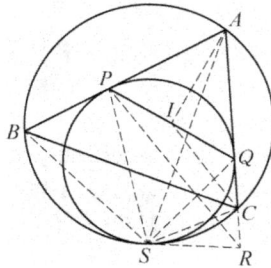

Figure 2.23

Let the point of tangency of the two circles be S, and join SB, SP, SA, SQ, SC. We know that SP bisects $\angle ASB$ and SQ bisects $\angle ASC$, so

$$\frac{BS}{BP} = \frac{AS}{AP} = \frac{AS}{AQ} = \frac{SC}{CQ}. \tag{1}$$

Extend QC to R such that $QC = CR$, and join SR. So, according to (1), we get

$$\frac{BS}{BP} = \frac{SC}{CR}.$$

Also $\angle PBS = \angle SCR$, and thus, $\triangle PBS \backsim \triangle SCR$. Then, $\angle BPS = \angle CRS$, and consequently, A, R, S, P are concyclic.

The segment IC is the middle line of $\triangle PQR$, so

$$\angle ACI = \angle ARP = \angle ASP$$

$$= \frac{1}{2}\angle ASB = \frac{1}{2}\angle ACB.$$

Thus, CI bisects $\angle ACB$.

Proof 2.   As shown in Figure 2.24, let the midpoint of PQ be I, the center of the big circle be O, the center of the small circle be T, and the radii of the two circles be R and r respectively.

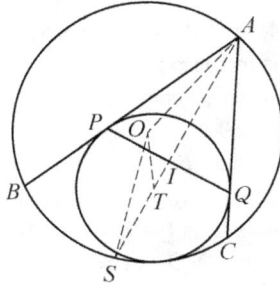

Figure 2.24

Let the extension of AI intersect $\odot O$ at point S. Then S is the midpoint of the arc $\overparen{BSC}$, and if it can be proved that $SI = BS = SC = 2R\sin\frac{A}{2}$, then I is the incenter of the circle (this familiar conclusion is left to the reader). In the following we prove that $SI = 2R\sin\frac{A}{2}$.

Join SO, OA, OT. By the power of a point and $OT = R - r$,

$$AT \cdot TS = (R + OT)(R - OT) = r(2R - r).$$

Also

$$AT = \frac{r}{\sin\frac{A}{2}},$$

$$TS = AS - AT,$$

and substitute it in to get

$$\frac{1}{\sin\frac{A}{2}}\left(AS - \frac{r}{\sin\frac{A}{2}}\right) = 2R - r.$$

Thus

$$AS - r\cot\frac{A}{2}\cos\frac{A}{2} = 2R\sin\frac{A}{2}.$$

Consequently, we see that

$$r\cot\frac{A}{2}\cos\frac{A}{2} = AI,$$

and so

$$SI = AS - AI = 2R\sin\frac{A}{2}.$$

【Score Situation】 This particular problem saw the following distribution of scores among contestants: no contestant scored 6 points, 36 contestants scored 5 points, 1 contestant scored 4 points, 1 contestant scored 3 points, 4 contestants scored 2 points, 2 contestants scored 1 point, and 4 contestants scored 0 point. The average score of this problem is 4.104, indicating that it was simple.

Among the top five teams in the team scores, the scores of this problem are as follows: the Romania team scored 40 points (with a total team score of 237 points), the United States team scored 39 points (with a total team score of 225 points), the United Kingdom team scored 36 points (with a total team score of 201 points), the Vietnam team scored 36 points (with a total team score of 200 points), and the Czechoslovakia team scored 40 points (with a total team score of 195 points).

The gold medal cutoff for this IMO was set at 35 points (with five contestants earning gold medals), the silver medal cutoff was 27 points (with 20 contestants earning silver medals), and the bronze medal cutoff was 22 points (with 38 contestants earning bronze medals).

In this IMO, only one contestant achieved a perfect score of 40 points, namely Mark Kleiman from the United States.

Problem 2.8 (IMO 25-4, proposed by Romania). Let $ABCD$ be a convex quadrilateral such that the line CD is a tangent to the circle with AB as a diameter. Prove that the line AB is tangent to a circle with CD as a diameter if and only if the lines BC and AD are parallel.

Proof 1. Denote the midpoint of AB as O. The circle $\odot O$ is tangent to CD at point E, the midpoint of CD is F, and $FG \perp AB$, with G the foot of the altitude (see Figure 2.25).

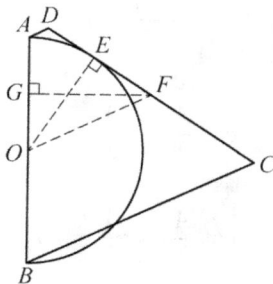

Figure 2.25

Sufficiency: If $AD \parallel BC$, then OF is the median line of a trapezoidal $ABCD$. Thus, $S_{\triangle AOF} = S_{\triangle DOF}$, and then $DF \cdot OE = GF \cdot AO$.

We know that $OE = AO$, so $DF = GF$, that is, the circle with diameter CD is tangent to AB.

Necessity: We can first reverse the above steps to get $S_{\triangle AOF} = S_{\triangle DOF}$, so $AD \parallel OF$, and then $BC \parallel AD$.

Proof 2. Before proving this problem, let's look at the following lemma.

Lemma. *Rays OX and OY intersect to form an acute angle θ, and A, B are the two points given on OX, in the order of O, A, B. A circle with diameter AB is denoted Γ (see Figure 2.26). Then the ray OY is tangent to the circle Γ if and only if*

$$\frac{AB}{OA} = \frac{2\sin\theta}{1 - \sin\theta}.$$

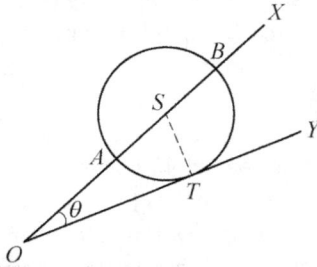

Figure 2.26

Proof of the Lemma. Let the tangent of circle Γ be OY with tangent point T and the midpoint of AB be S.

In the triangle $\triangle STO$, there holds

$$\sin\theta = \frac{ST}{OS} = \frac{SA}{SA + OA}.$$

Easily one can get

$$\frac{AB}{OA} = \frac{2AS}{OA} = \frac{2\sin\theta}{1 - \sin\theta}.$$

Conversely, suppose the equality

$$\frac{AB}{OA} = \frac{2\sin\theta}{1 - \sin\theta}$$

holds. Let α be the angle between ray OX and the tangent line of the circle Γ passing through point O. Since when θ is an acute angle, the function $f(\theta) = \frac{2\sin\theta}{1-\sin\theta}$ is monotonically increasing, so we get $\alpha = \theta$. This means that OY is the tangent of the circle Γ.

Below we prove the original problem.

If AB, CD intersect at the point O, then applying the above lemma separately for the two circles with diameters AB, CD, we get

$$\frac{AB}{OA} = \frac{DC}{OD}.$$

Note that $\frac{AB}{OA}$ and $\frac{DC}{OD}$ are equal if and only if BC is parallel to AD.

When AB, CD are parallel, the circle is tangent to the opposite line if and only if $AB = CD$. Then the quadrilateral $ABCD$ is a parallelogram, and BC is parallel to AD.

【Score Situation】 This particular problem saw the following distribution of scores among contestants: 82 contestants scored 7 points, 24 contestants scored 6 points, 3 contestants scored 5 points, 15 contestants scored 4 points, 11 contestants scored 3 points, 10 contestants scored 2 points, 9 contestants scored 1 point, and 38 contestants scored 0 point. The average score of this problem is 4.453, indicating that it was simple.

Among the top five teams in the team scores, the scores of this problem are as follows: the Soviet Union team scored 42 points (with a total team score of 235 points), the Bulgaria team scored 41 points (with a total team score of 203 points), the Romania team scored 42 points (with a total team score of 199 points), the Hungary team scored 42 points (with a total team score of 195 points), and the United States team scored 28 points (with a total team score of 195 points).

The gold medal cutoff for this IMO was set at 40 points (with 14 contestants earning gold medals), the silver medal cutoff was 26 points (with 35 contestants earning silver medals), and the bronze medal cutoff was 17 points (with 49 contestants earning bronze medals).

In this IMO, a total of eight contestants achieved a perfect score of 42 points.

Problem 2.9 (IMO 27-4, proposed by Iceland). Let A, B be adjacent vertices of a regular n-gon ($n \geq 5$) in a plane having centre at O. A triangle XYZ, which is congruent to and internally coincides with OAB, moves in the plane in such a way that Y and Z each trace out the whole boundary of the polygon, X remaining inside the polygon. Find the locus of X.

Solution. Let the radius of a regular n-sided circumscribed circle be R. As shown in Figure 2.27, from the symmetry of the regular polygon, just consider the locus of X when $\triangle XYZ$ moves from the position of $\triangle OAB$ to the position of $\triangle OBC$ (A, B, C are the three adjacent vertices of the regular polygon).

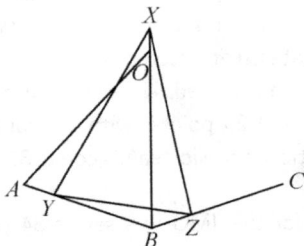

Figure 2.27

It is easy to know

$$\angle YXZ = \angle AOB = 180° - 2\angle OBA$$

$$= 180° - \angle ABC.$$

Thus, X, Y, B, Z are concyclic, so

$$\angle XBY = \angle XZY = \angle OBA.$$

Consequently, X, O, B are collinear.
By the law of sines,

$$BX = \frac{YZ}{\sin \angle ABC} \cdot \sin \angle XYB$$

$$\leq \frac{AB}{\sin \angle AOB} = \frac{2R \sin \frac{180°}{n}}{\sin \frac{360°}{n}}$$

$$= R \sec \frac{180°}{n}.$$

The equality holds only if $\angle XYB = \angle XZB = 90°$, in which case XB exactly bisects YZ vertically.

By the continuity and symmetry, the desired locus is a "star shape" composed of n rays centered on O, whose direction is opposite to the direction of the line from O to each vertex, and whose length is $R\left(\sec \frac{180°}{n} - 1\right)$.

【Score Situation】This particular problem saw the following distribution of scores among contestants: 58 contestants scored 7 points, 7 contestants scored 6 points, 24 contestants scored 5 points, 6 contestants scored 4 points, 8 contestants scored

3 points, 16 contestants scored 2 points, 41 contestants scored 1 point, and 50 contestants scored 0 point. The average score of this problem is 3.281, indicating that it was relatively straightforward.

Among the top five teams in the team scores, the scores of this problem are as follows: the Soviet Union team scored 39 points (with a total team score of 203 points), the United States team scored 38 points (with a total team score of 203 points), the Germany team scored 42 points (with a total team score of 196 points), the China team scored 29 points (with a total team score of 177 points), and the German Democratic Republic team scored 31 points (with a total team score of 172 points).

The gold medal cutoff for this IMO was set at 34 points (with 18 contestants earning gold medals), the silver medal cutoff was 26 points (with 41 contestants earning silver medals), and the bronze medal cutoff was 17 points (with 48 contestants earning bronze medals).

In this IMO, only three contestants achieved a perfect score of 42 points, namely Vladimir Roganov and Stanislav Smirnov from the Soviet Union, and Géza Kós from Hungary.

Problem 2.10 (IMO 37-2, proposed by Canada). Let P be a point inside triangle ABC such that

$$\angle APB - \angle ACB = \angle APC - \angle ABC.$$

Let D, E be the in-centres of triangles APB, APC respectively. Show that AP, BD, and CE meet at a point.

Proof. From the angular bisector theorem, it is obvious that we just need to prove that $\frac{AB}{BP} = \frac{AC}{CP}$.

The following is to prove the above formula.

As shown in Figure 2.28, make the circumcircle of $\triangle ABC$, extend AP to intersect the circle at point Q, and connect to form BQ, CQ. From the four-point concyclic and the problem conditions,

$$\angle PBQ = \angle APB - \angle AQB$$

$$= \angle APB - \angle ACB$$

$$= \angle APC - \angle ABC$$

$$= \angle APC - \angle AQC$$

$$= \angle PCQ.$$

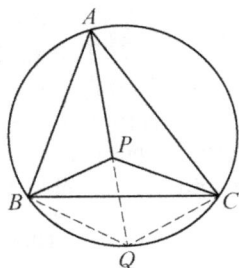

Figure 2.28

So, by the law of sines

$$\frac{BP}{CP} = \frac{PQ \cdot \frac{\sin \angle PQB}{\sin \angle PBQ}}{PQ \cdot \frac{\sin \angle PQC}{\sin \angle PCQ}}$$

$$= \frac{\sin \angle PQB}{\sin \angle PQC} = \frac{\sin \angle ACB}{\sin \angle ABC}$$

$$= \frac{AB}{AC}.$$

【Score Situation】This particular problem saw the following distribution of scores among contestants: 88 contestants scored 7 points, 2 contestants scored 6 points, 6 contestants scored 5 points, 11 contestants scored 4 points, 5 contestants scored 3 points, 10 contestants scored 2 points, 124 contestants scored 1 point, and 178 contestants scored 0 point. The average score of this problem is 2.031, indicating that it had a certain level of difficulty.

Among the top five teams in the team scores, the scores of this problem are as follows: the Romania team scored 39 points (with a total team score of 187 points), the United States team scored 28 points (with a total team score of 185 points), the Hungary team scored 30 points (with a total team score of 167 points), the Russia team scored 36 points (with a total team score of 162 points), and the United Kingdom team scored 14 points (with a total team score of 161 points).

The gold medal cutoff for this IMO was set at 28 points (with 35 contestants earning gold medals), the silver medal cutoff was 20 points (with 66 contestants earning silver medals), and the bronze medal cutoff was 12 points (with 99 contestants earning bronze medals).

In this IMO, only one contestant achieved a perfect score of 42 points, namely Ciprian Manolescu from Romania.

Problem 2.11 (IMO 45-1, proposed by Romania). Let ABC be an acute-angled triangle with $AB \neq AC$. The circle with diameter BC intersects the sides AB and AC at M and N respectively. Denote by O the midpoint of the side BC. The bisectors of the angles $\angle BAC$ and $\angle MON$ intersect at R. Prove that the circumcircles of the triangles BMR and CNR have a common point lying on the side BC.

Proof.   According to the famous Miquel theorem, it is only necessary to prove that A, M, R, N are concyclic.

The proof is as follows: as shown in Figure 2.29, make the circumscribed circle of triangle AMN, and let the point intersecting the angle bisector AK be R'. Then $MR' = NR'$.

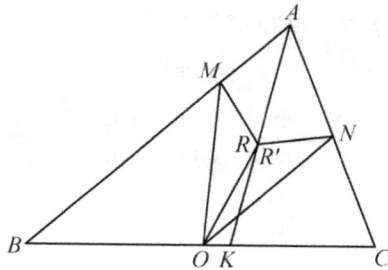

Figure 2.29

Also $MO = \frac{BC}{2} = NO$ and $OR' = OR'$, so $\triangle OMR' \cong \triangle ONR'$. Thus OR' bisects $\angle MON$.

Since $AB \neq AC$, the bisector of MON and AK do not coincide, and nor parallel, so there is a unique intersection point. Hence $R = R'$.

【Score Situation】 This particular problem saw the following distribution of scores among contestants: 191 contestants scored 7 points, 121 contestants scored 6 points, 2 contestants scored 5 points, 9 contestants scored 4 points, 19 contestants scored 3 points, 21 contestants scored 2 points, 29 contestants scored 1 point, and 94 contestants scored 0 point. The average score of this problem is 4.603, indicating that it was simple.

Among the top five teams in the team scores, the scores of this problem are as follows: the China team scored 37 points (with a total team score of 220 points), the United States team scored 40 points (with a total team score of 212 points), the Russia team scored 42 points (with a total team score of 205 points), the Vietnam team scored 38 points (with a total team score of 196 points), and the Bulgaria team scored 39 points (with a total team score of 194 points).

The gold medal cutoff for this IMO was set at 32 points (with 45 contestants earning gold medals), the silver medal cutoff was 24 points (with 78 contestants earning silver medals), and the bronze medal cutoff was 16 points (with 120 contestants earning bronze medals).

In this IMO, a total of four contestants achieved a perfect score of 42 points.

Problem 2.12 (IMO 46-5, proposed by Poland). Let $ABCD$ be a fixed convex quadrilateral with $BC = AD$ and BC be not parallel to AD. Let two variable points E and F lie on the sides BC and AD, respectively and satisfy $BE = DF$. The lines AC and BD meet at P, the lines BD and EF meet at Q, and the lines EF and AC meet at R. Prove that the circumcircles of the triangles PQR, as E and F vary, have a common point other than P.

Proof 1. As shown in Figure 2.30, let the perpendicular bisectors of AC and BD intersect at point O. It is shown below that when points E and F vary, the circumcircle of the triangle PQR passes through point O.

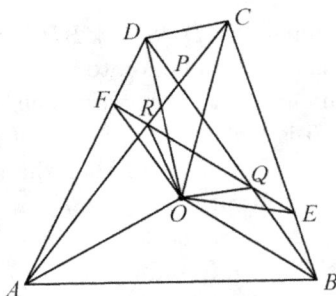

Figure 2.30

Since $OA = OC, OB = OD$, and $BC = AD$, we have $\triangle ODA \cong \triangle OBC$. In other words, $\triangle OBC$ can overlap with $\triangle ODA$ after rotating the degree of $\angle BOD$ around point O.

Since $BE = DF$, the previous rotation overlaps point E with point F, so $OE = OF$, and $\angle EOF = \angle BOD = \angle COA$.

Thus $\triangle EOF \backsim \triangle BOD \backsim \triangle COA$, so

$$\angle OEF = \angle OFE = \angle OBD = \angle ODB = \angle OCA = \angle OAC,$$

Then O, B, E, Q and O, E, C, R are concyclic, so

$$\angle OQB = \angle OEB = \angle ORC,$$

and thus the points P, Q, O, R are concyclic.

In summary, when the points E and F vary, the circumcircle of the triangle PQR passes through another fixed point O besides P.

Proof 2. As show in Figure 2.31, since BC and AD are not parallel, the circumcircles of $\triangle APD$ and $\triangle BPC$ are not tangent to each other. Otherwise, construct a common tangent line SS' through tangent point P.

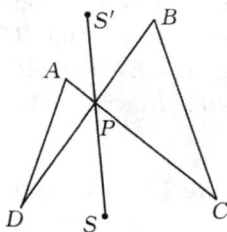

Figure 2.31

It follows from the equality $\angle DPS = \angle BPS'$ that $\angle BCP = \angle DAP$. Then $AD \parallel BC$, which is in contradiction to the condition.

Let the second common point of the circumcircles of triangles $\triangle APD$ and $\triangle BPC$ be O, which is fixed. Without loss of generality, let O be an interior point of $\triangle DPC$. We will prove that the circumcircle of $\triangle PQR$ passes through O as E and F vary.

As show in Figure 2.32, connect the lines $OA, OB, OC, OD, OE,$ OF, OP, OQ, OR. Since B, C, O, P and O, P, A, D are concyclic,

$$\angle OBC = \angle OPC, \quad \angle OPC = \angle ADO \Rightarrow \angle OBC = \angle ADO.$$

Similarly, $\angle OCB = \angle DPO = \angle DAO$.

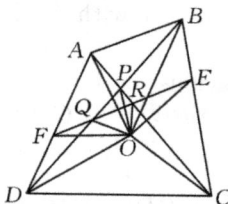

Figure 2.32

Together with $AD = BC$, we get $\triangle OBC \cong \triangle ODA$. So,

$$OB = OD, \angle OBE = \angle ODF.$$

Note that $BE = DF$, so the triangles OBE and ODF are congruent, giving $OE = OF$ and $OB = OD$.

The equalities $\angle FOE = \angle FOB + \angle BOE = \angle BOF + \angle FOD = \angle BOD$ imply that the triangles BOD and FOE are similar. This means $\angle EFO = \angle BDO$, that is $\angle QFO = \angle QDO$, so the points Q, F, D, O are concyclic. Thus $\angle RQO = \angle FDO$.

Since O, P, A, D are concyclic, $\angle FDO = \angle ADO = \angle RPO$, so $\angle RQO = \angle RPO$. The points O, R, P, Q are concyclic.

In summary, when the points E and F vary, the circumcircle of the triangle PQR passes through another fixed point O besides P.

【Score Situation】This particular problem saw the following distribution of scores among contestants: 125 contestants scored 7 points, 8 contestants scored 6 points, 6 contestants scored 5 points, 4 contestants scored 4 points, 16 contestants scored 3 points, 30 contestants scored 2 points, 36 contestants scored 1 point, and 288 contestants scored 0 point. The average score of this problem is 2.170, indicating that it had a certain level of difficulty.

Among the top five teams in the team scores, the scores of this problem are as follows: the China team scored 42 points (with a total team score of 235 points), the United States team scored 38 points (with a total team score of 213 points), the Russia team scored 35 points (with a total team score of 212 points), the Iran team scored 37 points (with a total team score of 201 points), and the South Korea team scored 42 points (with a total team score of 200 points).

The gold medal cutoff for this IMO was set at 35 points (with 42 contestants earning gold medals), the silver medal cutoff was 23 points (with 79 contestants earning silver medals), and the bronze medal cutoff was 12 points (with 128 contestants earning bronze medals).

In this IMO, a total of 16 contestants achieved a perfect score of 42 points.

Problem 2.13 (IMO 52-6, proposed by Japan). Let ABC be an acute triangle with circumcircle Γ. Let l be a tangent line to Γ, and let l_a, l_b, and l_c be the lines obtained by reflecting l in the lines BC, CA, and AB, respectively. Show that the circumcircle of the triangle determined by the lines l_a, l_b and l_c is tangent to the circle Γ.

Proof 1. As shown in Figure 2.33, let l intersect the lines l_a, l_b, l_c at the points L, M, N, and denote $l_a \cap l_b = F, l_b \cap l_c = D$, and $l_c \cap l_a = E$.

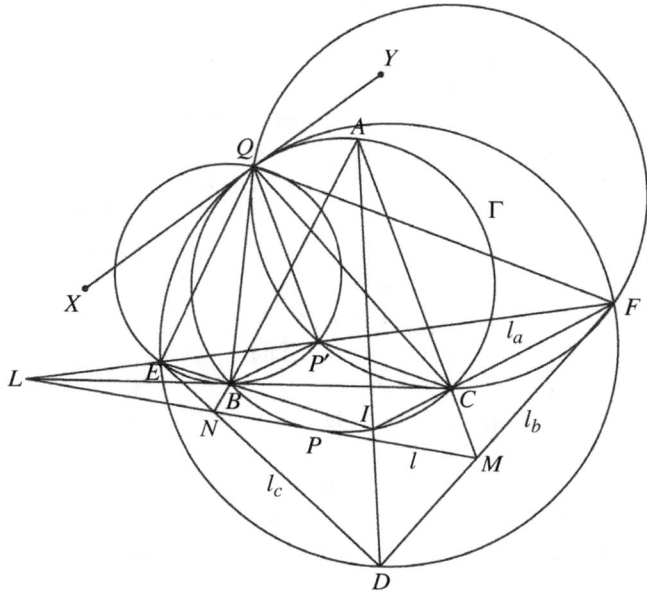

Figure 2.33

According to the known conditions, LB bisects $\angle ELM$ and NB bisects $\angle MNE$, so the point B is the excenter of $\triangle LNE$, hence EB bisects $\angle FED$. In the same way, DA bisects $\angle EDF$ and FC bisects $\angle EFD$. Thus, the three lines DA, EB, FC meet at a point I, which is the incenter of $\triangle DEF$.

Since A is the excenter of $\triangle DMN$,

$$\angle BAC = 180° - \frac{1}{2}\angle ENM - \frac{1}{2}\angle FMN$$

$$= 90° - \frac{1}{2}\angle NDM$$

$$= 90° - \frac{1}{2}\angle EDF.$$

And since $\angle BIC = \angle EIF = 90° + \frac{1}{2}\angle EDF = 180° - \angle BAC$, point I is on the circle Γ.

Let the symmetry point of point P with respect to line BC be P'. Then point P' is on line EF. Let another intersection point between the circumcircle of $\triangle P'CF$ and the circumcircle of $\triangle P'BE$ be Q. Then

$$\angle BQC = \angle BQP' + \angle CQP'$$
$$= \angle BEP' + \angle CFP'$$
$$= 180° - \angle BIC,$$

hence, point Q is on the circle Γ.

Since

$$\angle EQF = \angle EQP' + \angle FQP'$$
$$= \angle IBP' + \angle ICP'$$
$$= 360° - 2\angle BIC$$
$$= 180° - \angle EDF,$$

it follows that the point Q is also on the circumcircle c_1 of $\triangle DEF$.

Cross the point Q and make the tangent XY of the circle Γ. By symmetry, the circumcircle of $\triangle P'BC$ is tangent to the line EF at point P', from which $\angle FP'C = \angle P'BC$.

Hence,

$$\angle YQF = \angle YQC - \angle FQC$$
$$= \angle QBC - \angle FP'C$$
$$= \angle QBP' = \angle QEF,$$

and thus, the line XY is also tangent to the circle c_1.

Therefore, the circle c_1 is tangent to the circle Γ at point Q.

Proof 2. (Based on the Solution by Lin Chen). As shown in Figure 2.34, let P be the tangent point of l to Γ. Denote the symmetric points of P with respect to BC, CA, AB as P_a, P_b, P_c, respectively. These points are on circles Γ_a, Γ_b, and Γ_c and symmetric to Γ about, BC, CA and AB, respectively. Hence, l_a, l_b, and l_c are tangent to Γ_a, Γ_b, and Γ_c at points P_a, P_b, and P_c, respectively.

Denote $l_a \cap l_b = C', l_b \cap l_c = A'$, and $l_c \cap l_a = B'$.

Define the oriented angle of line m to line n by $\angle(m, n)$. The magnitude of $\angle(m, n)$ is the angle rotated anticlockwise from m to n.

We conclude that

(1) Points P_a, P_b, and P_c are collinear.

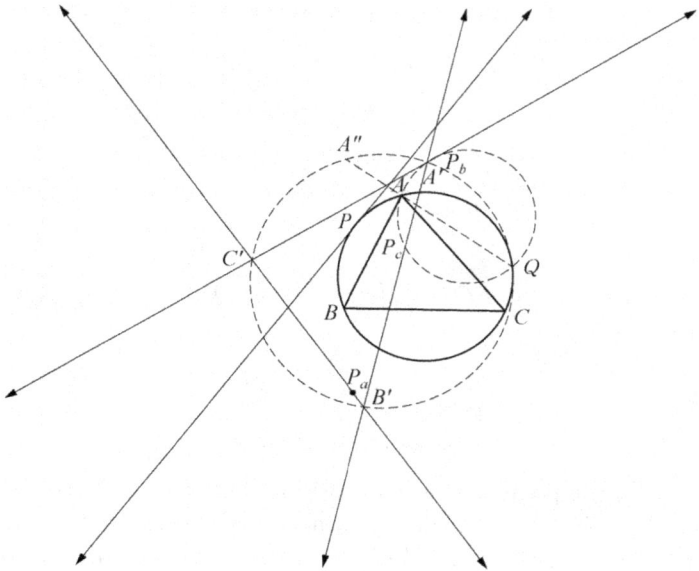

Figure 2.34

In fact, the midpoints of PP_a, PP_b, and PP_c are the pedal points of P to BC, CA, AB, respectively. The pedal points are collinear by Simson's theorem. So P_a, P_b, P_c are collinear.

(2) Denote the circumcircles of $\triangle A'P_bP_c, \triangle B'P_cP_a$, and $\triangle C'P_aP_b$ by Γ_1, Γ_2, and Γ_3, respectively and the circumcircle of $\triangle A'B'C'$ by Ω. Then the four circles $\Gamma_1, \Gamma_2, \Gamma_3$, and Ω meet at a point.

In fact, this is Miquel's theorem for the complete quadrilateral $A'P_cB'P_aC'P_b$. Denote the coincident point by Q.

(3) Points A, B, and C are on the circles Γ_1, Γ_2, and Γ_3, respectively.

In fact, the circles Γ_b and Γ_c intersect at point A, and $\widehat{AP_c} = \widehat{AP} = \widehat{AP_b}$. Rotate an angle of $\measuredangle(P_cA, P_bA)$ about point A, under which $\Gamma_c \to \Gamma_b$, $P_c \to P_b$, and tangent line $l_c \to l_b$. So, $\measuredangle(l_c, l_b) = \measuredangle(P_cA, P_bA)$ and $\measuredangle(l_c, l_b) = \measuredangle(P_cA', P_bA')$, which means the four points are concyclic. That is, point A is on circle Γ_1. Similarly, points B and C are on circles Γ_2 and Γ_3, respectively.

(4) Point Q is on circle Γ. By the definition of point Q,

$$\measuredangle(AQ, BQ) = \measuredangle(AQ, P_cQ) + \measuredangle(P_cQ, BQ)$$
$$= \measuredangle(AP_b, P_bP_c) + \measuredangle(P_cP_a, BP_a)$$

$$= \angle(AP_b, BP_a)$$
$$= \angle(AP_b, AC) + \angle(AC, BC) + \angle(BC, BP_a)$$
$$= \angle(AC, AP) + \angle(AC, BC) + \angle(BP, BC)$$
$$= 2\angle(AC, BC) - \angle(AP, BP) = \angle(AC, BC).$$

Hence, the four points A, B, C, and Q are concyclic, that is, point Q is on circle Γ.

(5) Circle Γ is tangent to circle Ω at point Q.

Let line QA intersect circle Ω at points Q and A''. Then

$$\angle(A''B', B'Q) = \angle(A''B', B'A') + \angle(A'B', B'Q)$$
$$= \angle(A''Q, QA') + \angle(A'B', B'Q)$$
$$= \angle(AQ, A'Q) + \angle(A'B', B'Q)$$
$$= \angle(AP_c, A'P_c) + \angle(A'B', B'Q)$$
$$= \angle(AB, BP_c) + \angle(A'B', B'Q)$$
$$= \angle(AB, BP_c) + \angle(P_cB', B'Q)$$
$$= \angle(AB, BP_c) + \angle(P_cB, BQ) = \angle(AB, BQ).$$

This shows that the degrees of $\overset{\frown}{A''Q}$ and $\overset{\frown}{AQ}$ are equal in circle Ω. Note that points A, Q, and A'' are collinear, and point Q is the same point on circles Ω and Γ. Therefore, the tangent lines of the two circles at point Q coincide. That is, the circles Ω and Γ are tangent at Q.

【Score Situation】 This particular problem saw the following distribution of scores among contestants: 6 contestants scored 7 points, no contestant scored 6 points, 3 contestants scored 5 points, no contestant scored 4 points, 2 contestants scored 3 points, 7 contestants scored 2 points, 102 contestants scored 1 point, and 443 contestants scored 0 point. The average score of this problem is 0.318, indicating that it was extremely difficult.

Among the top five teams in the team scores, the scores of this problem are as follows: the China team scored 9 points (with a total team score of 189 points), the United States team scored 5 points (with a total team score of 184 points), the Singapore team scored 9 points (with a total team score of 179 points), the Russia team scored 4 points (with a total team score of 161 points), and the Thailand team scored 3 points (with a total team score of 160 points).

The gold medal cutoff for this IMO was set at 28 points (with 54 contestants earning gold medals), the silver medal cutoff was 22 points (with 90 contestants earning silver medals), and the bronze medal cutoff was 16 points (with 137 contestants earning bronze medals).

In this IMO, only one contestant achieved a perfect score of 42 points, namely Lisa Sauermann from Germany.

Problem 2.14 (IMO 53-1, proposed by Greece). Given triangle ABC, a point J is the center of the excircle opposite the vertex A. This excircle is tangent to the side BC at M, and to the lines AB and AC at K and L, respectively. The lines LM and BJ meet at F, and the lines KM and CJ meet at G. Let S be the point of intersection of the lines AF and BC, and let T be the point of intersection of the lines AG and BC.

Prove that M is the midpoint of ST. (The excircle of ABC opposite the vertex A is the circle that is tangent to the line segment BC, to the ray AB beyond B, and to the ray AC beyond C).

Proof. As shown in Figure 2.35, let $\angle CAB = \alpha$, $\angle ABC = \beta$, $\angle BCA = \gamma$. Since AJ is the bisector of $\angle CAB$, we obtain $\angle JAK = \angle JAL = \frac{\alpha}{2}$. Since $\angle AKJ = \angle ALJ = 90°$, the points K, L are on the circle ω with diameter AJ.

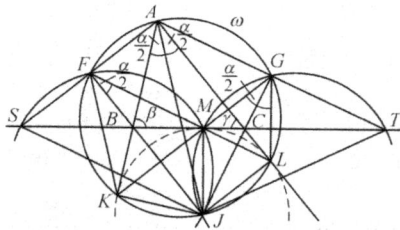

Figure 2.35

Since BJ is the bisector of $\angle KBM$, we have $\angle MBJ = 90° - \frac{\beta}{2}$. Similarly, $\angle CML = \frac{\gamma}{2}$ and $\angle MCJ = 90° - \frac{\gamma}{2}$. Consequently,

$$\angle LFJ = \angle MBJ - \angle BMF = \angle MBJ - \angle CML$$
$$= \left(90° - \frac{\beta}{2}\right) - \frac{\gamma}{2} = \frac{\alpha}{2}.$$

Hence, point F is on the circle ω. Similarly, point G is also on the circle ω. Since AJ is a diameter of circle ω, we have $\angle AFJ = \angle AGJ = 90°$.

The segments AB and BC are symmetric about the bisector of the external angle of $\angle ABC$. And by $AF \perp BF$ and $KM \perp BF$, we see that the segments SM and AK are symmetric about BF, that is $SM = AK$. Similarly, $TM = AL$.

And since $AK = AL$, we have $SM = TM$, namely, M is the midpoint of ST.

【Score Situation】This particular problem saw the following distribution of scores among contestants: 401 contestants scored 7 points, 2 contestants scored 6 points, 11 contestants scored 5 points, 16 contestants scored 4 points, 24 contestants scored 3 points, 15 contestants scored 2 points, 37 contestants scored 1 point, and 41 contestants scored 0 point. The average score of this problem is 5.625, indicating that it was simple.

Among the top five teams in the team scores, the scores of this problem are as follows: the South Korea team scored 42 points (with a total team score of 209 points), the China team scored 42 points (with a total team score of 195 points), the United States team scored 42 points (with a total team score of 194 points), the Russia team scored 42 points (with a total team score of 177 points), the Canada team scored 42 points (with a total team score of 159 points), and the Thailand team scored 42 points (with a total team score of 159 points).

The gold medal cutoff for this IMO was set at 28 points (with 51 contestants earning gold medals), the silver medal cutoff was 21 points (with 88 contestants earning silver medals), and the bronze medal cutoff was 14 points (with 137 contestants earning bronze medals).

In this IMO, only one contestant achieved a perfect score of 42 points, namely Jeck Lim from Singapore.

Problem 2.15 (IMO 54-3, proposed by Russia). Let the excircle of triangle ABC opposite the vertex A be tangent to the side BC at a point A_1. Define points B_1 on AC and C_1 on AB analogously, using the excircles opposite B and C, respectively. Suppose that the circumcenter of triangle $A_1 B_1 C_1$ lies on the circumcircle of triangle ABC. Prove that triangle ABC is right-angled.

The excircle of triangle ABC opposite the vertex A is the circle that is tangent to the line segment BC, to the ray AB beyond B, and to the ray AC beyond C. The excircles opposite B and C are similarly defined.

Proof. Denote the circumcircles of $\triangle ABC$ and $\triangle A_1 B_1 C_1$ by Ω and Γ, respectively. Let A_0 be the midpoint of arc BC on Ω containing point A. Points B_0 and C_0 are defined analogously. Let Q be the center of circle Γ.

Then Q is on Ω by the hypothesis of the problem. First, we give the following lemma.

Lemma. $A_0B_1 = A_0C_1$. *Points* A, A_0, B_1, *and* C_1 *are concyclic.*

Proof of the Lemma.   If points A_0 and A coincide, then $\triangle ABC$ is an isosceles triangle, and thus $AB_1 = AC_1$.

If points A_0 and A do not coincide, by the definition of A_0, we know that $A_0B = A_0C$. It is evident that

$$BC_1 = CB_1 \left(= \frac{1}{2}(b + c - a) \right),$$

and

$$\angle C_1BA_0 = \angle ABA_0 = \angle ACA_0 = \angle B_1CA_0.$$

Thus

$$\triangle A_0BC_1 \cong \triangle A_0CB_1.$$

So, $A_0B_1 = A_0C_1$.

Also, by (1), we have $\angle A_0C_1B = \angle A_0B_1C$. Hence $\angle A_0C_1A = \angle A_0B_1A$. Thus, points A, A_0, B_1 and C_1 are concyclic.

Obviously, points A_1, B_1, C_1 are on a semicircular arc of Γ, so $\triangle A_1B_1C_1$ is an obtuse-angled triangle. Without loss of generality, we may suppose that $\angle A_1B_1C_1$ is obtuse. Then points Q and B_1 are on different sides of A_1C_1. Obviously, so are points B and B_1. Hence, points Q and B are on the same side of A_1C_1.

Note that the perpendicular bisector of A_1C_1 intersects Γ at two points which are on different sides of A_1C_1. By the above argument, B_0 and Q are among the intersection points, and B_0 and Q are on the same side of A_1C_1. So B_0 is Q, as shown in Figure 2.36.

By the lemma, lines QA_0 and QC_0, are perpendicular bisectors of B_1C_1 and A_1B_1, respectively. Also A_0 and C_0 are midpoints of arcs CB and BA respectively. Thus,

$$\angle C_1B_0A_1 = \angle C_1B_0B_1 + \angle B_1B_0A_1 = 2\angle A_0B_0B_1 + 2\angle B_1B_0C_0$$

$$= 2\angle A_0B_0C_0 = 180° - \angle ABC.$$

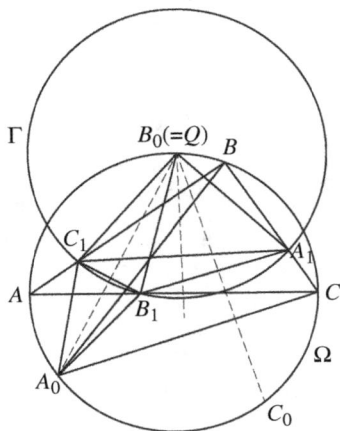

Figure 2.36

On the other hand, by the lemma once more,

$$\angle C_1 B_0 A_1 = \angle C_1 B A_1 = \angle ABC.$$

Hence, $\angle ABC = 180° - \angle ABC$, so $\angle ABC = 90°$. This completes the proof.

【Score Situation】 This particular problem saw the following distribution of scores among contestants: 41 contestants scored 7 points, 4 contestants scored 6 points, 3 contestants scored 5 points, no contestant scored 4 points, 16 contestants scored 3 points, 15 contestants scored 2 points, 10 contestants scored 1 point, and 438 contestants scored 0 point. The average score of this problem is 0.786, indicating that it was extremely difficult.

Among the top five teams in the team scores, the scores of this problem are as follows: the China team scored 30 points (with a total team score of 208 points), the South Korea team scored 26 points (with a total team score of 204 points), the Singapore team scored 14 points (with a total team score of 190 points), the Russia team scored 16 points (with a total team score of 187 points), and the North Korea team scored 21 points (with a total team score of 184 points).

The gold medal cutoff for this IMO was set at 31 points (with 45 contestants earning gold medals), the silver medal cutoff was 24 points (with 92 contestants earning silver medals), and the bronze medal cutoff was 15 points (with 141 contestants earning bronze medals).

In this IMO, no contestant achieved a perfect score of 42 points.

Problem 2.16 (IMO 55-3, proposed by Iran). Convex quadrilateral $ABCD$ has $\angle ABC = \angle CDA = 90°$. Point H is the foot of the altitude from A to BD. Points S and T lie on the sides AB and AD, respectively, such that H lies inside triangle SCT and

$$\angle CHS - \angle CSB = 90°, \angle THC - \angle DTC = 90°.$$

Prove that line BD is tangent to the circumcircle of triangle TSH.

Proof 1.   Suppose that the line passing C and perpendicular to line SC intersects line AB at point Q (see Figure 2.37). Then

$$\angle SQC = 90° - \angle BSC = 180° - \angle SHC.$$

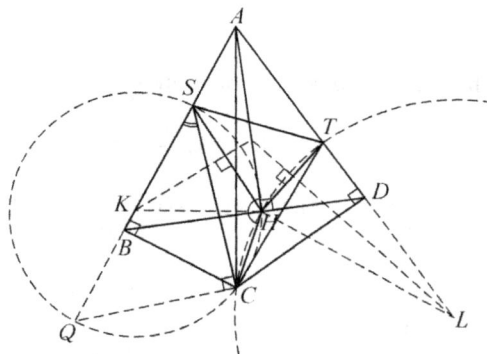

Figure 2.37

Hence, points C, H, S, and Q are concyclic with SQ its diameter. Therefore, the circumcenter K of $\triangle SHC$ is on line AB. In the same manner, the circumcenter L of $\triangle CHT$ is on line AD.

To show that line BD is tangent to the circumcircle of $\triangle TSH$, it suffices to show that the intersection point of perpendicular bisectors of HS and HT is on line AH. But the perpendicular bisectors are just the angle bisectors of $\angle AKH$ and $\angle ALH$, respectively. By the internal angle bisector theorem, it suffices to show that

$$\frac{AK}{KH} = \frac{AL}{LH}. \tag{1}$$

In the following, we give two proofs of (1).

(i) Let M be the intersection point of lines KL, HC (see Figure 2.38).

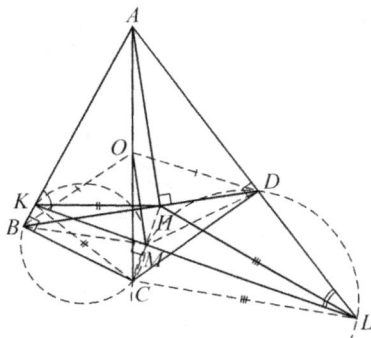

Figure 2.38

Since $KH = KC$ and $LH = LC$, points H and C are symmetric over line KL. Thus, M is the midpoint of HC. Let O be the circumcenter of the quadrilateral $ABCD$. Hence, O is the midpoint of AC and consequently $OM \parallel AH$. Therefore $OM \perp BD$. Further by $OB = OD$, we see that OM is the perpendicular bisector of BD. Thus $BM = DM$.

Since $CM \perp KL$, points B, C, M, K are concyclic with KC as its diameter. Similarly, points L, C, M, D are concyclic with LC as its diameter. So, by the law of sines,

$$\frac{AK}{AL} = \frac{\sin \angle ALK}{\sin \angle AKL} = \frac{DM}{CL} \cdot \frac{CK}{BM} = \frac{CK}{CL} = \frac{KH}{LH},$$

which gives (1).

(ii) If points A, H, C are collinear, then $AK = AL$ and $KH = LH$. Thus (1) holds. So, let ω be the circle passing A, H, C. Since points A, B, C, D are on a circle,

$$\angle BAC = \angle BDC = \angle 90° - \angle ADH = \angle DAH.$$

Let $N \neq A$ be the other intersection point of the circle ω and the bisector of $\angle CAH$. Then AN is also the bisector of $\angle BAD$. Since points H and C are symmetric with respect to line KL, and $HN = NC$, we see that point N and the center of ω are both on line KL. That is, the circle ω is the Apollonian circle of points K, L, and thus (1) holds.

Proof 2. Extend CB to E such that $CB = BE$. Extend CD to F such that $CD = DF$. Connect to form the line segment in Figure 2.39.

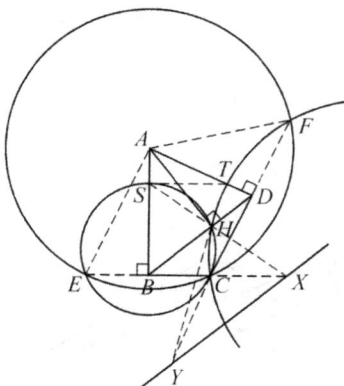

Figure 2.39

So, AB bisects CE vertically and AD bisects CF vertically. Thus $CS = ES$, $CT = FT$, and A is the circumcenter of $\triangle CEF$.

Since $BD \parallel EF$, $AH \perp BD$, so $AH \perp EF$. Then AH bisects EF vertically, so $HE = HF$.

And since

$$\angle CHS = 90° + \angle BSC = 180° - \angle SCB$$

$$= 180° - \angle SEB.$$

So, points S, E, C, H are concyclic, and consequently

$$XH \cdot XS = XC \cdot XE. \tag{2}$$

At the same time, since $CS = ES$ and S, H are on the same side of CE, we know that SH is the external angle bisector of $\angle CHE$.

Similarly, since

$$\angle CHT = 90° + \angle DTC = 180° - \angle TCD$$

$$= 180° - \angle TFD,$$

points T, F, C, H are concyclic, from which

$$YC \cdot YF = YH \cdot YT. \tag{3}$$

At the same time, since $CT = FT$ and T, H are on the same side of CF, we know that TH is the external angle bisector of $\angle CHF$.

According to the passage, C, H are on the same side of AB, so $CH < EH = FH$.

Let X be the intersection of lines EC, SH, and Y be the intersection of lines FC, TH. Then X is on the extension of line EC, and Y is on the extension of line FC. According to the external angle bisector theorem,

$$\frac{CX}{XE} = \frac{CH}{HE} = \frac{CH}{HF} = \frac{CY}{YF}.$$

So $XY \| EF$, and consequently $XY \perp AH$.

From (2) and (3), the line XY is the radical axis of the circumcircle of $\triangle CEF$ and the circumcircle of $\triangle TSH$ (let its circumcenter be O). Since $XY \perp AO$, we see that A, O, H are collinear. And $OH \perp XY$, $OH \perp BD$. Thus, BD is tangent to the circle O at point H.

Proof 3. Let $\angle THA = \alpha$, $\angle DAH = \beta$, and denote

$$AH = a, \quad BH = b, \quad DH = d, \quad DT = x.$$

Extend TH and DC at the point L. Take the point M on AH such that $MT \perp TH$; take the point N on AH such that $NS \perp SH$. Make $TR \perp AH$ at R; make $MU \perp AD$ at U; make $HV \perp AD$ at V. The next step is to prove that $M = N$.

According to $\angle DAH = \beta$, we know $\angle ADH = 90° - \beta$, $\angle BDC = \beta$. And at the same time,

$$\angle ABD = \angle ACD, \quad \angle AHB = \angle ADC.$$

Therefore, $\triangle ADC \backsim \triangle AHB$, and then $DC = b \cdot \frac{\sqrt{a^2+d^2}}{a}$, so

$$TC^2 = TD^2 + DC^2 = x^2 + \frac{b^2(a^2 + d^2)}{a^2}.$$

By the spread angle theorem

$$\frac{1}{d} = \frac{\cos \beta}{DL} + \frac{\sin \beta}{x},$$

so

$$DL = \frac{xd \cos \beta}{x - d \sin \beta}.$$

By the law of cosines,

$$TH^2 = x^2 + d^2 - 2xd \sin \beta.$$

Then

$$\frac{TL}{TH} = \frac{DL}{DH \sin(90° - \beta)} = \frac{x}{x - d \sin \beta},$$

so

$$TL \cdot TH = \frac{TL}{TH} \cdot TH^2$$

$$= (x^2 + d^2 - 2xd\sin\beta)\frac{x}{x - d\sin\beta}.$$

Because $\angle THC - \angle DTC = 90°$, we have $\angle THC = \angle TCL$, so $TC^2 = TL \cdot TH$, i.e.,

$$x^2 + \frac{b^2(a^2 + d^2)}{a^2} = (x^2 + d^2 - 2xd\sin\beta)\frac{x}{x - d\sin\beta}.$$

Hence,

$$x^2 + \frac{b^2(a^2 + d^2) - a^2 d^2}{a^2 d \sin\beta} x - \frac{b^2(a^2 + d^2)}{a^2} = 0, \tag{4}$$

as a result,

$$x = \frac{a^2 d^2 - a^2 b^2 - b^2 d^2 + \sqrt{k}}{2a^2 d \sin\beta} \tag{5}$$

(since T is on AD, we know that x should take the positive root), of which

$$k = a^4 b^4 + b^4 d^4 + a^4 d^4 + 2a^2 b^4 d^2 + 2a^2 b^2 d^4 - 2a^4 b^2 d^2.$$

Notice that the expression for k is symmetric about b and d.

$$\tan\alpha = \frac{TR}{RH} = \frac{(\sqrt{a^2 + d^2} - x)d}{ax},$$

$$\tan(\alpha + \beta) = \frac{DL}{DT} = \frac{\frac{xd\cos\beta}{x - d\cos\beta}}{x} = \frac{d\cos\beta}{x - d\cos\beta},$$

$$\frac{AM}{AH} = \frac{MU}{HV} = \frac{MU}{TV} \cdot \frac{TV}{VH} = \frac{MT}{TH} \cdot \frac{TV}{VH} = \frac{\tan\alpha}{\tan(\alpha + \beta)}$$

$$= \frac{(x - d\sin\beta)(\sqrt{a^2 + d^2} - x)}{ax\cos\beta}$$

$$= \frac{(\sqrt{a^2 + d^2} - x)(\sqrt{a^2 + d^2}x - d^2)}{a^2 x}$$

$$= \frac{1}{a^2}\left(-\sqrt{a^2 + d^2}x - \frac{d^2\sqrt{a^2 + d^2}}{x} + a^2 + 2d^2\right).$$

By formula (4),

$$\frac{1}{x} = \frac{a^2}{a^2b^2 + b^2d^2}x + \frac{a^2b^2 + b^2d^2 - a^2d^2}{(a^2b^2 + b^2d^2)d\sin\beta}.$$

Then

$$\frac{AM}{AH} = \frac{1}{a^2}\left(-\frac{a^2b^2 + b^2d^2 + a^2d^2}{a^2b^2 + b^2d^2}\sqrt{a^2 + d^2}x\right.$$

$$\left. -\frac{a^2b^2 + b^2d^2 - a^2d^2}{b^2} + a^2 + 2d^2\right).$$

Combining with formula (5), we get

$$\frac{AM}{AH} = \frac{1}{a^2}\left(-(a^2b^2 + b^2d^2 + a^2d^2)\frac{\sqrt{k}}{2a^2b^2d^2}\right.$$

$$\left. + \frac{a^4d^4 + a^4b^4 + b^4d^4 + 2a^2b^4d^2 + 2a^2b^2d^4}{2a^2b^2d^2}\right).$$

Since $\frac{AM}{AH}$ is symmetric about b and d, so $\frac{AM}{AH} = \frac{AN}{AH}$, i.e., $M = N$. Thus, points M, S, H, T are concyclic. This circumcenter is on AH, and $AH \perp BD$.

Hence, line BD is tangent externally to the circumcircle of $\triangle TSH$.

Note. This problem has an extension by the problem selection committee as follows.

For a convex quadrilateral $ABCD$, points S, T are on sides AB, AD, respectively. And point H is located in the inner part of $\triangle SCT$, satisfying $\angle BAC = \angle DAH, \angle CHS - \angle CSB = 90°$, and $\angle THC - \angle DTC = 90°$.

Then the circumcenter of $\triangle TSH$ is on AH and the circumcenter of $\triangle SCT$ is on AC.

【Score Situation】 This particular problem saw the following distribution of scores among contestants: 28 contestants scored 7 points, 4 contestants scored 6 points, no contestant scored 5 points, 3 contestants scored 4 points, 2 contestants scored 3 points, 1 contestant scored 2 points, 43 contestants scored 1 point, and 479 contestants scored 0 point. The average score of this problem is 0.505, indicating that it was extremely difficult.

Among the top five teams in the team scores, the scores of this problem are as follows: the China team scored 16 points (with a total team score of 201 points), the United States team scored 23 points (with a total team score of 193 points), the Chinese Taiwan team scored 28 points (with a total team score of 192 points), the Russia team scored 29 points (with a total team score of 191 points), and the Japan team scored 9 points (with a total team score of 177 points).

The gold medal cutoff for this IMO was set at 29 points (with 49 contestants earning gold medals), the silver medal cutoff was 22 points (with 113 contestants earning silver medals), and the bronze medal cutoff was 16 points (with 133 contestants earning bronze medals).

In this IMO, only three contestants achieved a perfect score of 42 points, namely Jiyang Gao from China, Alexander Gunning from Australia and Po-Sheng Wu from Chinese Taiwan.

Problem 2.17 (IMO 56-4, proposed by Greece). Triangle ABC has circumcircle Ω and circumcenter O. A circle Γ with center A intersects the segment BC at points D and E, such that B, D, E, C are all different and lie on line BC in this order. Let F and G be the points of intersection of Γ and Ω, such that A, F, B, C, G lie on Ω in this order. Let K be the second point of intersection of the circumcircle of triangle BDF and the segment AB Let L be the second point of intersection of the circumcircle of triangle CGE and the segment CA.

Suppose that the lines FK and GL are different and intersect at the point X. Prove that X lies on the line AO.

Proof. As shown in Figure 2.40, since $AF = AG$, and AO is the bisector of $\angle FAG$, so F, G are symmetric about the line AO. In order to prove that X is on AO, we just prove that $\angle AFK = \angle AGL$.

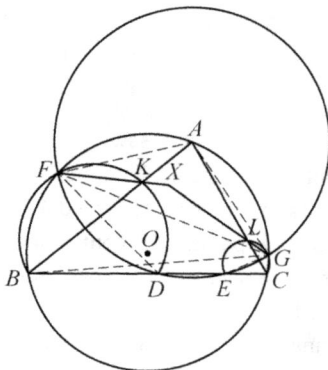

Figure 2.40

First of all,

$$\angle AFK = \angle DFG + \angle GFA - \angle DFK.$$

Since D, F, G, E are concyclic,

$$\angle DFG = \angle CEG.$$

Since A, F, B, G are concyclic,

$$\angle GFA = \angle GBA.$$

And since D, B, F, K are concyclic,

$$\angle DFK = \angle DBK.$$

Then,

$$\angle AFK = \angle CEG + \angle GBA - \angle DBK = \angle CEG - \angle CBG.$$

Since C, E, L, G are concyclic, $\angle CEG = \angle CLG$; and C, B, A, G are concyclic, so $\angle CBG = \angle CAG$. Hence

$$\angle AFK = \angle CLG - \angle CAG = \angle AGL.$$

【Score Situation】This particular problem saw the following distribution of scores among contestants: 351 contestants scored 7 points, 8 contestants scored 6 points, 1 contestant scored 5 points, 11 contestants scored 4 points, 18 contestants scored 3 points, 61 contestants scored 2 points, 36 contestants scored 1 point, and 91 contestants scored 0 point. The average score of this problem is 4.794, indicating that it was simple.

Among the top five teams in the team scores, the scores of this problem are as follows: the United States team scored 42 points (with a total team score of 185 points), the China team scored 42 points (with a total team score of 181 points), the South Korea team scored 42 points (with a total team score of 161 points), the North Korea team scored 42 points (with a total team score of 156 points), and the Vietnam team scored 42 points (with a total team score of 151 points).

The gold medal cutoff for this IMO was set at 26 points (with 39 contestants earning gold medals), the silver medal cutoff was 19 points (with 100 contestants earning silver medals), and the bronze medal cutoff was 14 points (with 143 contestants earning bronze medals).

In this IMO, only one contestant achieved a perfect score of 42 points, namely Zhuo Qun (Alex) Song from Canada.

Problem 2.18 (IMO 57-1, proposed by Belgium). Triangle BCF has a right angle at B. Let A be a point on line CF such that $FA = FB$, and

F lies between A and C. Point D is chosen such that $DA = DC$ and AC is the bisect or of $\angle DAB$. Point E is chosen such that $EA = ED$ and AD is the bisector of $\angle EAC$. Let M be the midpoint of CF. Let X be the point such that $AMXE$ is a parallelogram (where $AM \parallel EX$ and $AE \parallel MX$). Prove that lines BD, FX, ME are concurrent.

Proof. As shown in Figure 2.41, we have

$$\angle FAB = \angle FBA = \angle DAC = \angle DCA = \angle EAD = \angle EDA.$$

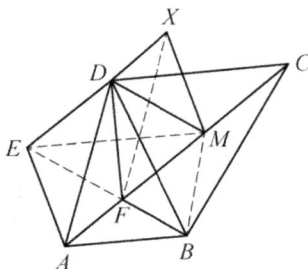

Figure 2.41

Denote their value by θ.

Since $\triangle ABF \backsim \triangle ACD$, we have $\frac{AB}{AC} = \frac{AF}{AD}$, so $\triangle ABC \backsim \triangle AFD$. And since $EA = ED$,

$$\angle AFD = \angle ABC = 90° + \theta = 180° - \frac{1}{2}\angle AED.$$

So F lies on the circle with centre E and radius EA. In particular, $EF = EA = ED$. And note that

$$\angle EFA = \angle EAF = 2\theta = \angle BFC.$$

Thus B, F, E are collinear.

Since $\angle EDA = \angle MAD$, we get $ED \parallel AM$, so E, D, X are collinear.

M is the midpoint on the hypotenuse of the right triangle CBF, so $MF = MB$.

In the isosceles triangles EFA and MFB, we have $\angle EFA = \angle MFB, AF = BF$.

So, they are congruent, implying $BM = AE = XM$ and

$$BE = BF + FE = AF + FM = AM = EX.$$

Therefore $\triangle EMB \cong \triangle EMX$.

And since $EF = ED$, we see that D and F are symmetric about the line EM; note that X and B are symmetric about EM as well. Hence the lines BD and XF are symmetric about the line EM.

From this, we conclude that BD, FX, ME are concurrent.

【Score Situation】 This particular problem saw the following distribution of scores among contestants: 391 contestants scored 7 points, 14 contestants scored 6 points, 35 contestants scored 5 points, 6 contestants scored 4 points, 9 contestants scored 3 points, 32 contestants scored 2 points, 63 contestants scored 1 point, and 52 contestants scored 0 point. The average score of this problem is 5.272, indicating that it was simple.

Among the top five teams in the team scores, the scores of this problem are as follows: the United States team scored 37 points (with a total team score of 214 points), the South Korea team scored 42 points (with a total team score of 207 points), the China team scored 42 points (with a total team score of 204 points), the Singapore team scored 42 points (with a total team score of 196 points), and the Chinese Taiwan team scored 42 points (with a total team score of 175 points).

The gold medal cutoff for this IMO was set at 29 points (with 44 contestants earning gold medals), the silver medal cutoff was 22 points (with 101 contestants earning silver medals), and the bronze medal cutoff was 16 points (with 135 contestants earning bronze medals).

In this IMO, a total of six contestants achieved a perfect score of 42 points.

Problem 2.19 (IMO 58-4, proposed by Luxembourg). Let R and S be different points on a circle Ω such that RS is not a diameter. Let l be the tangent line to Ω at R. Point T is such that S is the midpoint of the line segment RT. Point J is chosen on the shorter arc RS of Ω so that the circumcircle Γ of triangle JST intersects l at two distinct points. Let A be the common point of Γ and l that is closer to R. Line AJ meets Ω again at K. Prove that the line KT is tangent to Γ.

Proof. As shown in Figure 2.42, since R, K, S, J are concyclic, and S, J, A, T are also concyclic, so

$$\angle KRS = \angle KJS = \angle STA.$$

Since AR is tangent to Ω, we see that $\angle RKS = \angle TRA$. So $\triangle RKS \backsim \triangle TRA$, and then

$$\frac{RK}{RS} = \frac{TR}{TA}.$$

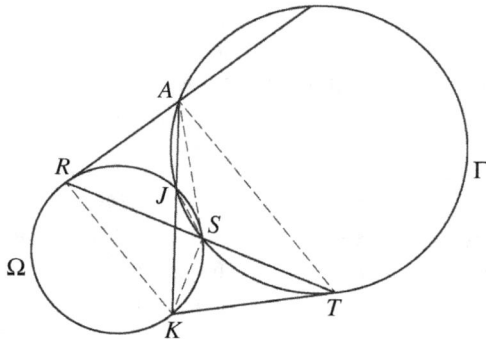

Figure 2.42

As S is the midpoint of $RT, RS = ST$. So

$$\frac{RK}{TS} = \frac{RT}{TA}.$$

Combing it with $\angle KRT = \angle STA$, we have $\triangle KRT \backsim \triangle STA$. Therefore $\angle SAT = \angle STK$, which means that line KT is tangent to circle Γ.

【Score Situation】 This particular problem saw the following distribution of scores among contestants: 394 contestants scored 7 points, 6 contestants scored 6 points, 4 contestants scored 5 points, 15 contestants scored 4 points, 14 contestants scored 3 points, 42 contestants scored 2 points, 93 contestants scored 1 point, and 47 contestants scored 0 point. The average score of this problem is 5.029, indicating that it was simple.

Among the top five teams in the team scores, the scores of this problem are as follows: the South Korea team scored 42 points (with a total team score of 170 points), the China team scored 42 points (with a total team score of 159 points), the Vietnam team scored 42 points (with a total team score of 155 points), the United States team scored 42 points (with a total team score of 148 points), and the Iran team scored 42 points (with a total team score of 142 points).

The gold medal cutoff for this IMO was set at 25 points (with 48 contestants earning gold medals), the silver medal cutoff was 19 points (with 90 contestants earning silver medals), and the bronze medal cutoff was 16 points (with 153 contestants earning bronze medals).

In this IMO, no contestant achieved a perfect score of 42 points.

Problem 2.20 (IMO 59-1, proposed by Greece). Let Γ be the circumcircle of acute-angled triangle ABC. Points D and E lie on segments

AB and AC, respectively, such that $AD = AE$. The perpendicular bisectors of BD and CE intersect the minor arcs AB and AC of Γ at points F and G, respectively. Prove that the lines DE and FG are parallel (or are the same line).

Proof. As shown in Figure 2.43, P, Q, R are the midpoints of minor arcs $\overset{\frown}{BC}, \overset{\frown}{CA}, \overset{\frown}{AB}$ on Γ, respectively, M, N are the midpoints of AB, AC, respectively, and O is the centre of circle Γ. Then O, M, R are collinear, and O, N, Q are collinear.

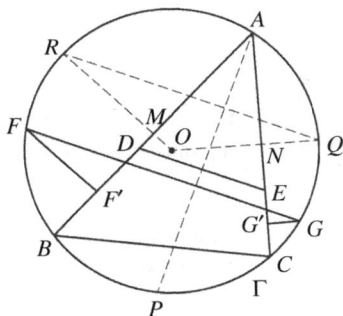

Figure 2.43

Since $AD = AE$, segment AP bisects $\angle BAC$, and then $AP \perp DE$. Furthermore,

$$\frac{1}{2}\overset{\frown}{AQ} + \frac{1}{2}\overset{\frown}{PBR} = \frac{B}{2} + \frac{A+C}{2} = 90°,$$

Hence, $QR \perp AP$ and $DE \| RQ$. Then we only need to prove that $FG \| RQ$, and that is equivalent to proving $\overset{\frown}{FR} = \overset{\frown}{GQ}$. Let F', G' be the midpoints of segments BD, CE, respectively. Then $FF' \perp AB$ and $GG' \perp AC$, so $FF' \| OR$ and $GG' \| OQ$. We have

$$F'M = BM - BF' = \frac{1}{2}(AB - BD) = \frac{AD}{2} = \frac{AE}{2} = G'N.$$

Then the distance between lines FF' and OR is equal to that between GG' and OQ. Therefore, $\overset{\frown}{FR} = \overset{\frown}{GQ}$. The proof is completed.

【Score Situation】 This particular problem saw the following distribution of scores among contestants: 381 contestants scored 7 points, 7 contestants scored 6 points, 7 contestants scored 5 points, 10 contestants scored 4 points, 15 contestants scored 3 points, 24 contestants scored 2 points, 54 contestants scored 1 point, and

96 contestants scored 0 point. The average score of this problem is 4.934, indicating that it was simple.

Among the top five teams in the team scores, the scores of this problem are as follows: the United States team scored 42 points (with a total team score of 212 points), the Russia team scored 42 points (with a total team score of 201 points), the China team scored 42 points (with a total team score of 199 points), the Ukraine team scored 42 points (with a total team score of 186 points), and the Thailand team scored 42 points (with a total team score of 183 points).

The gold medal cutoff for this IMO was set at 31 points (with 48 contestants earning gold medals), the silver medal cutoff was 25 points (with 98 contestants earning silver medals), and the bronze medal cutoff was 16 points (with 143 contestants earning bronze medals).

In this IMO, only two contestants achieved a perfect score of 42 points, namely Agnijo Banerjee from the United Kingdom and James Lin from the United States.

Problem 2.21 (IMO 60-2, proposed by Ukraine). In triangle ABC, point A_1 lies on side BC and point B_1 lies on side AC. Let P and Q be points on segments AA_1 and BB_1, respectively, such that PQ is parallel to AB. Let P_1 be a point on line PB_1, such that B_1 lies strictly between P and P_1, and $\angle PP_1C = \angle BAC$. Similarly, let Q_1 be a point on line QA_1, such that A_1 lies strictly between Q and Q_1, and $\angle CQ_1Q = \angle CBA$. Prove that points P, Q, P_1, Q_1 are concyclic.

Proof. As shown in Figure 2.44, let rays AA_1, BB_1 intersect the circumcircle of $\triangle ABC$ at A_2, B_2, respectively. Since $PQ \parallel AB$,

$$\angle A_2B_2B = \angle A_2AB = \angle A_2PQ,$$

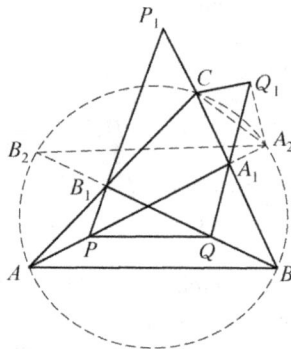

Figure 2.44

so A_2, B_2, P, Q are concyclic. And since

$$\angle CA_2A = \angle CBA = \angle CQ_1Q = \angle CQ_1A_1,$$

points C, A_1, A_2, Q_1 are concyclic. Hence,

$$\angle A_2Q_1Q = \angle A_2Q_1A_1 = \angle A_2CA_1 = \angle A_2CB$$

$$= \angle A_2AB = \angle A_2PQ.$$

This implies that Q_1, A_2, Q, P are concyclic. Similarly, we can prove that P_1, B_2, P, Q are concyclic. Therefore, P_1, Q_1, P, Q, A_2, B_2 all lie on the same circle, and the desired conclusion follows.

【Score Situation】 This particular problem saw the following distribution of scores among contestants: 97 contestants scored 7 points, 92 contestants scored 6 points, 3 contestants scored 5 points, 6 contestants scored 4 points, 6 contestants scored 3 points, 30 contestants scored 2 points, 135 contestants scored 1 point, and 251 contestants scored 0 point. The average score of this problem is 2.399, indicating that it had a certain level of difficulty.

Among the top five teams in the team scores, the scores of this problem are as follows: the China team scored 41 points (with a total team score of 227 points), the United States team scored 40 points (with a total team score of 227 points), the South Korea team scored 39 points (with a total team score of 226 points), the North Korea team scored 41 points (with a total team score of 187 points), and the Thailand team scored 35 points (with a total team score of 185 points).

The gold medal cutoff for this IMO was set at 31 points (with 52 contestants earning gold medals), the silver medal cutoff was 24 points (with 94 contestants earning silver medals), and the bronze medal cutoff was 17 points (with 156 contestants earning bronze medals).

In this IMO, a total of six contestants achieved a perfect score of 42 points.

Problem 2.22 (IMO 60-6, proposed by India). Let I be the incentre of acute triangle ABC with $AB \neq AC$. The incircle ω of ABC is tangent to sides BC, CA, and AB at D, E, and F, respectively. The line through D perpendicular to EF meets ω again at R. Line AR meets ω again at P. The circumcircles of triangles PCE and PBF meet again at Q. Prove that lines DI and PQ meet on the line through A perpendicular to AI.

Solution 1. As shown in Figure 2.45, let the exterior angle bisector of $\angle BAC$ and DI intersect at L. Then $AL \perp AI$. It suffices to show that L, Q, P are collinear. Let K be the second intersection of DL and ω, and let N be the midpoint of EF. We proceed with the following:

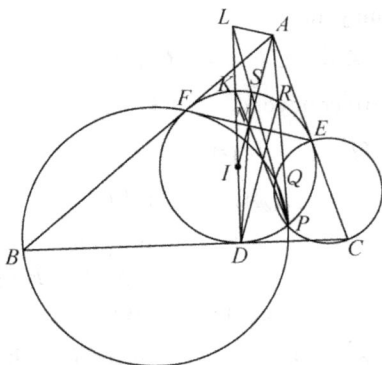

Figure 2.45

(i) The points K, N, P are collinear. Since $RFPE$ is a harmonic quadrilateral, EF bisects $\angle RNP$, and since

$$\frac{1}{2}\widehat{KF} + \frac{1}{2}\widehat{FD} = 90° = \frac{1}{2}\widehat{RE} + \frac{1}{2}\widehat{FD},$$

we have $\widehat{KF} = \widehat{RE}$, so K, R are symmetric with respect to AI. Hence $\angle KNF = \angle RNE = \angle PNE$, from which K, N, P are collinear.

(ii) The points L, S, P are collinear. Since A is the image of N under the inversion with respect to ω, and $LA \perp AI$, we see that AL is the polar of N with respect to ω. Let PS and DK intersect at L'. Then the polar of L' passes through N and the polar of N passes through L'. Hence L' lies on the line AL and $L' = L$. This shows that L, S, P are collinear.

Now it suffices to show that S, Q, P are collinear, as shown in Figure 2.46. We apply the notation of directed angles, in order to avoid dependence on the specific configuration of the points. Let $\angle(a, b)$ denote the smallest nonnegative angle α such that if we rotate the line a counterclockwise by α, then we obtain a line that is parallel to b. In the following argument, the arithmetic of directed angles is considered modulo π. Since B, F, Q, P are concyclic and C, E, Q, P are concyclic,

$$\angle(BQ, QC) = \angle(BQ, QP) + \angle(PQ, QC)$$

$$= \angle(BF, FP) + \angle(PE, EC)$$

$$= \angle(EF, EP) + \angle(FP, FE)$$

$$= \angle(FP, EP) = \angle(DF, DE)$$

$$= \angle(BI, IC).$$

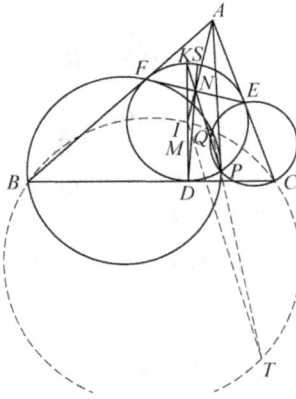

Figure 2.46

Hence, B, I, Q, C are concyclic. Let T be the second intersection of QP and $\odot(BIQC)$, and IT intersects DS at M.

(iii) The point M is the midpoint of DN. Note that

$$\angle(BI, IT) = \angle(BQ, QT) = \angle(BF, FP) = \angle(FK, KP).$$

And $FD \perp FK, FD \perp BI$, we have $FK \| BI$, so that $IT \| KNP$. Since I is the midpoint of DK, we see that M is also the midpoint of DN.

Finally, we show that S, P, T are collinear, and hence completing the proof. As shown in Figure 2.47, let F_1, E_1 be the midpoints of

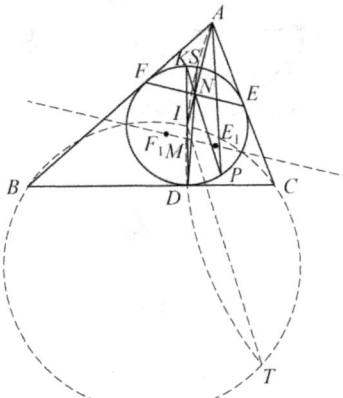

Figure 2.47

DF, DE. Since

$$DF_1 \cdot F_1F = DF_1^2 = BF_1 \cdot F_1I,$$

we see that F_1 lies on the radical axis of ω and $\odot(BIC)$. Similarly E_1 lies on the radical axis of ω and $\odot(BIC)$, so the line E_1F_1 is exactly the radical axis of ω and $\odot(BIC)$. Since M is the midpoint of DN, the point M lies on E_1F_1, so M has the same power to circles ω and $\odot(BIC)$. Therefore,

$$DM \cdot MS = IM \cdot MT,$$

and S, I, D, T are concyclic. Thus,

$$\angle(DS, ST) = \angle(DI, IT) = \angle(DK, KP) = \angle(DS, SP),$$

which is the same as saying S, P, T are collinear.

Solution 2 (By Jiajun Huang). As shown in Figure 2.48, let Ω be the circumcircle of $\triangle ABC$, and the midpoints of arcs $\overset{\frown}{BAC}$ and $\overset{\frown}{BC}$ are M_a and N_a, respectively. Let L be the intersection of the lines DL and AM_a. Note that since $AM_a \perp AN_a$ and A, I, N_a are collinear, $AL \perp AI$. Then it suffices to show that P, Q, L are collinear.

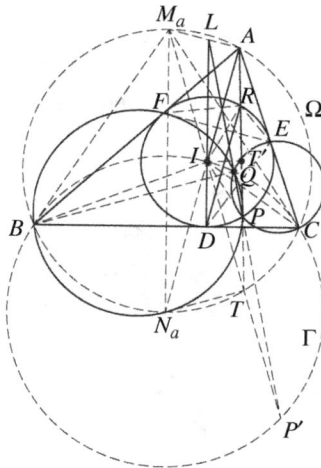

Figure 2.48

We construct circle ω with N_a as its center and IN_a as its radius. Then by the property of the incenter, B, C both lie on ω. Suppose line M_aI intersects Ω, ω again at T, P', respectively.

(i) The concave quadrilaterals $AERF$ and M_aCIB are similar.
First, we have $AE = AF$ and $M_aC = M_aB$, and

$$\angle(FA, AE) = \angle(BA, AC) = \angle(BM_a, M_aC),$$

so $\triangle EAF \backsim \triangle CM_aB$. Also

$$\angle(RE, EF) = 90° - \angle(DR, RE) = 90° - \angle(DE, EC) = \angle(IC, CB),$$

and similarly, $\angle(RF, FE) = \angle(IB, BC)$. Therefore, $AERF \backsim M_aCIB$.

(ii) A, P, T are collinear.
It follows by (i) that $\odot I$ and ω are corresponding circles in similar quadrilaterals $AERF$ and $ACIB$, and AR and M_aI are corresponding lines, so P, P' are corresponding points. Hence,

$$\angle(PA, AE) = \angle(P'M_a, M_aC) = \angle(TM_a, M_aC) = \angle(TA, AC),$$

Thus, A, P, T are collinear.

(iii) The points P, Q, P' are collinear. Since B, P, Q, F are concyclic,

$$\angle(BQ, QP) = \angle(BF, FP) = \angle(FR, RP)$$

and similarly, $\angle(PQ, QC) = \angle(PR, RE)$. Hence,

$$\angle(BQ, QC) = \angle(BQ, QP) + \angle(PQ, QC)$$
$$= \angle(FR, RP) + \angle(PR, RE)$$
$$= \angle(FR, RE) = \angle(BI, IC).$$

Then it follows that B, I, Q, C are concyclic, or equivalently, Q lies on ω. Consequently,

$$\angle(BQ, QP') = \angle(BI, IP') = \angle(BI, IM_a) = \angle(FR, RA)$$
$$= \angle(FR, RP) = \angle(BF, FP) = \angle(BQ, QP),$$

and therefore P, Q, P' are collinear.

(iv) The points P, Q, L are collinear. By Menelaus's theorem it suffices to show that

$$\frac{AP}{PT} \cdot \frac{TP'}{P'M_a} \cdot \frac{M_aL}{LA} = 1.$$

Construct the perpendicular line of PR through I, and let T' be its foot of altitude. Note that T and T' are corresponding points in quadrilaterals

$AERF$ and M_aCIB, since $N_aT \perp IP'$. This implies

$$\frac{TP'}{P'M_a} = \frac{T'P}{PA}.$$

In right triangles ITT' and M_aAN_a, we have

$$\angle(T'T, TI) = \angle(AT, TM_a) = \angle(AN_a, N_aM_a),$$

so $\triangle ITT' \sim \triangle M_aN_aA$. Combining with $AERF \sim M_aCIB$, we obtain that

$$\angle(IP, PT) = \angle(AR, RI) = \angle(M_aI, IN_a).$$

Thus, P, I are corresponding points in similar triangles ITT' and M_aN_aA. Combining with $IL \parallel M_aN_a$, we have

$$\frac{M_aL}{LA} = \frac{N_aI}{IA} = \frac{TP}{PT'}.$$

Thus,

$$\frac{AP}{PT} \cdot \frac{TP'}{P'M_a} \cdot \frac{M_aL}{LA} = \frac{AP}{PT} \cdot \frac{T'P}{PA} \cdot \frac{TP}{PT'} = 1,$$

which yields that P, Q, L are collinear.

Combining the above results, we obtain the desired conclusion.

Solution 3 (By Baiting Xie). As shown in Figure 2.49, on the circumcircle of $\triangle ABC$, let N and L be the midpoints of $\overparen{BC}$ and $\overparen{BAC}$,

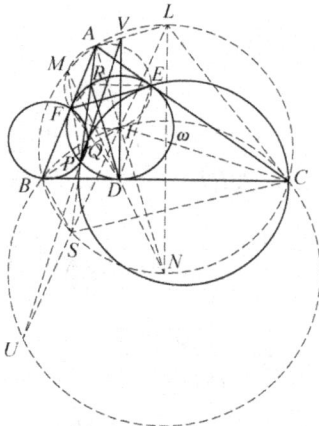

Figure 2.49

respectively. Construct circle $\odot N$ with radius IN, so that it passes through B, C. Let LI intersect $\odot N$ again at U, and intersect the circumcircle of $\triangle ABC$ again at S. Suppose the circumcircle of $\triangle AEF$ (denoted as Ω_1) and the circumcircle of $\triangle ABC$ (denoted as Ω) intersect at A, M. Let V be the intersection of DI, AL. We proceed to prove the following assertions:

(i) The points A, R, S are collinear.

Let $\angle BAC = 2\alpha$, $\angle ABC = 2\beta$, $\angle ACB = 2\gamma$. Note that $\angle BSI = \angle ISC$, and

$$\angle IBS = \angle IBC + \angle SBC = \beta + \angle CLI$$

$$= \gamma + (\beta - \gamma) + \angle CLI$$

$$= \angle ICA + \angle LCA + \angle CLI = \angle CIS,$$

so $\triangle IBS \backsim \triangle CIS$. Hence,

$$\frac{\sin \angle BAS}{\sin \angle CAS} = \frac{BS}{CS} = \frac{BS}{IS} \cdot \frac{IS}{CS} = \left(\frac{BI}{IC}\right)^2 = \left(\frac{\sin \gamma}{\sin \beta}\right)^2.$$

Since I is the center of ω, we see that $DR \perp EF$. Consequently $\angle EDR = \angle FDI = \beta, \angle FDR = \angle EDI = \gamma$. Thus,

$$\frac{\sin \angle BAR}{\sin \angle CAR} = \frac{\sin \angle AFR}{\sin \angle RFE} \cdot \frac{\sin \angle REF}{\sin \angle REA} = \left(\frac{\sin \angle FDR}{\sin \angle EDR}\right)^2$$

$$= \left(\frac{\sin \gamma}{\sin \beta}\right)^2 = \frac{\sin \angle BAS}{\sin \angle CAS}.$$

Therefore, A, R, S are collinear, and assertion (i) is proven.

(ii) The points U, P, Q are collinear.

Since

$$\angle AFE = \angle AEF = \angle LBC = \angle LCB = 90° - \alpha,$$

it follows that $\triangle AFE \sim \triangle LBC$ (rotationally), and M is the rotation center.

Now AI, LN are diameters of Ω_1 and Ω, respectively, so I, N are corresponding points in the pair of similar triangles, and thus ω and $\odot N$ are corresponding circles. Also $\angle FAR = \angle BAS = \angle BLS$, so AR, LS are corresponding lines, so the intersection points of AR and ω, namely R, P, should correspond to the intersection points of LS and $\odot N$, namely I, U.

Hence, $\angle BIU = \angle FRP = \angle BFP = \angle BQP$, and from

$$\angle BQC = \angle BAC + \angle ABQ + \angle ACQ = \angle BAC + \angle FPQ + \angle EPQ$$

$$= \angle BAC + \angle FPE = 2\alpha + \angle FDE = 90° + \alpha = \angle BIC,$$

we see that Q lies on $\odot N$. Therefore, $\angle BQU = \angle BIU = \angle BQP$, or equivalently, U, P, Q are collinear, and (ii) is proven.

(iii) The points A, M, P, V are concyclic.

From the rotational similarity of $\triangle AFE$ and $\triangle LBC$ in (ii) (where M is the rotation center), we obtain that $\triangle MFB \backsim \triangle MEC$, so $\frac{MB}{MC} = \frac{FB}{EC} = \frac{BD}{CD}$ and MD bisects $\angle BMC$. Thus, M, D, N are collinear.

Next, $DV \parallel LN$ implies that $\angle MDV = \angle MNL = 180° - \angle MAV$, so A, M, D, V are concyclic. Since $RD \perp EF, AN \perp EF$, we have $RD \parallel AN$, so

$$\angle PAN = \angle PRD = \angle PDB.$$

And

$$\angle MAN = \angle MAB + \alpha = \angle MCB + \angle CMN = \angle MDB,$$

so

$$\angle MAP = \angle MAN - \angle PAN = \angle MDB - \angle PDB = \angle MDP,$$

and A, M, P, D are concyclic.

Thus, A, M, P, D, V are concyclic, and (iii) is proven.

(iv) The points U, V, P, Q are concyclic.

By (iii) we have $\angle MPV = 180° - \angle MAV = 90° - \angle MAN = \angle MIA = \angle MFA$.

In the rotational similarity in (ii), the points M, F, P correspond to M, B, U, respectively. Hence,

$$\angle MPU = \angle MFB = 180° - \angle MFA = 180° - \angle MPV,$$

which means U, V, P are collinear. Combining with (ii), we conclude that U, V, P, Q are collinear, and (iv) is proven.

In combination, we derive the conclusion in the original problem.

【Score Situation】 This particular problem saw the following distribution of scores among contestants: 27 contestants scored 7 points, 3 contestants scored 6 points, no contestant scored 5 points, 1 contestant scored 4 points, no contestant scored 3 points, 7 contestants scored 2 points, 25 contestants scored 1 point, and 558 contestants scored 0 point. The average score of this problem is 0.403, indicating that it was extremely difficult.

Among the top five teams in the team scores, the scores of this problem are as follows: the China team scored 36 points (with a total team score of 227 points), the United States team scored 35 points (with a total team score of 227 points), the South Korea team scored 30 points (with a total team score of 226 points), the North Korea team scored 4 points (with a total team score of 187 points), and the Thailand team scored 21 points (with a total team score of 185 points).

The gold medal cutoff for this IMO was set at 31 points (with 52 contestants earning gold medals), the silver medal cutoff was 24 points (with 94 contestants earning silver medals), and the bronze medal cutoff was 17 points (with 156 contestants earning bronze medals).

In this IMO, a total of six contestants achieved a perfect score of 42 points.

Problem 2.23 (IMO 61-1, proposed by Poland). Consider a convex quadrilateral $ABCD$. A point P is in the interior of $ABCD$. The following ratio equalities hold:

$$\angle PAD : \angle PBA : \angle DPA = 1 : 2 : 3 = \angle CBP : \angle BAP : \angle BPC.$$

Prove that the following three lines meet at a point: the internal bisectors of angles $\angle ADP$ and $\angle PCB$ and the perpendicular bisector of segment AB.

Proof. As shown in Figure 2.50, let $\varphi = \angle PAD$ and $\psi = \angle CBP$. Then from the condition,

$$\angle PBA = 2\varphi, \angle DPA = 3\varphi, \angle BAP = 2\psi, \angle BPC = 3\psi.$$

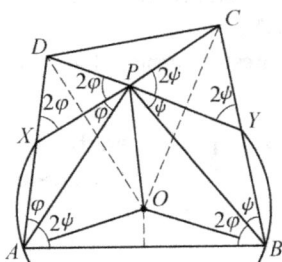

Figure 2.50

Suppose X is a point on AD such that $\angle XPA = \varphi$. Then

$$\angle PXD = \angle PAX + \angle XPA = 2\varphi = \angle DPA - \angle XPA = \angle DPA.$$

This implies that $DX = DP$ and the angle bisector of $\angle ADP$ is also the perpendicular bisector of XP. Similarly, if Y is a point on BC such that

$\angle BPY = \psi$, then the angle bisector of $\angle PCB$ is also the perpendicular bisector of PY. The problem then reduces to proving that the perpendicular bisectors of XP, PY, AB are concurrent. Note that

$$\angle AXP = 180° - \angle PXD = 180° - 2\varphi = 180° - \angle PBA.$$

So $AXPB$ is a cyclic quadrilateral, and X lies on the circumcircle of $\triangle APB$. Similarly, we can show that Y lies on the circumcircle of $\triangle APB$. Therefore, A, B, Y, P, X are concyclic, and the perpendicular bisectors of PX, PY, AB all pass through the center of this circle.

【Score Situation】 This particular problem saw the following distribution of scores among contestants: 451 contestants scored 7 points, 7 contestants scored 6 points, 3 contestants scored 5 points, 2 contestants scored 4 points, 5 contestants scored 3 points, 5 contestants scored 2 points, 26 contestants scored 1 point, and 117 contestants scored 0 point. The average score of this problem is 5.313, indicating that it was simple.

Among the top five teams in the team scores, the scores of this problem are as follows: the China team scored 42 points (with a total team score of 215 points), the Russia team scored 42 points (with a total team score of 185 points), the United States team scored 42 points (with a total team score of 183 points), the South Korea team scored 42 points (with a total team score of 175 points), and the Thailand team scored 42 points (with a total team score of 174 points).

The gold medal cutoff for this IMO was set at 31 points (with 49 contestants earning gold medals), the silver medal cutoff was 24 points (with 112 contestants earning silver medals), and the bronze medal cutoff was 15 points (with 155 contestants earning bronze medals).

In this IMO, only one contestant achieved a perfect score of 42 points, namely Jinmin Li from China.

Problem 2.24 (IMO 63-4, proposed by Slovakia). Let $ABCDE$ be a convex pentagon such that $BC = DE$. Assume that there is a point T inside $ABCDE$ with $TB = TD, TC = TE$, and $\angle ABT = \angle TEA$. Let line AB intersect lines CD and CT at points P and Q, respectively. Assume that the points P, B, A, Q occur on their line in that order. Let line AE intersect lines CD and DT at points R and S, respectively. Assume that the points R, E, A, S occur on their line in that order. Prove that the points P, S, Q, R lie on a circle.

Proof 1.   From the given conditions, it follows that $BC = DE$, $CT = ET$, and $TB = TD$. Hence $\triangle TBC$ and $\triangle TDE$ are congruent. In particular $\angle BTC = \angle DTE$. In $\triangle TBQ$ and $\triangle TES$, we have $\angle TBQ = \angle SET$

and

$$\angle QTB = 180° - \angle BTC = 180° - \angle DTE = \angle ETS.$$

Hence, they are similar triangles, so $\angle TSE = \angle BQT$ and

$$\frac{TD}{TQ} = \frac{TB}{TQ} = \frac{TE}{TS} = \frac{TC}{TS}.$$

We see that $TD \cdot TS = TC \cdot TQ$ and C, D, Q, S are concyclic. (An alternative approach is to use similar triangles $\triangle TCS$ and $\triangle TDQ$ to derive $\angle CQD = \angle CSD$.)

Now, $\angle DCQ = \angle DSQ, \angle RPQ = \angle RCQ - \angle PQC = \angle DSQ - \angle DSR = \angle RSQ$ and thus P, Q, R, S are concyclic.

Proof 2. As shown in Proof 1, we find that $\triangle TBC$ and $\triangle TDE$ are congruent. Let DT, BA, and CT, EA meet at V and W, respectively, as shown in Figure 2.51.

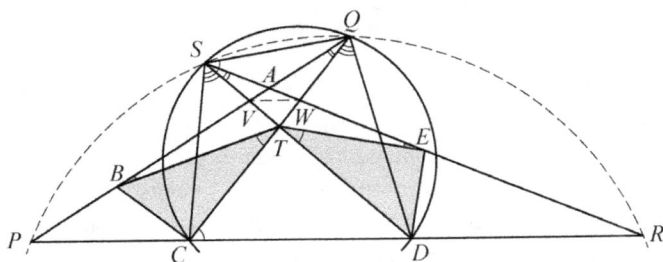

Figure 2.51

In $\triangle BCQ$ and $\triangle DES$, we find

$$\angle VSW = \angle DSE = 180° - \angle SED - \angle EDS$$
$$= 180° - \angle AET - \angle TED - \angle EDT$$
$$= 180° - \angle TBA - \angle TCB - \angle CBT$$
$$= 180° - \angle QCB - \angle CBQ$$
$$= \angle BQC = \angle VQW,$$

and hence V, S, Q, W are concyclic. In particular, $\angle WVQ = \angle WSQ$. Since

$$\angle VTB = 180° - \angle BTC - \angle CTD = 180° - \angle CTD - \angle DTE = \angle ETW,$$

and given that $\angle TBV = \angle WET$, we arrive at $\triangle VTB \sim \triangle WTE$. Hence,

$$\frac{VT}{WT} = \frac{BT}{ET} = \frac{DT}{CT},$$

and $CD \parallel VW$.

Finally, the conclusion follows from $\angle RPQ = \angle WVQ = \angle WSQ = \angle RSQ$. Thus, P, Q, R, S are concyclic.

Note. Proof 1 for C, D, Q, S to be concyclic is from $\triangle TBQ \sim \triangle TES$. Notice that $\triangle TBD \backsim \triangle TCE$, together with $\triangle TBQ \sim \triangle TES$, we can get similar quadrilaterals (concave) $TBDQ \backsim TECS$. Then $\triangle TDQ \sim \triangle TCS$ and C, D, Q, S lie on a circle.

Another approach for $CD \parallel VW$ in Proof 2 is similar to the first proof, that is, we can find that V, S, Q, W are concyclic. Moreover, $\triangle TBC$ and $\triangle TDE$ are congruent, and hence $\triangle TBD \backsim \triangle TCE$. Since they are isosceles triangles, $\angle TBD = \angle TEC, \angle TDB = \angle TCE$. It follows that $\triangle VBD \sim \triangle WCE$, and BT, ET divide $\angle VBD, \angle WEC$, respectively at the same ratio. Consequently, $\frac{VT}{TD} = \frac{WT}{TC}$ and $CD \parallel VW$.

【Score Situation】 This particular problem saw the following distribution of scores among contestants: 433 contestants scored 7 points, 1 contestant scored 6 points, 5 contestants scored 5 points, 10 contestants scored 4 points, 17 contestants scored 3 points, 18 contestants scored 2 points, 31 contestants scored 1 point, and 74 contestants scored 0 point. The average score of this problem is 5.467, indicating that it was simple.

Among the top five teams in the team scores, the scores of this problem are as follows: the China team scored 42 points (with a total team score of 252 points), the South Korea team scored 42 points (with a total team score of 208 points), the United States team scored 42 points (with a total team score of 207 points), the Vietnam team scored 42 points (with a total team score of 196 points), and the Romania team scored 42 points (with a total team score of 194 points).

The gold medal cutoff for this IMO was set at 34 points (with 44 contestants earning gold medals), the silver medal cutoff was 29 points (with 101 contestants earning silver medals), and the bronze medal cutoff was 23 points (with 140 contestants earning bronze medals).

In this IMO, a total of 10 contestants achieved a perfect score of 42 points.

2.2.3 *Quantity relation problems*

Problem 2.25 (IMO 23-5, proposed by the Netherlands). The diagonals AC and CE of a regular hexagon $ABCDEF$ are divided by two inner points M and N, respectively, so that $\frac{AM}{AC} = \frac{CN}{CE} = r$. Determine r if B, M, N are collinear.

Solution 1. As shown in Figure 2.52, connect to form DN. It is easy to know that $AM = CN$ and $\triangle AMB \cong \triangle CND$. So,

$$\angle BND = \angle 3 + \angle 4 = \angle 1 + \angle 4 = \angle 2 + \angle 4$$
$$= 180° - \angle ACE = 120°.$$

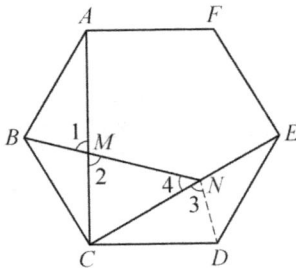

Figure 2.52

Make a circle with C as the center and BC as the radius, in which $\angle BCD = 120°, CB = CD$. Since the inscribed angle of the minor arc BD is $60°$, the inscribed angle of the major arc BD is $120°$. Point N satisfies this requirement, so point N is on the circle, and $CN = CD$.

Thus,

$$r = \frac{CN}{CE} = \frac{CD}{CE} = \frac{\sqrt{3}}{3}.$$

Solution 2. Let the side length of the regular hexagon $ABCDEF$ be 1. As shown in Figure 2.53, connect to form BE intersecting AC at point G. Clearly, $BE = 2$, $GB = \frac{1}{2}$, so,

$$CE = \sqrt{3}, \quad CN = \sqrt{3}r, \quad CM = EN = \sqrt{3}(1 - r),$$
$$GM = \frac{1}{2}AC - CM = \frac{\sqrt{3}}{2} - \sqrt{3}(1 - r) = \frac{\sqrt{3}}{2}(2r - 1).$$

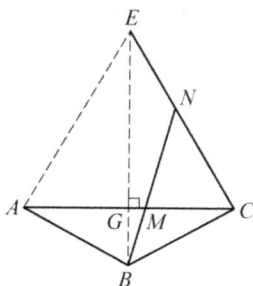

Figure 2.53

By the condition and Menelaus's theorem,

$$\frac{EB}{BG} \cdot \frac{GM}{MC} \cdot \frac{CN}{NE} = 1,$$

so we get the equation

$$\frac{2r-1}{2(1-r)} \cdot \frac{r}{1-r} = \frac{1}{4}.$$

Hence,

$$r = \frac{\sqrt{3}}{3}.$$

【Score Situation】 This particular problem saw the following distribution of scores among contestants: 82 contestants scored 7 points, 7 contestants scored 6 points, 3 contestants scored 5 points, 3 contestants scored 4 points, 1 contestant scored 3 points, 3 contestants scored 2 points, 3 contestants scored 1 point, and 17 contestants scored 0 point. The average score of this problem is 5.504, indicating that it was simple.

Among the top five teams in the team scores, the scores of this problem are as follows: the Germany team scored 28 points (with a total team score of 145 points), the Soviet Union team scored 28 points (with a total team score of 137 points), the German Democratic Republic team scored 28 points (with a total team score of 136 points), the United States team scored 27 points (with a total team score of 136 points), and the Vietnam team scored 28 points (with a total team score of 133 points).

The gold medal cutoff for this IMO was set at 37 points (with 10 contestants earning gold medals), the silver medal cutoff was 30 points (with 20 contestants earning silver medals), and the bronze medal cutoff was 21 points (with 31 contestants earning bronze medals).

In this IMO, only three contestants achieved a perfect score of 42 points, namely Bruno Haible from Germany, Grigori Perelman from the Soviet Union, and Lê Tự Quốc Thắng from Vietnam.

Problem 2.26 (IMO 35-2, proposed by Armenia-Australia). An isosceles triangle ABC is given with $AB = AC$. Suppose that:

(i) M is the midpoint of BC and O is a point on the line AM such that OB is perpendicular to AB;
(ii) Q is an arbitrary point on the segment BC different from B and C;
(iii) E lies on the line AB and F lies on the line AC such that E, Q, F are all distinct and collinear.

Prove that OQ is perpendicular to EF if and only if $QE = QF$.

Proof. Suppose that $OQ \perp EF$, as shown in Figure 2.54. Connect to form OE, OF, OC. Then O is outside $\triangle ABC$ and
$$\angle ACO = \angle ABO = \angle EQO = 90°.$$

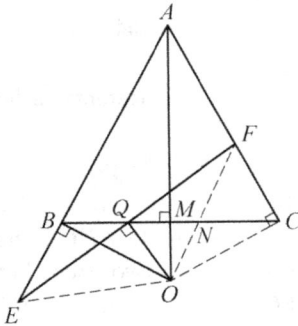

Figure 2.54

So, B, E, O, Q are concyclic, and F, Q, O, C are also concyclic.
Thus, $\angle QEO = \angle QBO = \angle QCO = \angle QFO$.
Also $\triangle OEF$ is an isosceles triangle, so by $OQ \perp EF$, we get $QE = QF$.

Note. In fact BQC is a Simson line of $\triangle AEF$.
Conversely, if $EQ = FQ$, then make $FN \parallel AE$. As shown in the figure, the point N is on BC, from which $FN = BE$.
And $\angle FNC = \angle ABC = \angle ACB$, so $FN = FC$. From the above formula, $FC = BE$. Meanwhile, $BO = CO, \angle EBO = 90° = \angle FCO$, so $\triangle EBO \cong \triangle FCO$.

Thus, $EO = FO$. Combining with the $EQ = FQ$, we see that $OQ \perp EF$.

【Score Situation】This particular problem saw the following distribution of scores among contestants: 227 contestants scored 7 points, 12 contestants scored 6 points, 14 contestants scored 5 points, 17 contestants scored 4 points, 26 contestants scored 3 points, 12 contestants scored 2 points, 25 contestants scored 1 point, and 52 contestants scored 0 point. The average score of this problem is 5.003, indicating that it was simple.

Among the top five teams in the team scores, the scores of this problem are as follows: the United States team scored 42 points (with a total team score of 252 points), the China team scored 42 points (with a total team score of 229 points), the Russia team scored 42 points (with a total team score of 224 points), the Bulgaria team scored 42 points (with a total team score of 223 points), and the Hungary team scored 42 points (with a total team score of 221 points).

The gold medal cutoff for this IMO was set at 40 points (with 30 contestants earning gold medals), the silver medal cutoff was 30 points (with 64 contestants earning silver medals), and the bronze medal cutoff was 19 points (with 98 contestants earning bronze medals).

In this IMO, a total of 22 contestants achieved a perfect score of 42 points.

Problem 2.27 (IMO 38-2, proposed by the United Kingdom).
Angle A is the smallest one in triangle ABC. The points B and C divide the circumcircle of the triangle into two arcs. Let U be an interior point of the arc between B and C which does not contain A. The perpendicular bisectors of AB and AC meet the line AU at V and W, respectively. The lines BV and CW meet at T. Show that $AU = TB + TC$.

Proof.   By the property of $\angle A$, we know that $\angle WCA = \angle WAC < \angle A \le \angle C$. Then, extend CW intersecting at a point of the circumcircle $\overset{\frown}{AB}$, denoted as S. As shown in Figure 2.55, connect to form BS, BU.

We know that $AV = BV, AW = CW$.

Thus

$$\angle S = \angle A = \angle BAU + \angle CAU$$
$$= \angle VBA + \angle SCA$$

$$= \angle VBA + \angle SBA$$

$$= \angle SBV.$$

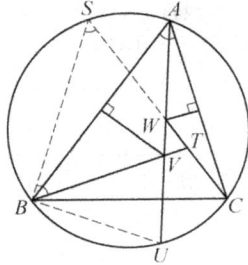

Figure 2.55

Then, $ST = BT, TB + TC = SC$.

Accordingly, the remaining thing is just to prove $AU = SC$. Let the radius of the circumcircle be R. Then

$$AU = 2R\sin\angle ABU = 2R\sin(\angle B + \angle CBU)$$

$$= 2R\sin(\angle B + \angle CAU) = 2R\sin(\angle B + \angle ACS)$$

$$= 2R\sin(\angle B + \angle ABS) = 2R\sin\angle SBC$$

$$= SC.$$

【Score Situation】 This particular problem saw the following distribution of scores among contestants: 236 contestants scored 7 points, no contestant scored 6 points, 9 contestants scored 5 points, 7 contestants scored 4 points, 4 contestants scored 3 points, 16 contestants scored 2 points, 24 contestants scored 1 point, and 164 contestants scored 0 point. The average score of this problem is 3.898, indicating that it was relatively straightforward.

Among the top five teams in the team scores, the scores of this problem are as follows: the China team scored 42 points (with a total team score of 223 points), the Hungary team scored 42 points (with a total team score of 219 points), the Iran team scored 42 points (with a total team score of 217 points), the Russia team scored 42 points (with a total team score of 202 points) and the United States team scored 37 points (with a total team score of 202 points).

The gold medal cutoff for this IMO was set at 35 points (with 39 contestants earning gold medals), the silver medal cutoff was 25 points (with 70 contestants earning silver medals), and the bronze medal cutoff was 15 points (with 122 contestants earning bronze medals).

In this IMO, a total of four contestants achieved a perfect score of 42 points.

Problem 2.28 (IMO 39-1, proposed by Luxembourg). In convex quadrilateral $ABCD$, the diagonals AC and BD are perpendicular and the opposite sides AB and DC are not parallel. Suppose that a point P, where the perpendicular bisectors of AB and DC meet, is inside $ABCD$. Prove that $ABCD$ is a cyclic quadrilateral if and only if the triangles ABP and CDP have equal areas.

Proof. As shown in Figure 2.56, let BA and CD intersect at point Q after extension, AC and BD intersect at point G, the points E, F are the midpoints of AB, CD, respectively, and EG intersects CD at point H after extension.

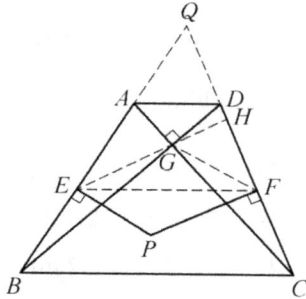

Figure 2.56

If A, B, C, D are concyclic, by $AE = EG$ and $AG \perp BG$, it follows that

$$\angle HGC = \angle AGE = \angle GAE = \angle BDC.$$

Clearly $EH \perp CD$, so $EG \parallel PF$. Similarly, $GF \parallel PE$. Thus, the quadrilateral $PFGE$ is a parallelogram.

So,

$$S_{\triangle ABP} = \frac{1}{2}AB \cdot PE = \frac{1}{2}AB \cdot GF = \frac{1}{4}AB \cdot CD$$

$$= \frac{1}{2}PF \cdot CD = S_{\triangle CDP}.$$

The necessity is thus proved.

On the other hand, if $S_{\triangle ABP} = S_{\triangle CDP}$, then

$$\frac{EG}{GF} = \frac{AB}{CD} = \frac{PF}{EP}.$$

And

$$\angle P = 180° - \angle Q = 180° - (90° - \angle ABD - \angle ACD)$$

$$= 90° + \angle EGB + \angle FGC = \angle EGF.$$

Consequently, $\triangle GEF \cong \triangle PFE$, and $EF = EF$, so $\triangle GEF \cong \triangle PFE$. Then the quadrilateral $PFGE$ is a parallelogram. Thus, $GE \perp CD$, which means that

$$\angle CDB = 90° - \angle DGH$$

$$= 90° - \angle EGB = 90° - \angle ABG$$

$$= \angle BAC,$$

so, A, B, C, D are concyclic. The sufficiency is also proved.

【Score Situation】 This particular problem saw the following distribution of scores among contestants: 103 contestants scored 7 points, 5 contestants scored 6 points, 23 contestants scored 5 points, 27 contestants scored 4 points, 41 contestants scored 3 points, 115 contestants scored 2 points, 16 contestants scored 1 point, and 89 contestants scored 0 point. The average score of this problem is 3.205, indicating that it was relatively straightforward.

Among the top five teams in the team scores, the scores of this problem are as follows: the Iran team scored 42 points (with a total team score of 211 points), the Bulgaria team scored 36 points (with a total team score of 195 points), the Hungary team scored 31 points (with a total team score of 186 points), the United States team scored 36 points (with a total team score of 186 points), and the Chinese Taiwan team scored 40 points (with a total team score of 184 points).

The gold medal cutoff for this IMO was set at 31 points (with 37 contestants earning gold medals), the silver medal cutoff was 24 points (with 66 contestants earning silver medals), and the bronze medal cutoff was 14 points (with 102 contestants earning bronze medals).

In this IMO, only one contestant achieved a perfect score of 42 points, namely Omid Amini from Iran.

Problem 2.29 (IMO 42-5, proposed by Israel). In a triangle ABC, let AP bisect $\angle BAC$, with P on BC, and let BQ bisect $\angle ABC$, with Q

on CA. It is known that $\angle BAC = 60°$ and $AB + BP = AQ + QB$. What are possible angles of triangle ABC?

Solution. As shown in Figure 2.57, extend AB to E such that $BP = BE$. Denote $\angle ABC = \angle B, \angle ACB = \angle C$.

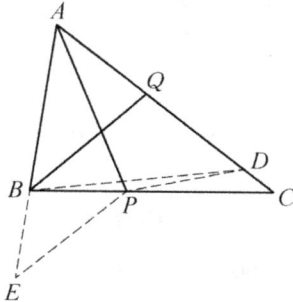

Figure 2.57

Find a point D on AC or its extension line such that $AD = AE$. Then the condition implies that $QD = QB$.

Since $\triangle APD \cong \triangle APE$,

$$\angle ADP = \angle E = \frac{1}{2}\angle B,$$

while

$$\angle QDB = \frac{1}{2}\angle AQB$$

$$= \frac{1}{2}(\angle QBC + \angle C)$$

$$= \frac{1}{4}\angle B + \frac{1}{2}\angle C.$$

The following is discussed for different cases.

(1) When points C and D coincide, $\angle B = 80°, \angle C = 40°$, and it is easy to verify that

$$AB + BP = AC = AQ + QB.$$

(2) When point D is in CQ,

$$\angle BDP = \angle ADP - \angle ADB$$

$$= \frac{1}{2}\angle B - \left(\frac{1}{4}\angle B + \frac{1}{2}\angle C\right) = \frac{1}{4}\angle B - \frac{1}{2}\angle C.$$

Meanwhile

$$\angle DPC = \angle ADP - \angle C$$

$$= \frac{1}{2}\angle B - \angle C = 2\angle BDP.$$

Thus

$$BP = DP.$$

By the law of sines for $\triangle ABP$ and $\triangle ADP$, we can get $\sin \angle B = \sin \angle ADP$.

Since $AD = AB + BP > AB, \angle B = \angle ADP$ is impossible, so A, B, P, D are concyclic. Thus $\angle DPC = 60°$, i.e., $\frac{1}{2}\angle B - \angle C = 60°$.

And $\angle B + \angle C = 120°$, so $\angle C = 0°$ is impossible.

(3) When point D is on the extension line AC, there is still $PB = DP$, and it can be seen that A, B, P, D are concyclic, but point P is inside $\triangle ABD$, which is impossible.

In summary, there can only be $\angle B = 80°, \angle C = 40°$.

【Score Situation】 This particular problem saw the following distribution of scores among contestants: 82 contestants scored 7 points, 5 contestants scored 6 points, 16 contestants scored 5 points, 17 contestants scored 4 points, 67 contestants scored 3 points, 145 contestants scored 2 points, 48 contestants scored 1 point, and 93 contestants scored 0 point. The average score of this problem is 2.729, indicating that it had a certain level of difficulty.

Among the top five teams in the team scores, the scores of this problem are as follows: the China team scored 42 points (with a total team score of 225 points), the Russia team scored 39 points (with a total team score of 196 points), the United States team scored 33 points (with a total team score of 196 points), the South Korea team scored 39 points (with a total team score of 185 points), and the Bulgaria team scored 35 points (with a total team score of 185 points).

The gold medal cutoff for this IMO was set at 30 points (with 39 contestants earning gold medals), the silver medal cutoff was 20 points (with 81 contestants earning silver medals), and the bronze medal cutoff was 11 points (with 122 contestants earning bronze medals).

In this IMO, a total of four contestants achieved a perfect score of 42 points, namely Liang Xiao and Zhiqiang Zhang from China, and Reid Barton and Gabriel Carroll from the Soviet Union.

Problem 2.30 (IMO 50-4, proposed by Belgium). Let ABC be a triangle with $AB = AC$. The angle bisectors of $\angle CAB$ and $\angle ABC$ meet the sides BC and CA at D and E, respectively. Let K be the incentre of triangle ADC. Suppose that $\angle BEK = 45°$. Find all possible values of $\angle CAB$.

Solution 1. Since CK is the angle bisector of $\angle ACB$, the symmetric point F of the point E with respect to CK is on the side of BC. Connect to form IF, and by symmetry $\angle IFK = 45°$. Connect to form DK, and since DK is the angle bisector of $\angle ADC$, so $\angle IDK = 45°$. Thus, there are two kinds of situations:

(1) As shown in Figure 2.58, if the points F and D do not coincide, then the four points I, D, F, K are concyclic, so $\angle IKF = 180° - \angle IDF = 90°$.

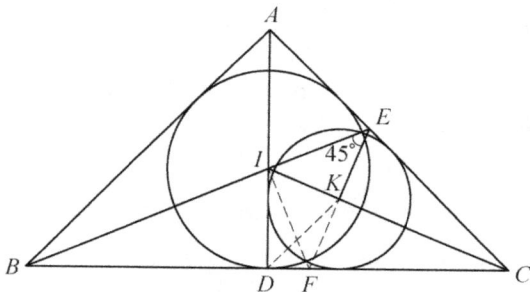

Figure 2.58

By symmetry, $\angle IKE = 90°$, and then $\angle EIK = 45°$, so

$$45° = \angle EIK = \angle IBC + \angle ICB = \frac{1}{2}\angle ABC + \frac{1}{2}\angle ACB.$$

Thus $\angle ABC + \angle ACB = 90°$, from which $\angle CAB = 90°$.

(2) As shown in Figure 2.59, if points F and D coincide, then $\angle IEC = \angle IDC = 90°$, which means that the bisector of $\angle ABC$ is also the altitude of side AC, so $AB = BC$. The triangle $\triangle ABC$ is an equilateral one, and consequently $\angle CAB = 60°$.

When $\angle CAB = 90°$ or $\angle CAB = 60°$, it is easy to verify that $\angle BEK = 45°$.

In summary, all possible values of $\angle CAB$ are $60°$ and $90°$.

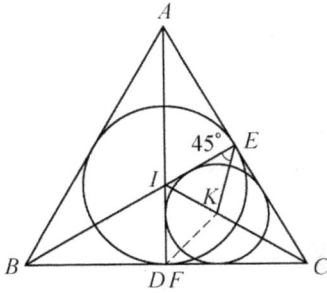

Figure 2.59

Solution 2. Since AD, BE are the angle bisectors of $\angle CAB, \angle ABC$, their intersection point is the incenter I. Connect to form CI. Then CI is the angle bisector of $\angle ACB$. Since K is the incenter of $\triangle ADC$, the point K lies on the segment CI.

Let $\angle BAC = \alpha$. Since $AB = AC, AD \perp BC$, we have

$$\angle ABC = \angle ACB = 90^\circ - \frac{\alpha}{2}.$$

As BI and CI bisect $\angle ABC$ and $\angle ACB$ respectively,

$$\angle ABI = \angle IBC = \angle ACI = \angle ICB = 45^\circ - \frac{\alpha}{4}.$$

Thus,

$$\angle EIC = \angle IBC + \angle ICB = 90^\circ - \frac{\alpha}{2},$$

$$\angle IEC = \angle BAE + \angle ABE = 45^\circ + \frac{3\alpha}{4}.$$

So

$$\frac{IK}{KC} = \frac{S_{\triangle IEK}}{S_{\triangle EKC}}$$

$$= \frac{\frac{1}{2} IE \cdot EK \cdot \sin \angle IEK}{\frac{1}{2} EC \cdot EK \cdot \sin \angle KEC}$$

$$= \frac{\sin 45^\circ}{\sin \frac{3\alpha}{4}} \cdot \frac{IE}{EC}$$

$$= \frac{\sin 45^\circ}{\sin \frac{3\alpha}{4}} \cdot \frac{\sin \left(45^\circ - \frac{\alpha}{4} \right)}{\sin \left(90^\circ - \frac{\alpha}{2} \right)}.$$

On the other hand, since K is the incenter of $\triangle ADC$, the segment DK bisects $\angle IDK$.

It follows from the property of the angle bisector that

$$\frac{IK}{KC} = \frac{ID}{DC} = \tan \angle ICD = \frac{\sin \left(45° - \frac{\alpha}{4}\right)}{\cos \left(45° - \frac{\alpha}{4}\right)}.$$

Thus,

$$\frac{\sin 45°}{\sin \frac{3\alpha}{4}} \cdot \frac{\sin \left(45° - \frac{\alpha}{4}\right)}{\sin \left(90° - \frac{\alpha}{2}\right)} = \frac{\sin \left(45° - \frac{\alpha}{4}\right)}{\cos \left(45° - \frac{\alpha}{4}\right)}.$$

Removing the denominators, we have

$$\sin 45° \cdot \cos \left(45° - \frac{\alpha}{4}\right) = \sin \frac{3\alpha}{4} \cos \frac{\alpha}{2}.$$

By the trigonometric formula,

$$\sin \left(90° - \frac{\alpha}{4}\right) + \sin \frac{\alpha}{4} = \sin \frac{5\alpha}{4} + \sin \frac{\alpha}{4},$$

that is,

$$\sin \left(90° - \frac{\alpha}{4}\right) = \sin \frac{5\alpha}{4}.$$

Since $0° < \alpha < 180°, \sin(90° - \frac{\alpha}{4}) > 0$, we have $\sin \frac{5\alpha}{4} > 0$, i.e., $0° < \frac{5\alpha}{4} < 180°$. It follows that either

$$90° - \frac{\alpha}{4} = \frac{5\alpha}{4} \Rightarrow \alpha = 60°,$$

or

$$90° - \frac{\alpha}{4} = 180° - \frac{5\alpha}{4} \Rightarrow \alpha = 90°.$$

When $\alpha = 60°$, it is easy to see that $\triangle IEC \cong \triangle IDK$, so $\triangle IEK \cong \triangle IDK$, and therefore

$$\angle BEK = \angle IDK = 45°.$$

When $\alpha = 90°$,

$$\angle EIC = 90° - \frac{\alpha}{2} = 45° = \angle KDC.$$

Since $\angle ICE = \angle DCK$, the triangles $\triangle ICE$ and $\triangle DCK$ are similar, which implies that

$$IC \cdot KC = DC \cdot EC.$$

It follows that $\triangle IDC$ and $\triangle EKC$ are similar, so $\angle EKC = \angle IDC = 90°$, and hence

$$\angle BEK = 180° - \angle EIK - \angle EKI = 45°.$$

Combining the above arguments, we conclude that all possible values of $\angle CAB$ are $60°$ and $90°$.

【Score Situation】This particular problem saw the following distribution of scores among contestants: 100 contestants scored 7 points, 52 contestants scored 6 points, 69 contestants scored 5 points, 17 contestants scored 4 points, 23 contestants scored 3 points, 37 contestants scored 2 points, 79 contestants scored 1 point, and 188 contestants scored 0 point. The average score of this problem is 2.915, indicating that it had a certain level of difficulty.

Among the top five teams in the team scores, the scores of this problem are as follows: the China team scored 42 points (with a total team score of 221 points), the Japan team scored 33 points (with a total team score of 212 points), the Russia team scored 42 points (with a total team score of 203 points), the South Korea team scored 40 points (with a total team score of 188 points), and the North Korea team scored 39 points (with a total team score of 183 points).

The gold medal cutoff for this IMO was set at 32 points (with 49 contestants earning gold medals), the silver medal cutoff was 24 points (with 98 contestants earning silver medals), and the bronze medal cutoff was 14 points (with 135 contestants earning bronze medals).

In this IMO, only two contestants achieved a perfect score of 42 points, namely Dongyi Wei from China and Makoto Soejima from Japan.

Problem 2.31 (IMO 51-4, proposed by Poland). Let P be a point inside a triangle ABC. The lines AP, BP, and CP intersect the circumcircle Γ of triangle ABC again at points K, L, and M respectively. The tangent to Γ at C intersects the line AB at S. Suppose that $SC = SP$. Prove that $MK = ML$.

Proof. Without loss of generality, we assume that $CA > CB$. Then S lies on the extension of AB. Suppose that the line SP intersects the circumcircle of the triangle ABC at E, F, as shown in Figure 2.60. By assumption and the power of a point,

$$SP^2 = SC^2 = SB \cdot SA,$$

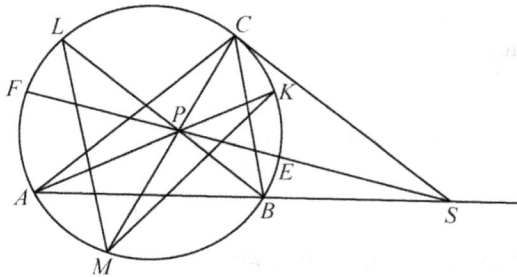

Figure 2.60

and hence $\frac{SP}{SB} = \frac{SA}{SP}$. Then $\triangle PSA \backsim \triangle BSP$ and $\angle BPS = \angle SAP$.
Since $2\angle BPS = \widehat{BE} + \widehat{LF}$, $2\angle SAP = \widehat{BE} + \widehat{EK}$, we have

$$\widehat{LF} = \widehat{EK}. \tag{1}$$

It follows from $\angle SPC = \angle SCP$ that $\widehat{EC} + \widehat{MF} = \widehat{EC} + \widehat{EM}$, and
therefore

$$\widehat{MF} = \widehat{EM}. \tag{2}$$

From (1) and (2),

$$\widehat{MFL} = \widehat{MF} + \widehat{FL} = \widehat{ME} + \widehat{EK} = \widehat{MEK},$$

and thus $MK = ML$.

【Score Situation】This particular problem saw the following distribution of scores
among contestants: 365 contestants scored 7 points, 2 contestants scored 6 points,
4 contestants scored 5 points, 2 contestants scored 4 points, 47 contestants scored
3 points, 10 contestants scored 2 points, 2 contestants scored 1 point, and 84
contestants scored 0 point. The average score of this problem is 5.345, indicating
that it was simple.

Among the top five teams in the team scores, the scores of this problem are as
follows: the China team scored 42 points (with a total team score of 197 points),
the Russia team scored 42 points (with a total team score of 169 points), the United
States team scored 42 points (with a total team score of 168 points), the South
Korea team scored 42 points (with a total team score of 156 points), the Thailand
team scored 42 points (with a total team score of 148 points), and the Kazakhstan
team scored 42 points (with a total team score of 148 points).

The gold medal cutoff for this IMO was set at 27 points (with 47 contestants
earning gold medals), the silver medal cutoff was 21 points (with 103 contestants

earning silver medals), and the bronze medal cutoff was 15 points (with 115 contestants earning bronze medals).

In this IMO, only one contestant achieved a perfect score of 42 points, namely Zipei Nie from China.

Problem 2.32 (IMO 62-4, proposed by Poland). Let Γ be a circle with center I, and $ABCD$ a convex quadrilateral such that each of the segments AB, BC, CD, and DA is tangent to Γ. Let Ω be the circumcircle of the triangle AIC. The extension of BA beyond A meets Ω at X, and the extension of BC beyond C meets Ω at Z. The extensions of AD and CD beyond D meet Ω at Y and T, respectively.

Prove that $AD + DT + TX + XA = CD + DY + YZ + ZC$.

Proof 1. As shown in Figure 2.61, notice that I is the intersection of the external angle bisector of $\angle TCZ$ and the circumcircle Ω of $\triangle TCZ$. Thus, I is the midpoint of the arc TCZ, so $IT = IZ$. Likewise, I is the midpoint of the arc YAX, namely $IX = IY$. Let O be the centre of Ω. Points X, T are the symmetric points of Y, Z with respect to IO, respectively, and hence $XT = YZ$.

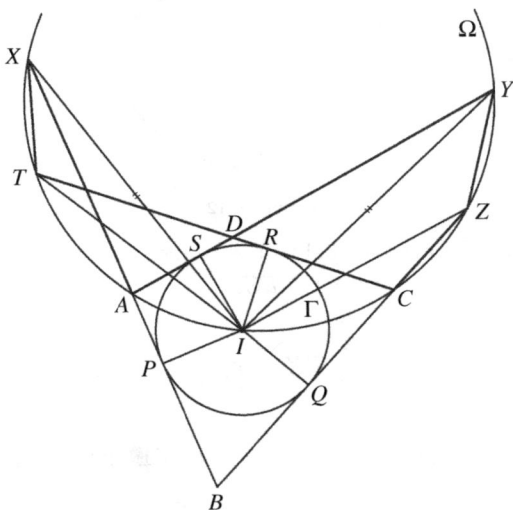

Figure 2.61

Let the inscribed circle of $ABCD$ touch the four sides at P, Q, R, S, respectively. Since $IP = IS$ and $IX = IY$, the right triangles IXP and

IYS are congruent. Likewise, the right triangles IRT and IQZ are congruent. They imply $XP = YS$ and $RT = QZ$.

Finally, from $AS = AP$, $CQ = RC$, and $SD = DR$, we obtain

$$AD + DT + TX + XA = TX + XP + RT = YZ + SY + QZ$$
$$= CD + DY + YZ + ZC,$$

as desired.

Proof 2.   As shown in Figure 2.62, since AI, CI are the external angle bisectors of $\angle XAY$, $\angle ZCT$, respectively, I is the midpoint of $\overset{\frown}{XAY}$, as well as the midpoint of $\overset{\frown}{TCZ}$.

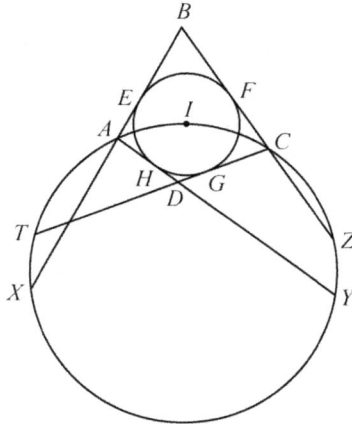

Figure 2.62

Both XY and TZ are parallel to the tangent line of the circumcircle of $\triangle AIC$ at point I, and hence $XY \parallel TZ$.

Then,

$$TX = YZ. \tag{1}$$

And since

$$AD + DT + XA = AH + HD + DT + XA$$
$$= AE + GD + DT + XA$$
$$= XA + AE + TD + DG$$
$$= XE + TG,$$

similarly, $CD + DY + ZC = YH + ZF$.

According to the Pythagorean theorem,

$$XE = \sqrt{XI^2 - EI^2} = \sqrt{YI^2 - HI^2} = YH.$$

In the same manner, $TG = ZF$, so

$$AD + DT + XA = CD + DY + ZC. \tag{2}$$

From (1) and (2), the conclusion holds.

【Score Situation】 This particular problem saw the following distribution of scores among contestants: 309 contestants scored 7 points, 5 contestants scored 6 points, 1 contestant scored 5 points, 12 contestants scored 4 points, 2 contestants scored 3 points, 39 contestants scored 2 points, 33 contestants scored 1 point, and 218 contestants scored 0 point. The average score of this problem is 3.817, indicating that it was relatively straightforward.

Among the top five teams in the team scores, the scores of this problem are as follows: the China team scored 42 points (with a total team score of 208 points), the Russia team scored 42 points (with a total team score of 183 points), the South Korea team scored 42 points (with a total team score of 172 points), the United States team scored 42 points (with a total team score of 165 points), and the Canada team scored 42 points (with a total team score of 151 points).

The gold medal cutoff for this IMO was set at 24 points (with 52 contestants earning gold medals), the silver medal cutoff was 19 points (with 103 contestants earning silver medals), and the bronze medal cutoff was 12 points (with 148 contestants earning bronze medals).

In this IMO, only one contestant achieved a perfect score of 42 points, namely Yichuan Wang from China.

2.3 Summary

In the first 64 IMOs, there are 32 problems on basic properties of circles and four points on a circle as depicted in Figure 2.63, which can be broadly categorized into three types.

Problems 2.1–2.6 focus on "existence problems;" among these six problems, the one with the lowest average score is Problem 2.1 (IMO 1-6), proposed by Czechoslovakia. Problems 2.7–2.24 deal with "position structure problems;" among these 18 problems, the one with the lowest average

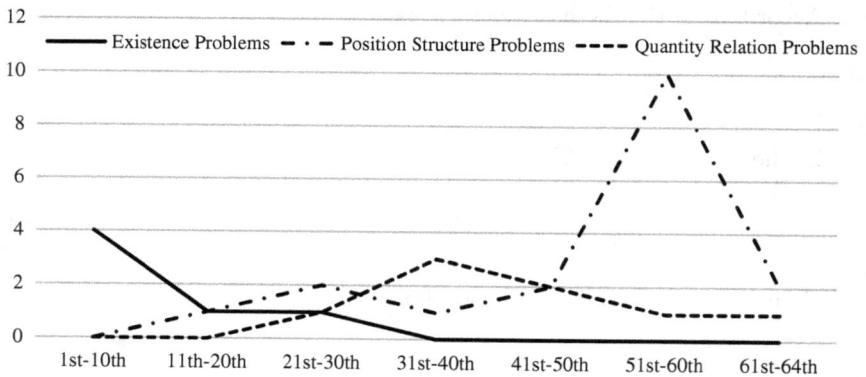

Figure 2.63 Numbers of Basic Properties of Circles and Four Points on a Circle Problems in the First 64 IMOs

score is Problem 2.13 (IMO 52-6), proposed by Japan. Problems 2.25–2.32 are about "quantity relation problems;" among these eight problems, the one with the lowest average score is Problem 2.29 (IMO 42-5), proposed by Israel. Furthermore, Problem 2.15 (IMO 54-3), proposed by Russia, Problem 2.16 (IMO 55-3), proposed by Iran, and Problem 2.22 (IMO 60-6), proposed by India all had extremely low average scores.

These problems were proposed by 23 countries and regions, with Poland contributing the most, totaling four problems. Greece proposed three problems, while Belgium, the Netherlands, Czechoslovakia, Luxembourg, and Romania each contributed two problems.

From Table 2.2, it can be observed that in the first 64 IMOs, there were four problems with an average score of 0–1 point; no problem with an average score of 1–2 points; seven problems with an average score of 2–3 points; seven problems with an average score of 3–4 points; fourteen problems with an average score above 4 points. Overall, basic properties of circles and four points on a circle problems were relatively simple.

In the 24th to 64th IMOs, there were a total of 24 basic properties of circles and four points on a circle problems. Among these, four had an average score of 0–1 point; no problem had an average score of 1–2 points; five had an average score of 2–3 points; four had an average score of 3–4 points;

Table 2.2 Score Details of Basic Properties of Circles and Four Points on a Circle Problems in the First 64 IMOs

Problem	2.1	2.2	2.3	2.4	2.5	2.6	2.7
Full points	7.000	5.000	7.000	7.000	6.000	7.000	6.000
Average score	2.556	2.800	3.000	4.176	3.758	3.807	4.104
Top five mean					4.375	5.575	4.775
6th–15th mean						3.43	3.92
16th–25th mean							
Problem number in IMO	1-6	2-7	3-5	4-5	14-2	21-3	20-4
Proposing country	Czechoslovakia	The German Democratic Republic	Czechoslovakia	Bulgaria	Netherlands	Soviet Union	United States

Problem	2.8	2.9	2.10	2.11	2.12	2.13	2.14
Full points	7.000	7.000	7.000	7.000	7.000	7.000	7.000
Average score	4.453	3.281	2.031	4.603	2.170	0.318	5.625
Top five mean	6.500	5.967	4.900	6.533	6.467	0.867	7.000
6th–15th mean	5.5	4.28	4.32	6.52	5.57	1.09	6.85
16th–25th mean	4.07	2.17	2.48	6.1	3.73	0.27	6.8
Problem number in IMO	25-4	27-4	37-2	45-1	46-5	52-6	53-1
Proposing country	Romania	Iceland	Canada	Romania	Poland	Japan	Greece

(Continued)

IMO Problems, Theorems, and Methods: Geometry

Table 2.2 (*Continued*)

Problem	2.15	2.16	2.17	2.18	2.19	2.20	2.21
Full points	7.000	7.000	7.000	7.000	7.000	7.000	7.000
Average score	0.786	0.505	4.794	5.272	5.029	4.934	2.399
Top five mean	3.567	3.500	7.000	6.833	7.000	7.000	6.533
6th–15th mean	1.82	1.6	6.93	6.72	6.64	6.95	4.97
16th–25th mean	1.48	0.73	6.47	6.92	6.52	6.8	4.61
Problem number in IMO	54-3	55-3	56-4	57-1	58-4	59-1	60-2
Proposing country	Russia	Iran	Greece	Belgium	Luxembourg	Greece	Ukraine

Problem	2.22	2.23	2.24	2.25	2.26	2.27	2.28
Full points	7.000	7.000	7.000	7.000	7.000	7.000	7.000
Average score	0.403	5.313	5.467	5.504	5.003	3.898	3.205
Top five mean	4.200	7.000	7.000	6.950	7.000	6.833	6.167
6th–15th mean	1.15	6.92	6.77	6.83	6.45	6.43	5.12
16th–25th mean	0.41	6.77	6.8	5.03	5.97	4.72	4.25
Problem number in IMO	60-6	61-1	63-4	23-5	35-2	38-2	39-1
Proposing country	India	Poland	Slovakia	Netherlands	Armenia and Australia	United Kingdom	Luxembourg

Problem	2.29	2.30	2.31	2.32
Full points	7.000	7.000	7.000	7.000
Average score	2.729	2.915	5.345	3.817
Top five mean	6.267	6.533	7.000	7.000
6th–15th mean	4.23	5.67	6.80	6.57
16th–25th mean	3.47	4.73	6.8	6.3
Problem number in IMO	42-5	50-4	51-4	62-4
Proposing country	Israel	Belgium	Poland	Poland

Note. Top five mean = Total score of the top five teams/Total number of contestants from the top five teams,
6th–15th mean = Total score of the 6th–15th teams/Total number of contestants from the 6th–15th teams,
16th–25th mean = Total score of the 16th–25th teams/Total number of contestants from the 16th–25th teams.

eleven had an average score above 4 points. Further analysis of the problem numbers of these 24 basic properties of circles and four points on a circle problems, as shown in Table 2.3, reveals that these problems frequently appeared as the 1st/4th problem. A majority type of these problems was position structure problems, totaling 17 problems, followed by quantity relation problems, totaling seven problems, and existence problems did not appear.

Table 2.3 Numbers of Basic Properties of Circles and Four Points on a Circle Problems in the 24th-64th IMOs

Basic Properties of Circles and Four Points on a Circle Problems	Problem Number			Numbers of Problems in the First 64 IMOs
	1, 4	**2, 5**	**3, 6**	
Existence problems	0	0	0	6
Position structure problems	10	3	4	18
Quantity relation problems	4	3	0	8
Total	14	6	4	32

From the 25th to 64th IMOs, basic properties of circles and four points on a circle problems are arranged in order of their average scores, from left to right, and a scatter plot of the scoring details is presented, as shown in Figure 2.64.

From Table 2.2 and Figure 2.64, it can be observed that the average score of the top five teams is generally about 0.5 to 4 points higher than the average score of the problem. However, in simpler problems, the difference in average scores between the top five teams, 6th-15th teams, and 16th–25th teams is typically not significant, as seen in problems such as Problem 2.11 (IMO 45-1), Problem 2.14 (IMO 53-1), Problem 2.18 (IMO 57-1), etc.

From Table 2.2 and Figure 2.64, it can also be observed that after the 30th IMO, the average scores of the 6th–15th teams and 16th–25th teams also tend to be higher than the average score of the problem. This phenomenon is due to the smaller number of participating teams in early

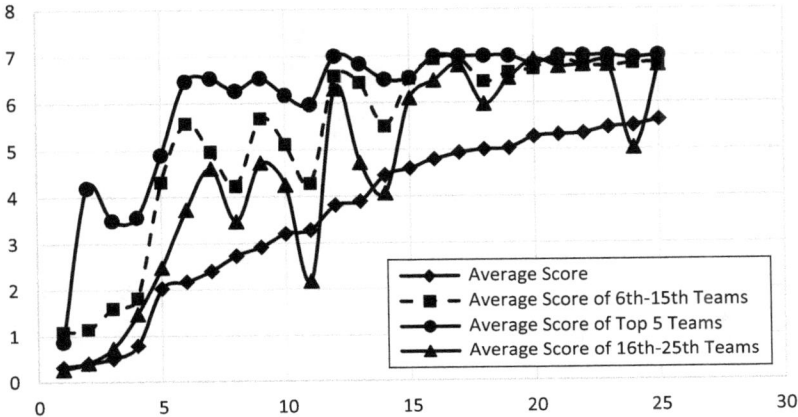

Figure 2.64 Scoring Details of Basic Properties of Circles and Four Points on a Circle Problems in the 25th-64th IMOs

Note. Average score of top five teams = Total score of the top five teams/Total number of contestants from the top five teams,

Average score of 6th–15th teams = Total score of the 6th–15th teams/Total number of contestants from the 6th–15th teams,

Average score of 16th–25th teams = Total score of the 16th–25th teams/Total number of contestants from the 16th–25th teams.

IMOs. It was not until the 30th IMO in 1989 that the number of participating teams exceeded 50. In addition, the appearance of extremely difficult geometry problems can have a depressing effect on the scores of many contestants, thus affecting the average score of the problem.

Chapter 3

Power of a Point, Radical Axis, and Radical Center

In a plane, a given point P and a given circle O of radius r determine an extremely important invariant, which is the "power of the point with respect to the circle": $PO^2 - r^2$, or "power of a point" for short. When point P is inside the circle O, on the circle, or outside the circle, its values are negative, zero, or positive, respectively.

When point P is outside the circle O, the power of point P is the square of the tangent length from point P to circle O.

This invariant is guaranteed by the intersecting chords theorem and the tangent-secant theorem, and it is also the basis of the inversion transformation.

The intersecting chords theorem, secants theorem, and tangent-secant theorem are collectively called the power of a point theorem. The power of a point theorem describes quantitative relationships between segments in the case of specific position relations, which is essentially to express a length relation between segments by an algebraic proportion relation. Therefore, the power of a point theorem itself can be regarded as a conversion between geometry and algebra. In the process of geometric proofs, a proper use of the power of a point theorem can transform geometry problems into algebraic problems, so as to solve geometry problems succinctly.

The locus of points with equal powers of two non-concentric circles is a straight line perpendicular to the line connecting the centers of the two circles, called the radical axis of the two circles. The radical axis is a basic straight line that communicates the relationship between circles. Because the radical axis and coaxial circles are easy to construct and calculate, many analytic geometry problems involving circles can be solved simply and clearly by using the knowledge of a radical axis.

In particular, when two circles intersect, the radical axis is the line through the two points of intersection. When two circles are tangent, the radical axis is the common tangent line through the point of tangency. When two circles are separate or inclusive, the radical axis does not intersect either circle.

When any two of three circles are not concentric, the radical axes of these three circles either intersect at one point (called the radical center of three circles) or are parallel (the radical center is at infinity). There is "Monge's theorem" about the radical center, which is also quite beautiful and useful.

In the first 64th IMOs, there had been a total of 12 problems related to the power of a point, radical axis, and radical center, accounting for 9.8% of all geometry problems. These problems are mainly categorized into two types: (1) position structure problems, totaling nine problems; (2) quantity relation problems, totaling three problems. The statistical distribution of these two types of problems in the previous IMOs is presented in Table 3.1.

It can be seen that the problems related to the power of a point, radical axis, and radical center began to appear after the 21st IMO, and accounted

Table 3.1   Numbers of Power of a Point, Radical Axis and Radical Center Problems in the First 64 IMOs

Content	Session							
	1–10	11–20	21–30	31–40	41–50	51–60	61–64	Total
Position structure problems	0	0	1	2	1	2	3	9
Quantity relation problems	0	0	0	0	2	1	0	3
Geometry problems	29	18	18	15	20	17	6	123
Percentage of power of a point, radical axis, and radical center problems among geometry problems	0.0%	0.0%	5.6%	13.3%	15.0%	17.6%	50%	9.8 %

for the highest proportion of geometry problems in the 51st to 64th IMOs, about 19.0%. Using the knowledge of the power of a point, radical axis, and radical center, we can flexibly communicate geometry and algebra, which is very helpful to deal with position structure problems and quantity relation problems in geometry.

Therefore, this chapter will be divided into three parts. The first part introduces the theorems related to circle powers, such as Monge's theorem, the power theorem of Casey, etc., and through some examples to expound the application of these theorems.

The second part revolves around two types of problems: "position structure problems" and "quantity relation problems." These problems are presented in chronological order, and some problems include various solutions and generalizations.

It is important to note that for each problem, the solutions are followed by information on the scores, including the number of contestants in each score range, the average score, and the scores of the top five teams. However, early IMOs often lacked information on contestant scores, so the number of contestants in each score range only represents the counted number of contestants, and some problems lack scores of the top five teams.

The third part provides a brief summary of this chapter.

3.1 Related Properties, Theorems, and Methods

3.1.1 *Power of a point*

(1) *Power of a point theorem*

Theorem 3.1 (Secants Theorem). Through a point P outside a circle, make two secant lines PAB and PCD to the circle. Then

$$PA \cdot PB = PC \cdot PD.$$

Theorem 3.2 (Tangent-Secant Theorem). Through a point P outside a circle, make the tangent line PT (T is the tangent point) and the secant PAB to the circle. Then

$$PT^2 = PA \cdot PB.$$

Theorem 3.3 (Intersecting Chords Theorem). Let two chords AB, CD in a circle intersect at point P. Then

$$PA \cdot PB = PC \cdot PD.$$

Note. The above three theorems all have inverse theorems.

(2) *The concept of power of a point*

Let P be a point in the plane of $\odot O$ and the radius of $\odot O$ be r. Define the power of the point P with respect to $\odot O$ as

$$OP^2 - r^2,$$

denoted $p(P, \odot O)$.

    If a line is drawn through point P and intersects $\odot O$ at two points A, B, then

$$p(P, \odot O) = \overline{PA} \cdot \overline{PB}.$$

3.1.2 *Radical axis and radical center*

(1) *The concept of radical axis*

The locus of a point to which the powers of two circles (non-concentric circles) are equal is a line, which is called the radical axis of the two circles.

    The radical axis of the two circles is perpendicular to the line connecting the centers of the two circles

    Obviously, there exist the following conclusions:

(i) The radical axis of two intersecting circles is the line connecting their two intersection points;

(ii) The radical axis of two tangent circles is the common tangent line at their tangent point;

(iii) The radical axis of two separate circles is the line through the midpoints of their two external common tangent segments and two internal common tangent segments.

(2) *Monge's theorem (radical center theorem)*

The radical axes of three circles (whose centers do not coincide with each other) are parallel to each other or intersect at one point.

    If the radical axes intersect at one point, the point is called the radical center of the three circles. (see Figures 3.1 and 3.2).

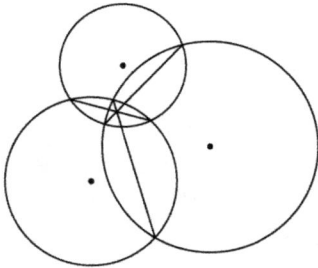

Figure 3.1

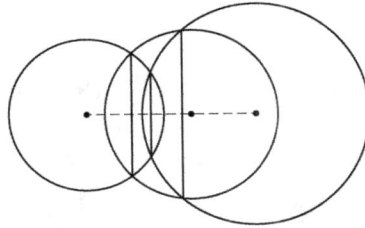

Figure 3.2

(3) *The power theorem of Casey*

The locus of a point to which the ratio of the powers of the two circles is constant ($\neq 1$) is a circle, which is coaxial with the two circles (that is, the radical axis between each pair is the same).

(4) *Jacobi's theorem and Brocard's theorem*

As shown in Figure 3.3, the quadrilateral $ABCD$ is inscribed in $\odot O$. Lines AB, CD intersect at point P, lines AD, BC intersect at point Q, and diagonal lines AC, BD intersect at point R.

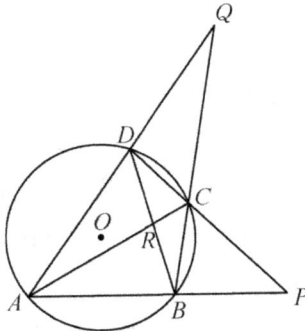

Figure 3.3

Theorem 3.4 (Jacobi's Theorem). Denote the powers of points P, Q, R to $\odot O$ as $p(P, \odot O), p(Q, \odot O), p(R, \odot O)$, respectively. Then

$$PQ^2 = p(P, \odot O) + p(Q, \odot O),$$

$$PR^2 = p(P, \odot O) + p(R, \odot O),$$
$$QR^2 = p(Q, \odot O) + p(R, \odot O).$$

Theorem 3.5 (Brocard's Theorem). Four points O, P, Q, R form an orthocentric system, that is, using any three points in O, P, Q, R as the vertices of a triangle, its orthocenter must be the fourth point.

Example 3.1. As shown in Figure 3.4, $\triangle ABC$ is inscribed in $\odot O$ and line AO intersects BC at point P. Both AE and AD are tangent to $\odot O$ and satisfy $PD \perp AB, PE \perp AC$. Let the lines BD, CE intersect at point Q. Prove: $PQ \perp BC$.

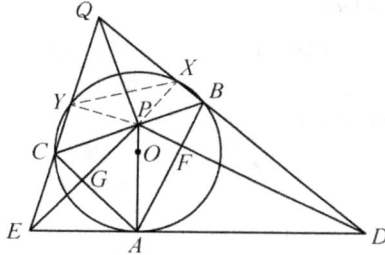

Figure 3.4

Proof. Let DQ and EQ intersect $\odot O$ at points $X(\neq B)$ and $Y(\neq C)$, respectively.

Combining the similarity and the tangent-secant theorem, we know that

$$DF \cdot DP = DA^2 = DB \cdot DX,$$

so, the four points B, X, P, F are concyclic, and then

$$\angle DXP = \angle DFB = 90°,$$

i.e., $PX \perp DQ$. Similarly, $PY \perp EQ$.

Thus, the four points P, X, Q, Y are concyclic. Since B, C, X, Y are concyclic,

$$\angle QPX = \angle QYX = \angle QBC = \angle QBP,$$

and then

$$\angle BPQ = \angle PXQ = 90°.$$

Hence $PQ \perp BC$.

Example 3.2. As shown in Figure 3.5, the tangency points of the incircle $\odot I$ of $\triangle ABC$ on the sides BC, CA, AB are D, E, F, respectively. Construct a parallel line passing through point A and parallel to EF, intersecting rays DE, DF at points P, Q, respectively. Prove: $\angle PIQ$ is an acute angle.

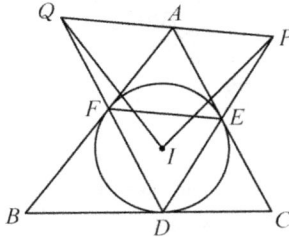

Figure 3.5

Proof.   We just have to prove that $IP^2 + IQ^2 > PQ^2$.

Let the radius of $\odot I$ be r. By the definition of the power of a point,

$$IP^2 = p(P, \odot I) + r^2 = PD \cdot PE + r^2. \tag{1}$$

Since $EF \parallel PQ$,

$$\angle CED = \angle DFE = \angle DQA,$$

so A, E, D, Q are concyclic, and then

$$PD \cdot PE = PA \cdot PQ. \tag{2}$$

Combining equations (1) and (2), we get

$$IP^2 = PA \cdot PQ + r^2.$$

Similarly, $IQ^2 = QA \cdot QP + r^2$.

Therefore,

$$IP^2 + IQ^2 = PA \cdot PQ + QA \cdot QP + 2r^2 = PQ^2 + 2r^2 > PQ^2,$$

from which, $\angle PIQ < 90°$.

Example 3.3. As shown in Figure 3.6, AD, BE, CF are the altitudes in a scalene triangle ACB, and their perpendicular feet are D, E, F, respectively. Lines AB, DE intersect at point P and lines BC, EF intersect at point Q. Let O, H be the circumcenter and the orthocenter of $\triangle ABC$ respectively. Prove: $OH \perp PQ$.

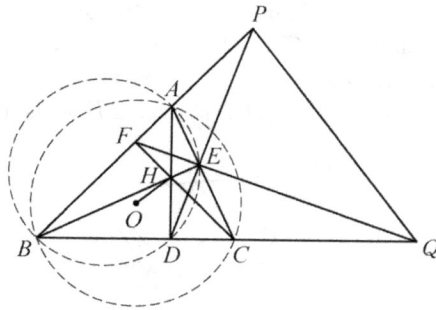

Figure 3.6

Proof.   Obviously, A, B, D, E are concyclic, and the circle has AB as its diameter, denoted as ω.

In the inscribed quadrilateral $ABDE$, by Jacobi's theorem,

$$PH^2 = p(P, \omega) + p(H, \omega) = PA \cdot PB - HB \cdot HE.$$

Similarly,

$$QH^2 = QB \cdot QC - HB \cdot HE.$$

Then,

$$PH^2 - QH^2 = PA \cdot PB - QB \cdot QC = p(P, \odot O) - p(Q, \odot O) = OP^2 - OQ^2.$$

By the equivalent difference of power of lines theorem, $OH \perp PQ$.

Example 3.4. As shown in Figure 3.7, in $\triangle ABC$, points D, E, F are on the sides BC, CA, AB respectively, and satisfy $EF \parallel BC$. The line

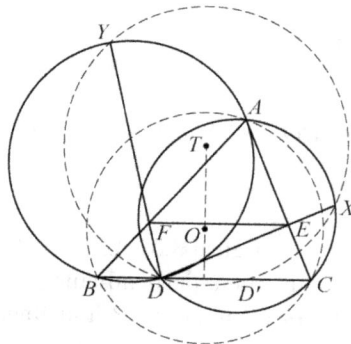

Figure 3.7

DE intersects the circumcircle of $\triangle ACD$ at another point X, and the line DF intersects the circumcircle of $\triangle ABD$ at another point Y. Let the symmetric point of D with respect to the midpoint of the side BC be D'. Prove: the four points D, D', X, Y are concyclic.

Proof. Construct the circumcircle $\odot O$ of $\triangle ABC$ and the circumcircle $\odot T$ of $\triangle DXY$. We only need to prove that the point D' is on $\odot T$.

Note that the powers of point E to $\odot O, \odot T$ are $\overline{EA} \cdot \overline{EC}$ and $\overline{ED} \cdot \overline{EX}$, respectively. Since points A, D, C, X are concyclic, the two previous values are equal. Thus, the powers of point E to $\odot O, \odot T$ are equal.

Similarly, the powers of point F to $\odot O, \odot T$ are also equal, so the line EF is the radical axis of circles $\odot O, \odot T$. Furthermore, by the properties of the radical axis, $EF \perp OT$.

By the condition $EF \parallel BC$, so $OT \perp BC$.

Notice again that $BD = CD'$, so point O is on the perpendicular bisector of DD', and then point T is also on the perpendicular bisector of DD'. Thus $TD = TD'$. That means point D' is on $\odot T$.

To sum up, points D, D', X, Y are concyclic.

3.2 Problems and Solutions

3.2.1 *Position structure problems*

Problem 3.1 (IMO 26-5, proposed by the Soviet Union). A circle with centre O passes through the vertices A and C of triangle ABC and intersects the segments AB, BC again at distinct points K, N, respectively. The circumscribed circles of the triangles ABC and KBN intersect at exactly two distinct points B and M. Prove that angle OMB is a right angle.

Proof 1. As shown in Figure 3.8, three circles intersect each other, and the lines AC, KN, and BM are their three radical axes. By the condition (point K is different from N; point B is different from M), we know that the three radical axes must meet at a point, that is, the radical center P, which is equal for the powers of the three circles.

It is easy to know $\angle PMN = \angle BKN = \angle NCA$, so M, N, C, P are concyclic. Thus,

$$BM \cdot BP = BN \cdot BC = BO^2 - r^2, \tag{1}$$

$$PM \cdot PB = PN \cdot PK = PO^2 - r^2. \tag{2}$$

Here r is the radius of the circle O.

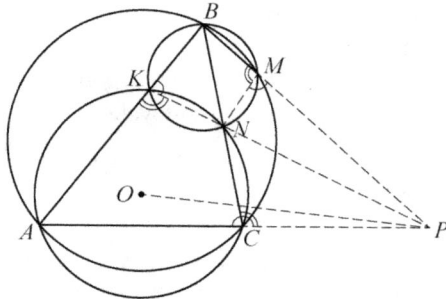

Figure 3.8

By (1)–(2),

$$BO^2 - PO^2 = BP(BM - PM)$$
$$= (BM + PM)(BM - PM)$$
$$= BM^2 - PM^2,$$

and thus, $OM \perp BP$.

Proof 2.   As shown in Figure 3.9, let the centers of the circumcircles of $\triangle ABC, \triangle BKN$ be O_1, O_2, respectively. Connect to form $KN, O_1B, O_1O, OO_2, O_1O_2, BO_2$, and MO_2.

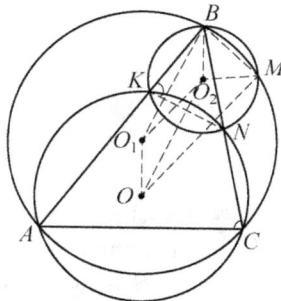

Figure 3.9

There is a familiar conclusion that $BO_1 \perp KN$, because

$$\angle O_1BA = 90° - \frac{1}{2}\angle BO_1A$$
$$= 90° - \angle BCA$$
$$= 90° - \angle BKN.$$

Similarly,

$$BO_2 \perp AC.$$

Since $OO_2 \perp KN$, so $BO_1 \parallel OO_2$, and similarly, $BO_2 \parallel O_1O$. So, the quadrilateral BO_1OO_2 is a parallelogram. And then O_1O_2 perpendicularly bisects BM, and consequently:

(i) $MO_2 = BO_2 = O_1O$;
(ii) $\angle OO_1O_2 = \angle BO_2O_1 = \angle MO_2O_1$.

Therefore, the quadrilateral O_1OMO_2 is an isosceles trapezoid. So, by $OM \parallel O_1O_2$, we have $O_1O_2 \perp BM$, from which $\angle OMB = 90°$.

Proof 3. As shown in Figure 3.10, connect to form KN. Extend MN to intersect $\odot O$ at another point D, and connect to form AD.

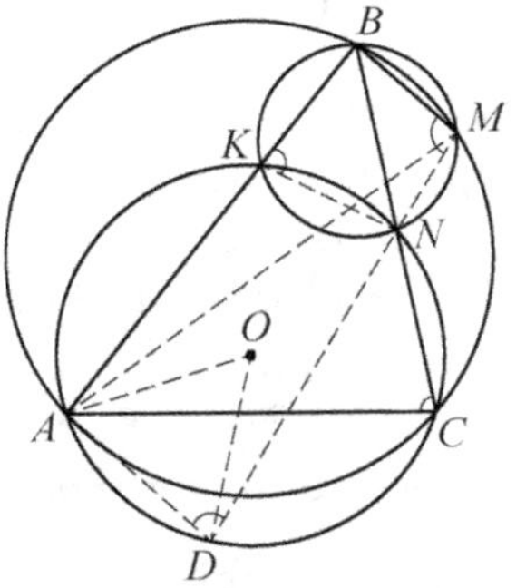

Figure 3.10

It is obvious that $\angle MDA = \angle BKN = 180° - \angle BMD$.

Thus, $AD \parallel BM$.

If we can prove that $MO \perp AD$, then the conclusion holds immediately.

It is easy to know $OA = OD$, that is, point O is on the perpendicular bisector of AD. If we can prove that M is also on the perpendicular bisector of AD, then $MO \perp AD$. So, the problem turns to a proof that $AM = DM$.

In the following, we will prove it. According to a series of concyclic points, we can see that:

$$\angle MAD = \angle MAC + \angle DAC$$

$$= \angle MBC + \angle DNC$$

$$= \angle MBN + \angle MNB$$

$$= 180° - \angle BMN$$

$$= \angle MDA.$$

Thus, $AM = DM$, so $\angle OMB = 90°$.

【Score Situation】 This particular problem saw the following distribution of scores among contestants: 35 contestants scored 7 points, 1 contestant scored 6 points, 5 contestants scored 5 points, 7 contestants scored 4 points, 7 contestants scored 3 points, 20 contestants scored 2 points, 21 contestants scored 1 point, and 113 contestants scored 0 point. The average score of this problem is 1.847, indicating that it was relatively challenging.

Among the top five teams in the team scores, the scores of this problem are as follows: the Romania team scored 32 points (with a total team score of 201 points), the United States team scored 10 points (with a total team score of 180 points), the Hungary team scored 23 points (with a total team score of 168 points), the Bulgaria team scored 33 points (with a total team score of 165 points), and the Vietnam team scored 28 points (with a total team score of 144 points).

The gold medal cutoff for this IMO was set at 34 points (with 14 contestants earning gold medals), the silver medal cutoff was 22 points (with 35 contestants earning silver medals), and the bronze medal cutoff was 15 points (with 52 contestants earning bronze medals).

In this IMO, only two contestants achieved a perfect score of 42 points, namely Géza Kós from Hungary and Daniel Tătaru from Romania.

Problem 3.2 (IMO 36–1, proposed by Bulgaria). Let A, B, C, and D be four distinct points on a line, in that order. The circles with diameters AC and BD intersect at points X and Y. The line XY meets BC at a point Z. Let P be a point on the line XY different from Z. The line CP intersects the circle with diameter AC at the points C and M, and the line BP intersects the circle with diameter BD at the points B and N. Prove that the lines AM, DN, and XY are concurrent.

Proof.   As shown in Figure 3.11, let lines AM and DN intersect line XY at points Q and Q', respectively.

Clearly $\angle AMC = 90°$. Thus $\triangle QAZ \backsim \triangle CPZ$, so

$$QZ = \frac{CZ \cdot AZ}{PZ} = \frac{XZ \cdot YZ}{PZ}.$$

Similarly,

$$Q'Z = \frac{BZ \cdot DZ}{PZ} = \frac{XZ \cdot YZ}{PZ}.$$

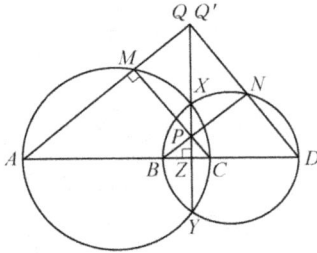

Figure 3.11

Hence, $QZ = Q'Z$, so Q and Q' coincide. This means that the three lines AM, DN, XY are concurrent.

【Score Situation】This particular problem saw the following distribution of scores among contestants: 239 contestants scored 7 points, 38 contestants scored 6 points, 4 contestants scored 5 points, 7 contestants scored 4 points, 18 contestants scored 3 points, 21 contestants scored 2 points, 38 contestants scored 1 point, and 47 contestants scored 0 point. The average score of this problem is 5.056, indicating that it was simple.

Among the top five teams in the team scores, the scores of this problem are as follows: the China team scored 42 points (with a total team score of 236 points), the Romania team scored 39 points (with a total team score of 230 points), the Russia team scored 42 points (with a total team score of 227 points), the Vietnam team scored 42 points (with a total team score of 220 points), and the Hungary team scored 38 points (with a total team score of 210 points).

The gold medal cutoff for this IMO was set at 37 points (with 30 contestants earning gold medals), the silver medal cutoff was 29 points (with 71 contestants earning silver medals), and the bronze medal cutoff was 19 points (with 100 contestants earning bronze medals).

In this IMO, a total of 14 contestants achieved a perfect score of 42 points.

Problem 3.3 (IMO 40-5, proposed by Russia). Two circles Γ_1 and Γ_2 are contained inside a circle Γ, and are tangent to Γ at distinct points M and N, respectively. The circle Γ_1 passes through the centre of Γ_2. The line passing through the two intersection points of Γ_1 and Γ_2 meets Γ at A and B. The lines MA and MB meet Γ_1 at C and D, respectively. Prove that CD is tangent to Γ_2.

Proof 1. As shown in Figure 3.12, let the centers of the three circles be $\Gamma, \Gamma_1, \Gamma_2$ and the radii be r, r_1, r_2, respectively. Join points to form $\Gamma_2\Gamma$, $\Gamma_1\Gamma_2$, $\Gamma\Gamma_1 M$, and $\Gamma_2 M$.

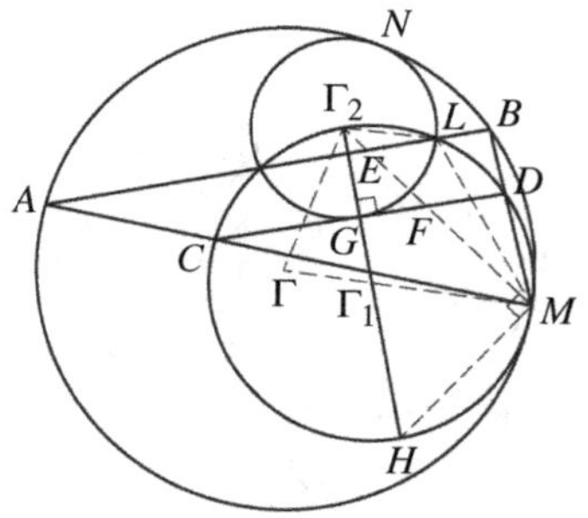

Figure 3.12

Since $\Gamma\Gamma_2 = r - r_2, \Gamma_1\Gamma_2 = r_1, \Gamma\Gamma_1 = r - r_1, \Gamma_1M = r_1$, by the law of cosines,

$$\frac{\Gamma_2M^2 - 2r_1^2}{2r_1^2} = \cos\angle\Gamma_2\Gamma_1\Gamma = \frac{r_1^2 + (r-r_1)^2 - (r-r_2)^2}{2r_1(r-r_1)},$$

so

$$\Gamma_2M^2 = \frac{r_1r_2(2r-r_2)}{r-r_1}.$$

And let $M\Gamma_2$ intersect AB, CD at E, F, respectively.

Since the homothetic center of $\odot\Gamma_1$ and $\odot\Gamma$ is M, it is easy to know that $CD \parallel AB$.

Connect to get the segment $\Gamma_2\Gamma_1$, and we know that $\Gamma_2\Gamma_1 \perp AB$, so $\Gamma_2\Gamma_1 \perp CD$. Extend $\Gamma_2\Gamma_1$ to intersect the circle Γ_1 at point H. Then connect to get MH. Let $\Gamma_2\Gamma_1$ intersect CD at point G. Then the four points G, H, M, F are concyclic. Thus

$$\Gamma_2G \cdot \Gamma_2H = \Gamma_2F \cdot \Gamma_2M.$$

Let one point of intersection between $\odot\Gamma_1$ and $\odot\Gamma_2$ be L. From $\angle\Gamma_2LA = \angle LM\Gamma_2$, it is easy to prove that

$$\Gamma_2E \cdot \Gamma_2M = \Gamma_2L^2 = r_2^2.$$

Thus,

$$EM = \Gamma_2M - \Gamma_2E = \Gamma_2M - \frac{r_2^2}{\Gamma_2M}.$$

While

$$\frac{EF}{EM} = 1 - \frac{FM}{EM} = 1 - \frac{MD}{MB} = 1 - \frac{r_1}{r},$$

so

$$EF = \frac{r - r_1}{r} \cdot \frac{\Gamma_2 M^2 - r_2^2}{\Gamma_2 M},$$

$$\Gamma_2 F = \Gamma_2 E + EF$$

$$= \frac{r_2^2}{\Gamma_2 M} + \frac{r - r_1}{r} \cdot \frac{\Gamma_2 M^2 - r_2^2}{\Gamma_2 M},$$

and then,

$$\Gamma_2 F \cdot \Gamma_2 M = r_2^2 + \frac{r - r_1}{r}(\Gamma_2 M^2 - r_2^2)$$

$$= r_2^2 + \frac{r_1 r_2 (2r - r_2) - (r - r_1) r_2^2}{r}$$

$$= 2 r_1 r_2.$$

Also $\Gamma_2 H = 2 r_1$, so, $\Gamma_2 G = r_2$. Thus, CD is tangent to the circle Γ_2.

Proof 2. Construct tangent lines through points M, N, respectively and tangent to the circle Γ, and let them intersect at point P. Since the radical axes of the three circles intersect at one point, P is on the extension line of AB (if three radical axes are parallel, readers can handle it themselves). As shown in Figure 3.13, construct the radius $\Gamma_2 R$ of $\odot \Gamma_2$ to be perpendicular to AB, noting that point R is on $\odot \Gamma_2$. Now it is not known whether point R is on CD (obviously if it is on CD, then the conclusion holds).

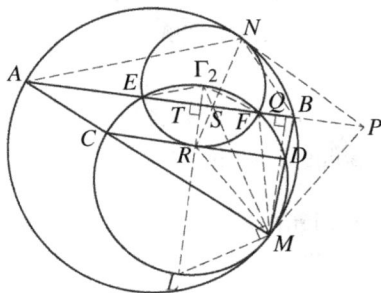

Figure 3.13

Let $\odot \Gamma_1$, and $\odot \Gamma_2$ intersect at points E, F. Then $\Gamma_2 R$ and EF intersect at points T.

Since $\odot \Gamma_1$ and $\odot \Gamma$ are homothetic, and the homothetic center is M, it is easy to know that $\frac{MC}{MA} = \frac{MD}{MB}$, so $CD \parallel AB$. Thus $\widehat{CE} = \widehat{FD}$ and $\widehat{\Gamma_2 E} = \widehat{\Gamma_2 F}$, and they add up to $\widehat{\Gamma_2 C} = \widehat{\Gamma_2 D}$, so $M\Gamma_2$ bisects $\angle AMB$. Let $M\Gamma_2$ and AB intersect at point S.

Similarly, NR bisects $\angle ANB$, and let NR and AB intersect at point S'. By the similarity of triangles,

$$\frac{AS}{SB} = \frac{AM}{BM} = \frac{AP}{PM} = \frac{AP}{PN} = \frac{AN}{BN} = \frac{AS'}{S'B},$$

and then S and S' coincide.

In addition, we can easily get $PS = PM = PN$.

Extend $\Gamma_2 R$ to intersect $\odot\Gamma_1$ at point L. Connect to form $\Gamma_2 E, \Gamma_2 F, FM$. It is easy to know

$$\Gamma_2 R^2 = \Gamma_2 S \cdot \Gamma_2 M = \Gamma_2 T \cdot \Gamma_2 L.$$

Let $\angle RSM = \angle RMQ = \theta, \angle \Gamma_2 ST = \angle PSM = \alpha$, where $MQ \perp AP$ and Q is the perpendicular foot.

Let the radii of $\odot\Gamma, \odot\Gamma_1, \odot\Gamma_2$ be r, r_1, r_2, respectively. Then

$$\frac{r_1}{r} = \frac{\Gamma_2 L}{2PM \tan \frac{\angle NPM}{2}}$$

$$= \frac{\Gamma_2 R^2}{2\Gamma_2 T \cdot PM \tan \theta}$$

$$= \frac{\Gamma_2 S \cdot \Gamma_2 M \cdot \cos \alpha}{2\Gamma_2 S \sin \alpha \cdot \frac{SM}{2} \tan \theta}$$

$$= \frac{TQ}{QM \tan \theta} = \frac{TQ}{QM} \cdot \frac{QM - TR}{TQ}$$

$$= 1 - \frac{TR}{QM} = 1 - \frac{AC}{AM} = 1 - \frac{BD}{BM},$$

which proves that the point R is on CD.

Proof 3. As shown in Figure 3.14, construct a common tangent line ST of the two small circles, which is tangent to $\odot\Gamma_1$ and $\odot\Gamma_2$ at points F, E, respectively.

First, we know that ST is the chord of $\odot\Gamma$. Then NE and MF intersect at the midpoint of $\overparen{ST}$, and let it be G.

In addition, the four points E, N, M, F are concyclic. Then

$$GE \cdot GN = GF \cdot GM.$$

This means that point G has the equal powers with respect to $\odot\Gamma_1$ and $\odot\Gamma_2$. Thus point G is on AB, that is, point A coincides with point G, and then point F coincides with point C.

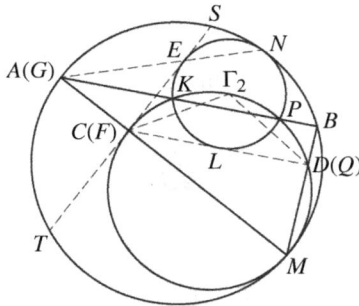

Figure 3.14

Construct the tangent line CL of $\odot\Gamma_2$ and extend to intersect $\odot\Gamma_1$ at point Q. Then

$$\angle\Gamma_2 QC = \angle\Gamma_2 CS = \angle\Gamma_2 CQ,$$

so $\overset{\frown}{\Gamma_2 C} = \overset{\frown}{\Gamma_2 Q}$.

As shown in Figure 3.14, let circles $\odot\Gamma_1$ and $\odot\Gamma_2$ intersect at two points K, P, where point K is close to CS. By $\overset{\frown}{\Gamma_2 K} = \overset{\frown}{\Gamma_2 P}$, we get $\overset{\frown}{KC} = \overset{\frown}{PQ}$.

So $CQ \parallel AB$, while $CD \parallel AB$. Then point D coincides with point Q, which means that CD is tangent to the $\odot\Gamma_2$.

【Score Situation】This particular problem saw the following distribution of scores among contestants: 47 contestants scored 7 points, 31 contestants scored 6 points, 7 contestants scored 5 points, 5 contestants scored 4 points, 18 contestants scored 3 points, 44 contestants scored 2 points, 103 contestants scored 1 point, and 195 contestants scored 0 point. The average score of this problem is 1.811, indicating that it was relatively challenging.

Among the top five teams in the team scores, the scores of this problem are as follows: the Russia team scored 37 points (with a total team score of 182 points), the China team scored 33 points (with a total team score of 182 points), the Vietnam team scored 40 points (with a total team score of 177 points), the Romania team scored 32 points (with a total team score of 173 points), and the Bulgaria team scored 25 points (with a total team score of 170 points).

The gold medal cutoff for this IMO was set at 28 points (with 38 contestants earning gold medals), the silver medal cutoff was 19 points (with 70 contestants earning silver medals), and the bronze medal cutoff was 12 points (with 118 contestants earning bronze medals).

In this IMO, no contestant achieved a perfect score of 42 points.

Problem 3.4 (IMO 49-1, proposed by Russia). An acute-angled triangle ABC has orthocenter H. The circle passing through H with center the midpoint of BC intersects the line BC at A_1 and A_2. Similarly, the circle passing through H with center the midpoint of CA intersects the line CA at B_1 and B_2, and the circle passing through H with center the midpoint of AB intersects the line AB at C_1 and C_2. Show that $A_1, A_2, B_1, B_2, C_1, C_2$ lie on a circle.

Proof 1.   As shown in Figure 3.15, let B_0, C_0 be the midpoints of CA, AB, respectively. Denote A' as the other intersection of the circle centered at B_0 which passes through H and the circle centered at C_0 which passes through H. We know $A'H \perp C_0 B_0$. Since $B_0 C_0$ are the midpoints of CA, AB, respectively, so $B_0 C_0 \parallel BC$. Therefore $A'H \perp BC$, and then A' lies on the segment AH.

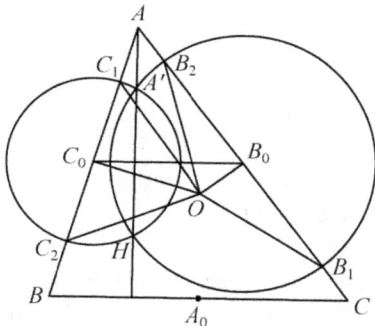

Figure 3.15

By the tangent-secant theorem,

$$AC_1 \cdot AC_2 = AA' \cdot AH = AB_1 \cdot AB_2,$$

so B_1, B_2, C_1, C_2 all lie on a circle.

Denote O as the intersection of perpendicular bisectors of $B_1 B_2, C_1 C_2$. Then O is the circumcenter of quadrilateral $B_1 B_2 C_1 C_2$, as well as the circumcenter of $\triangle ABC$. Thus $OB_1 = OB_2 = OC_1 = OC_2$.

Similarly, $OA_1 = OA_2 = OB_1 = OB_2$.

Therefore, the six points $A_1, A_2, B_1, B_2, C_1, C_2$ are all on the same circle, whose center is O, and the radius is OA_1.

Proof 2 (By Dongyi Wei). As shown in Figure 3.16, let O be the circumcenter of the triangle ABC, and D, E, F are the midpoints of

BC, CA, AB respectively. The segment BH intersects DF at point P, and then $BH \perp DF$. By the Pythagorean theorem,

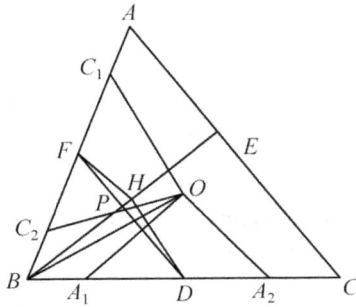

Figure 3.16

$$BF^2 - FH^2 = BP^2 - PH^2 = BD^2 - DH^2, \qquad (1)$$

$$BO^2 - A_1O^2 = BD^2 - A_1D^2 = BD^2 - DH^2. \qquad (2)$$

Similarly,

$$BO^2 - C_2O^2 = BF^2 - FH^2, \qquad (3)$$

By (1), (2), (3), $A_1O = C_2O$. It is obvious that $A_1O = A_2O, C_1O = C_2O$. Thus

$$A_1O = A_2O = C_1O = C_2O.$$

Similarly, $A_1O = A_2O = B_1O = B_2O$.

Therefore, the six points $A_1, A_2, B_1, B_2, C_1, C_2$ are all on the same circle, whose center is O.

【Score Situation】 This particular problem saw the following distribution of scores among contestants: 321 contestants scored 7 points, 7 contestants scored 6 points, 44 contestants scored 5 points, 8 contestants scored 4 points, 5 contestants scored 3 points, 17 contestants scored 2 points, 74 contestants scored 1 point, and 59 contestants scored 0 point. The average score of this problem is 4.979, indicating that it was simple.

Among the top five teams in the team scores, the scores of this problem are as follows: the China team scored 42 points (with a total team score of 217 points), the Russia team scored 42 points (with a total team score of 199 points), the United States team scored 42 points (with a total team score of 190 points), the South

Korea team scored 41 points (with a total team score of 188 points), and the Iran team scored 42 points (with a total team score of 181 points).

The gold medal cutoff for this IMO was set at 31 points (with 47 contestants earning gold medals), the silver medal cutoff was 22 points (with 100 contestants earning silver medals), and the bronze medal cutoff was 15 points (with 120 contestants earning bronze medals).

In this IMO, only three contestants achieved a perfect score of 42 points, namely Xiaosheng Mu and Dongyi Wei from China, and Alex Zhai from the United States.

Problem 3.5 (IMO 54-4, proposed by Thailand). Let $\triangle ABC$ be an acute-angled triangle with orthocenter H, and let W be a point on the side BC, lying strictly between B and C. Two points M and N are the feet of the altitudes from B and C, respectively. Denote by ω_1 the circumcircle of $\triangle BWN$, and let X be a point on ω_1 such that WX is a diameter of ω_1. Analogously, denote by ω_2 the circumcircle of $\triangle CWM$, and let Y be a point on ω_2 such that WY is a diameter of ω_2. Prove that X, Y, and H are collinear.

Proof.   As shown in Figure 3.17, let AL be the altitude to side BC, and Z be another intersection point of circles ω_1 and ω_2 different from W. We show that points X, Y, Z, and H are collinear.

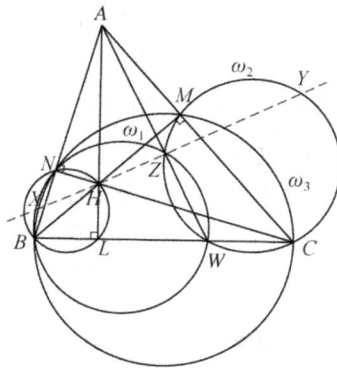

Figure 3.17

Points B, C, M, and N are concyclic (denote the circle by ω_3) since $\angle BNC = \angle BMC = 90°$. WZ, BN, and CM intersect at a point since they are the radical axis of ω_1 and ω_2, ω_1 and ω_3, and ω_2 and ω_3, respectively. And since BN and CM intersect at A, the segment WZ passes through A.

Since $\angle WZX = \angle WZY = 90°$, the segments WX and WY are diameters of ω_1 and ω_2, respectively. Thus, points X and Y are on the line l perpendicular from Z to WZ.

So, points B, L, H, and N are concyclic since $\angle BNH = \angle BLH = 90°$. By the power of a point,

$$AL \cdot AH = AB \cdot AN = AW \cdot AZ. \tag{1}$$

If point H is on the line AW, then H and Z coincide. Otherwise, by (1),

$$\frac{AZ}{AH} = \frac{AL}{AW}.$$

Thus, $\triangle AHZ \backsim \triangle AWL$, and consequently, $\angle HZA = \angle WLA = 90°$. So, point H is also on the line l.

【Score Situation】 This particular problem saw the following distribution of scores among contestants: 385 contestants scored 7 points, 9 contestants scored 6 points, 5 contestants scored 5 points, 2 contestants scored 4 points, 14 contestants scored 3 points, 14 contestants scored 2 points, 16 contestants scored 1 point, and 82 contestants scored 0 point. The average score of this problem is 5.442, indicating that it was simple.

Among the top five teams in the team scores, the scores of this problem are as follows: the China team scored 41 points (with a total team score of 208 points), the South Korea team scored 42 points (with a total team score of 204 points), the Singapore team scored 42 points (with a total team score of 190 points), the Russia team scored 42 points (with a total team score of 187 points), and the North Korea team scored 42 points (with a total team score of 184 points).

The gold medal cutoff for this IMO was set at 31 points (with 45 contestants earning gold medals), the silver medal cutoff was 24 points (with 92 contestants earning silver medals), and the bronze medal cutoff was 15 points (with 141 contestants earning bronze medals).

In this IMO, no contestant achieved a perfect score of 42 points.

Problem 3.6 (IMO 56-3, proposed by Ukraine). Let ABC be an acute triangle with $AB > AC$. Let Γ be its circumcircle, H its orthocenter, and F the foot of the altitude from A. Let M be the midpoint of BC, let Q be the point on Γ such that $\angle HQA = 90°$, and let K be the point on Γ such that $\angle HKQ = 90°$. Assume that the points A, B, C, K, and Q are all different, and lie on Γ in this order. Prove that the circumcircles of triangles KQH and FKM are tangent to each other.

Proof. As shown in Figure 3.18, extend QH to intersect Γ at A'. Since $\angle AQH = 90°$, the segment AA' is a diameter of Γ.

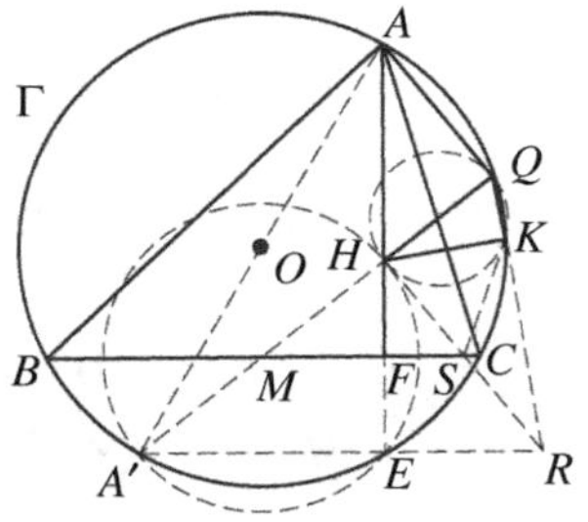

Figure 3.18

Since $A'B \perp AB$, so $A'B \parallel CH$. Similarly, $A'C \parallel BH$, so $BA'CH$ is a parallelogram, and M is the midpoint of $A'H$.

Extend AF to intersect Γ at E. Since $A'E \perp AE$, so $A'E \parallel BC$, and MF is the midsegment of $\triangle HA'E$ with F the midpoint of HE.

Let R be the intersection of $A'E$ and QK. By the power of a point,

$$RK \cdot RQ = RE \cdot RA'.$$

The circumcircles ω_1 of $\triangle HKQ$ and ω_2 of $\triangle HEA'$ have diameters HQ and HA', respectively. They are externally tangent at H, and then the point R has the same power to them, so R is on the radical axis of ω_1 and ω_2, so RH is their common tangent line, and $RH \perp A'Q$.

Let S be the intersection of the lines MF and HR. Then S is the midpoint of RH. Since $\triangle RHK$ is a right triangle, and S is the midpoint of its hypotenuse RH, so $SH = SK$. And because SH is tangent to ω_1, we see that SK is also tangent to ω_1. In the right triangle SHM, the segment HF is the altitude on the hypotenuse, from which

$$SF \cdot SM = SH^2 = SK^2,$$

Hence SK is also tangent to the circumcircle of $\triangle KMF$. Therefore SK is tangent to both circumcircles of $\triangle KQH$ and $\triangle FKM$ at the point K and these two circles are tangent at K as well.

【Score Situation】 This particular problem saw the following distribution of scores among contestants: 30 contestants scored 7 points, 1 contestant scored 6 points, no contestant scored 5 points, 3 contestants scored 4 points, 1 contestant scored 3 points, 12 contestants scored 2 points, 122 contestants scored 1 point, and 408 contestants scored 0 point. The average score of this problem is 0.653, indicating that it was extremely difficult.

Among the top five teams in the team scores, the scores of this problem are as follows: the United States team scored 31 points (with a total team score of 185 points), the China team scored 12 points (with a total team score of 181 points), the South Korea team scored 18 points (with a total team score of 161 points), the North Korea team scored 17 points (with a total team score of 156 points), and the Vietnam team scored 23 points (with a total team score of 151 points).

The gold medal cutoff for this IMO was set at 26 points (with 39 contestants earning gold medals), the silver medal cutoff was 19 points (with 100 contestants earning silver medals), and the bronze medal cutoff was 14 points (with 143 contestants earning bronze medals).

In this IMO, only one contestant achieved a perfect score of 42 points, namely Zhuo Qun Alex Song from Canada.

Problem 3.7 (IMO 62-3, proposed by Ukraine). Let D be an interior point of an acute triangle ABC with $AB > AC$ so that $\angle DAB = \angle CAD$. A point E on the segment AC satisfies $\angle ADE = \angle BCD$, a point F on the segment AB satisfies $\angle FDA = \angle DBC$, and a point X on the line AC satisfies $CX = BX$. Let O_1 and O_2 be the circumcenters of the triangles ADC and EXD, respectively. Prove that the lines BC, EF, and O_1O_2 are concurrent.

Proof 1. As shown in Figure 3.19, let Q be the isogonal conjugate of point D in $\triangle ABC$. Since $\angle DAB = \angle CAD$, point Q is on AD, and $\angle QBA = \angle DBC = \angle FDA$. It follows that Q, D, F, B are concyclic. Similarly, Q, D, E, C are concyclic. Then
$$AF \cdot AB = AD \cdot AQ = AE \cdot AC,$$
and therefore B, F, E, C concyclic.

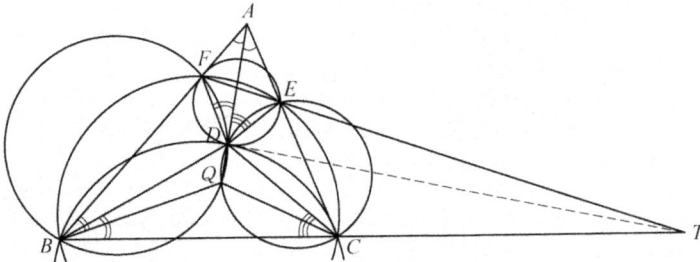

Figure 3.19

Lemma. *Let the extensions of BC and FE beyond C and E meet at T. Then $TD^2 = TB \cdot TC = TF \cdot TE$.*

Proof of the Lemma.   First, $\odot DEF$ and $\odot BDC$ are tangent to each other, since

$$\angle BDF = \angle AFD - \angle ABD$$

$$= (180° - \angle FAD - \angle FDA) - (\angle ABC - \angle DBC)$$

$$= 180° - \angle FAD - \angle ABC = 180° - \angle DAE - \angle FEA$$

$$= \angle FED + \angle ADE = \angle FED + \angle DCB.$$

Next, since B, C, E, F are concyclic, the powers of T with respect to $\odot BDC$ and $\odot DEF$ are equal. Therefore, the radical axis through T is the common tangent at D, so $TD^2 = TB \cdot TC = TF \cdot TE$. The lemma is proved.

As shown in Figure 3.20, let the line TA and $\odot ABC$ intersect at another point M. Notice that B, C, E, F are concyclic, A, M, C, B are concyclic, and by the lemma,

$$TM \cdot TA = TF \cdot TE = TB \cdot TC = TD^2,$$

implying that A, M, E, F are concyclic.

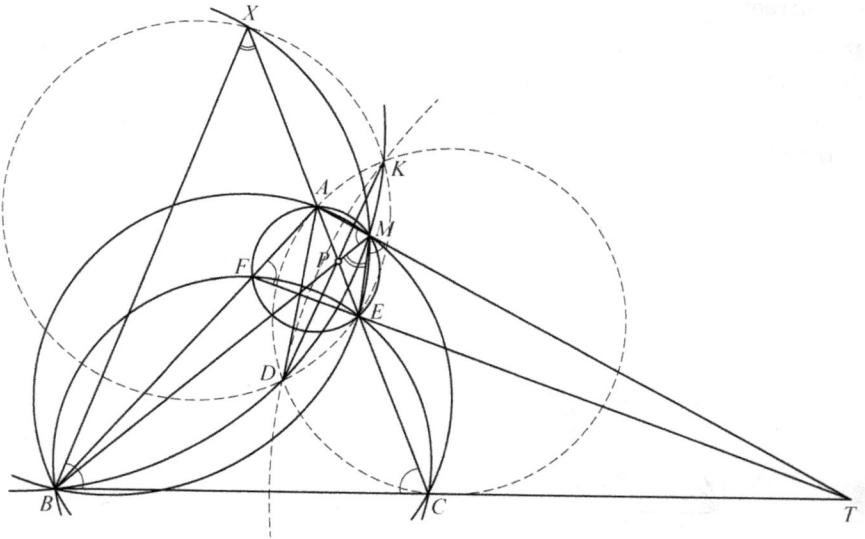

Figure 3.20

Consider the inversion transformation centered at T and with radius TD: the point M is mapped to A and B is mapped to C, i.e., $\odot MBD$ is mapped to $\odot ADC$. As their common point D lies on the inversion circle, so does the other common point K, namely $TK = TD$. It follows that T and the centers of $\odot KDE$ and $\odot ADC$ all lie on the perpendicular bisector of KD.

As O_1 is the circumcenter of $\triangle ADC$, it remains to prove D, K, E, X lie on a circle, whose center is O_2.

Observe that BM, DK, and AC are the radical axes of pairs from $\odot ABCM$, $\odot ACDK$, and $\odot BMDK$, respectively; they are concurrent at P. In addition, M lies on $\odot AEF$. Thus,

$$\measuredangle(EX, XB) = \measuredangle(CX, XB) = \measuredangle(XC, BC) + \measuredangle(BC, BX)$$

$$= 2\measuredangle(AC, CB) = \measuredangle(AC, CB) + \measuredangle(EF, FA)$$

$$= \measuredangle(AM, BM) + \measuredangle(EM, MA) = \measuredangle(EM, BM).$$

This implies that M, E, X, B are concyclic, and hence

$$PE \cdot PX = PM \cdot PB = PK \cdot PD.$$

Therefore, D, K, E, X are concyclic.

Proof 2. As shown in Figure 3.21, let the intersection point of lines BC and EF be P. Extend AD to intersect the circumcircle of $\triangle BCD$ at point D'. Then

$$\angle AD'C = \angle DBC = \angle ADF.$$

By the same argument, $\angle AD'B = \angle ADE$.

Combined with $\angle BAD = \angle CAD$, we know that quadrilateral $AEDF$ is similar to quadrilateral $ABD'C$, so

$$\frac{AE}{AF} = \frac{AB}{AC},$$

and thus, B, C, E, F are concyclic.

Let the tangent line of the circumcircle of $\triangle DEF$ at point D be α, and the tangent lines of the circumcircle of the quadrilateral $BDCD'$ at point D, D' be β, β', respectively. Then

$$\measuredangle(\alpha, AD) = \measuredangle(AD, \beta') = \measuredangle(\beta, AD),$$

so $\alpha \parallel \beta$, and then $\alpha = \beta$, that is, the circumcircle of $\triangle DEF$ is tangent to the circumcircle of the quadrilateral $BDCD'$.

For the circumcircle of $\triangle DEF$, the circumcircle of quadrilateral $BDCD'$, and the circumcircle of quadrilateral $BCEF$, it is known by Monge's theorem that DP is tangent to the circumcircle of $\triangle BCD$.

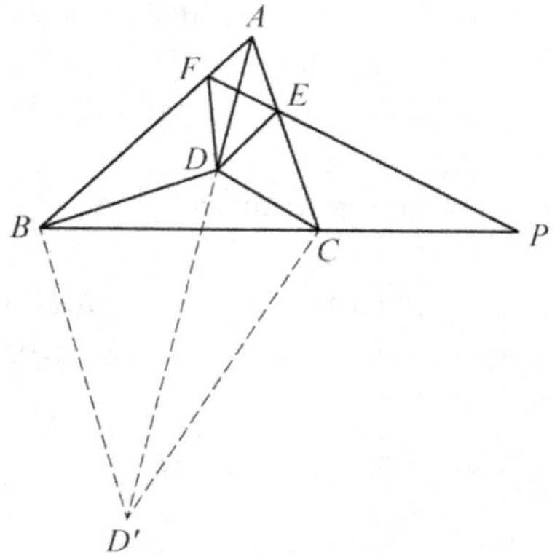

Figure 3.21

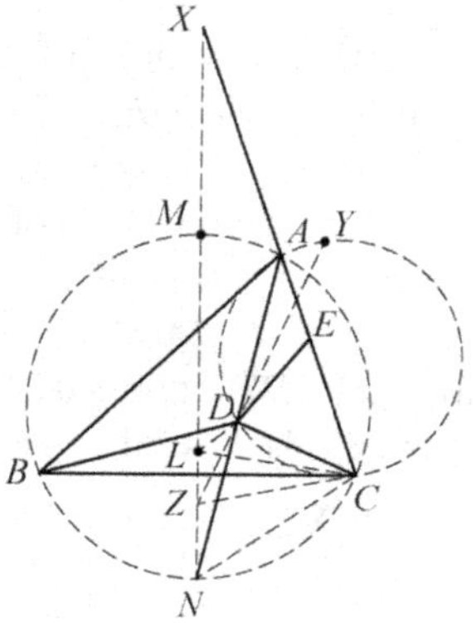

Figure 3.22

As shown in Figure 3.22, let the midpoints of $\overset{\frown}{BC}$ and $\overset{\frown}{BAC}$ on the circumcircle of $\triangle ABC$ be N and M respectively.

Crossing point C makes the tangent line of the circumcircle of $\triangle ACD$, which intersects MN at point Z. Let ZD and the circumcircle of $\triangle ACD$ intersect at another point Y.

In the following we prove that Y is on the circumcircle of $\triangle DEX$. Take a point L on the line ZX such that $ZL \cdot ZX = ZD \cdot ZY = ZC^2$. Then X, L, D, Y are concyclic, and $\triangle ZLC \backsim \triangle ZCX$, so consequently $\angle ZLC = \angle XCZ = \angle NDC$. Among them $\angle XCZ = \angle NDC$, because ZC is tangent to the circumcircle of $\triangle ACD$.

Then N, L, D, C are concyclic, and thus

$$\angle DLX = \angle DCN = \angle DCB + \angle BCN$$

$$= \angle ADE + \angle DAC = \angle DEC,$$

so X, L, D, E are concyclic. Furthermore, the five points X, L, D, E, Y are concyclic.

As shown in Figure 3.23, in order to prove that O_1O_2, BC, and EF have a common point, we just prove that O_1, O_2, P are collinear, that is, we prove that $PD = PY$, where D and Y are the two intersection points of the circumcircle of $\triangle ADC$ and the circumcircle of $\triangle EXD$.

Since

$$ZB^2 = ZC^2 = ZD \cdot ZY,$$

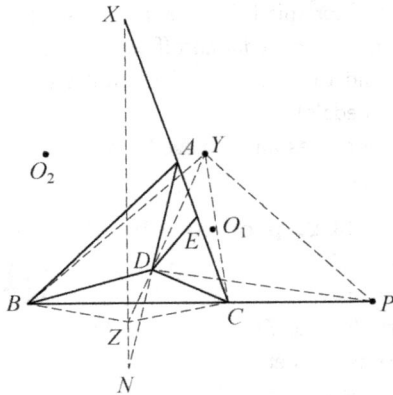

Figure 3.23

so $\triangle ZBD \backsim \triangle ZYB$ and $\triangle ZCD \backsim \triangle ZYC$. Then

$$\frac{BD}{BY} = \sqrt{\frac{DZ}{YZ}} = \frac{CD}{CY}.$$

Consequently, $\frac{BD}{CD} = \frac{BY}{CY}$. Since

$$\frac{BY}{CY} = \frac{BD}{CD} = \sqrt{\frac{BP}{CP}},$$

we have $\triangle PCY \backsim \triangle PYB$. Hence

$$PY = \sqrt{PC \cdot PB} = PD.$$

The original proposition is proved.

【Score Situation】 This particular problem saw the following distribution of scores among contestants: 15 contestants scored 7 points, no contestant scored 6 points, no contestant scored 5 points, 1 contestant scored 4 points, 1 contestant scored 3 points, 4 contestants scored 2 points, 110 contestants scored 1 point, and 488 contestants scored 0 point. The average score of this problem is 0.372, indicating that it was extremely difficult.

Among the top five teams in the team scores, the scores of this problem are as follows: the China team scored 24 points (with a total team score of 208 points), the Russia team scored 21 points (with a total team score of 183 points), the South Korea team scored 10 points (with a total team score of 172 points), the United States team scored 17 points (with a total team score of 165 points), and the Canada team scored 4 points (with a total team score of 151 points).

The gold medal cutoff for this IMO was set at 24 points (with 52 contestants earning gold medals), the silver medal cutoff was 19 points (with 103 contestants earning silver medals), and the bronze medal cutoff was 12 points (with 148 contestants earning bronze medals).

In this IMO, only one contestant achieved a perfect score of 42 points, namely Yichuan Wang from China.

Problem 3.8 (IMO 64-2, proposed by Portugal). Let ABC be an acute-angled triangle with $AB < AC$. Let Ω be the circumcircle of ABC. Let S be the midpoint of the arc BC of Ω containing A. The perpendicular from A to BC meets BS at D and meets Ω again at $E \neq A$. The line through D parallel to BC meets line BE at L. Denote the circumcircle of triangle BDL by ω. Let ω meet Ω again at $P \neq B$. Prove that the line tangent to ω at P meets line BS on the internal angle bisector of $\angle BAC$.

Proof 1.　As shown in Figure 3.24, let S' be the midpoint of arc BC of circle Ω, so SS' is the diameter of Ω, and AS' is the angle bisector of $\angle BAC$. Let the line tangent to ω at P again intersect Ω at the point $Q(Q \neq P)$. Then $\angle SQS' = 90°$. Next, it will be proved that the corresponding sides of $\triangle APD$ and $\triangle SQS'$ are parallel, so as to draw the segments of the corresponding vertices, that is, the angle bisector of $\angle BAC$, the line tangent to ω at P, and line DS. Then the three lines are concurrent at the (reverse) homothetic center.

Since A, P, B, E are concyclic and D, P, L, B are concyclic,

$$\angle PAD = \angle PAE = 180° - \angle EBP = \angle PBL = \angle PDL = 90° - \angle ADP.$$

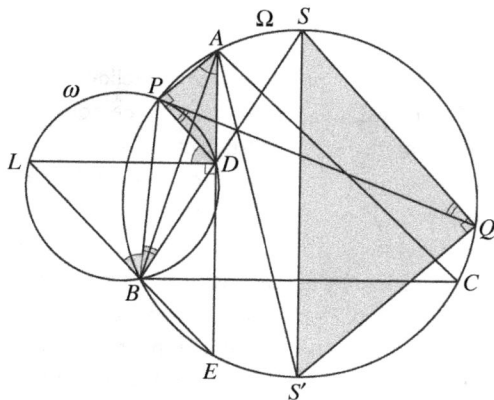

Figure 3.24

Thus, $AP \perp DP$. Now we can see:

Both lines ADE and SS' are perpendicular to BC, so $AD \parallel SS'$.

The segment PQ is tangent to the circle ω at P, so

$$\angle DPQ = \angle DBP = \angle SBP = \angle SQP,$$

from which it follows that $PD \parallel QS$.

Finally, since $AP \perp PD \parallel QS \perp S'Q$, $AP \parallel S'Q$. Therefore, the corresponding sides of $\triangle APD$ and $\triangle SQS'$ are parallel. This completes the proof.

Proof 2. As shown in Figure 3.25, let S' be the middle point of the inferior arc BC, which is the antipodal point of S. Thus $AES'S$ is an isosceles trapezoid and $\angle S'BS = \angle S'PS = 90°$. Let segments AE, PS' intersect at point T, and $AP, S'B$ intersect at point M. We need to prove that the points L, P, S are collinear, and the points T and M are on the circle ω.

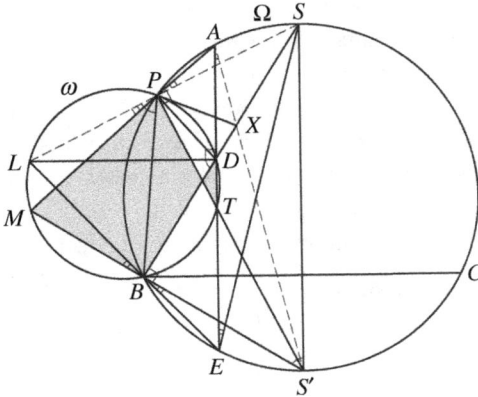

Figure 3.25

Since

$$\angle LPB = \angle LDB = 90° - \angle BDE = 90° - \angle BSS' = \angle SS'B = 180° - \angle BPS,$$

we get that L, P, S are indeed collinear.

And SS' is the diameter of Ω, so the lines LPS and PTS' are perpendicular. We also have $LD \parallel BC \perp AE$, so $\angle LDT = \angle LPT = 90°$. Therefore $T \in \omega$.

And since $\angle LPM = \angle SPA = \angle SEA = \angle EAS' = \angle EBS' = \angle LBM$, the points M, B, P, L are concyclic. Therefore $M \in \omega$.

Now let X be the intersection of the line BDS and the line tangent to ω at P. Applying Pascal's theorem to the degenerate hexagon $PPMBDT$ inscribed in the circle ω ($PPMBDT$ need not be a convex hexagon), and it follows that the three points $PP \cap BD = X$, $PM \cap DT = A$, $MB \cap TP = S'$ are collinear. Hence X is on the line AS', which is the angle bisector of $\angle BAC$.

【Score Situation】This particular problem saw the following distribution of scores among contestants: 215 contestants scored 7 points, 6 contestants scored 6 points, 7 contestants scored 5 points, 20 contestants scored 4 points, 62 contestants scored 3 points, 6 contestants scored 2 points, 100 contestants scored 1 point, and 202 contestants scored 0 point. The average score of this problem is 3.162, indicating that it was relatively straightforward.

Among the top five teams in the team scores, the scores of this problem are as follows: the China team scored 42 points (with a total team score of 240 points), the United States team scored 42 points (with a total team score of 222 points), the South Korea team scored 40 points (with a total team score of 215 points), the Romania team scored 42 points (with a total team score of 208 points), and the Canada team scored 29 points (with a total team score of 183 points).

The gold medal cutoff for this IMO was set at 32 points (with 54 contestants earning gold medals), the silver medal cutoff was 25 points (with 90 contestants earning silver medals), and the bronze medal cutoff was 18 points (with 170 contestants earning bronze medals).

In this IMO, a total of 5 contestants achieved a perfect score of 42 points.

Problem 3.9 (IMO 64-6, proposed by the United States). Let ABC be an equilateral triangle. Let A_1, B_1, C_1 be interior points of ABC such that $BA_1 = A_1C$, $CB_1 = B_1A$, $AC_1 = C_1B$, and

$$\angle BA_1C + \angle CB_1A + \angle AC_1B = 480°.$$

Let BC_1 and CB_1 meet at A_2, let CA_1 and AC_1 meet at B_2, and let AB_1 and BA_1 meet at C_2. Prove that if triangle $A_1B_1C_1$ is scalene, then the three circumcircles of triangles AA_1A_2, BB_1B_2, and CC_1C_2 all pass through two common points (*Note*: a scalene triangle is one where no two sides have equal length.).

Proof 1.   As shown in Figure 3.26, denote the three circumcircles of $\triangle AA_1A_2, \triangle BB_1B_2, \triangle CC_1C_2$ as $\delta_A, \delta_B, \delta_C$, respectively. We need to find two different points that have equal powers with respect to $\delta_A, \delta_B, \delta_C$.

Lemma 1. *The point A_1 is the circumcenter of $\triangle A_2BC$, the point B_1 is the circumcenter of $\triangle B_2CA$, and the point C_1 is the circumcenter of $\triangle C_2AB$.*

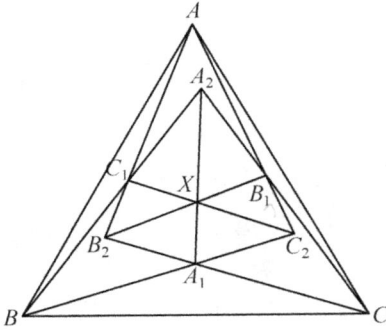

Figure 3.26

Proof of Lemma 1. We only prove the first one, and the rest is of the same. Since A_1 lies on the perpendicular bisector of BC and is within $\triangle A_2BC$, we simply need to prove that $\angle BA_1C = 2\angle BA_2C$. This can be obtained from the following equality:

$$\angle BA_2C = \angle A_2BA + \angle BAC + \angle A_2CA$$
$$= \frac{1}{2}((180° - \angle AC_1B) + (180° - \angle AB_1C)) + 60°$$
$$= 240° - \frac{1}{2}(480° - \angle BA_1C)$$
$$= \frac{1}{2}\angle BA_1C.$$

So, $\angle B_1B_2C_1 = \angle B_1B_2A = \angle B_1AB_2 = \angle C_1AC_2 = \angle C_1C_2A = \angle C_1C_2B_1$.

Thus, the four points B_1, B_2, C_1, C_2 are concyclic. Similarly, A_1, A_2, C_1, C_2 and A_1, A_2, B_1, B_2 are concyclic, respectively. And since

$$\angle B_2A_1C_2 + \angle A_2C_1B_2 + \angle A_2B_1C_2 = 480° \neq 360°,$$

the hexagon $A_1B_2C_1A_2B_1C_2$ is not a cyclic polygon. Therefore, by the radical center theorem, A_1A_2, B_1B_2, and C_1C_2 are concurrent (denoted as point X), and the powers of point X with respect to $\delta_A, \delta_B, \delta_C$ are equal.

Let the intersection of the circumcircle of $\triangle A_2BC$ and δ_A be $A_3(\neq A_2)$. Define B_3 and C_3 similarly.

Lemma 2. *The four points B, B_3, C, C_3 are concyclic.*

Proof of Lemma 2. As shown in Figure 3.27, using the directed angle, we get

$$\angle BC_3C = \angle BC_3C_2 + \angle C_2C_3C$$

$$= \angle BAC_2 + \angle C_2C_1C$$

$$= 90° + \angle(C_1C, AC_2) + \angle C_2C_1C \quad \text{(since } C_1C \perp AB\text{)}$$

$$= 90° + \angle C_1C_2B_1.$$

Similarly, we have $\angle CB_3B = 90° + \angle B_1B_2C_1$. Therefore, by the four points B_1, B_2, C_1, C_2 on a circle,

$$\angle BB_3C = 90° + \angle C_1B_2B_1 = 90° + \angle C_1C_2B_1 = \angle BC_3C.$$

This completes the proof.

Similarly, CAC_3A_3 and ABA_3B_3 are cyclic quadrilateral. However, the hexagon $AC_3BA_3CB_3$ is not a cyclic polygon, because otherwise A, B_2, C, B_3 would be concyclic, which would mean that B_2 is on $\odot ABC$, but B_2 is inside $\triangle ABC$, so this is impossible. Thus, using the radical center theorem for these three circles, we can get the concurrent point of AA_3, BB_3, CC_3 (denoted as point Y), and the powers of point Y with respect to $\delta_A, \delta_B, \delta_C$ are equal.

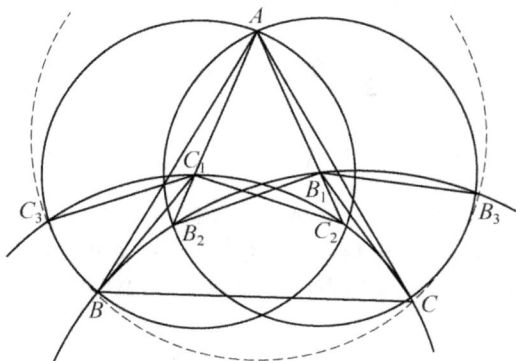

Figure 3.27

In addition, some positional relationships need to be discussed. Let O be the center of $\triangle ABC$. We have

$$\angle BA_1C = 480° - \angle CB_1A - \angle AC_1B > 480° - 180° - 180° = 120°.$$

So A_1 is inside $\triangle BOC$. For B_1 and C_1, we have similar results. Thus, $\triangle BA_1C$, $\triangle AB_1C$, and $\triangle AC_1B$ have non-overlapping interiors. It follows that $A_1B_2C_1A_2B_1C_2$ is a convex hexagon, so X lies on the segment A_1A_2 and then is inside δ_A.

Since A_1 is the circumcenter of $\triangle A_2BC$, we have $A_1A_2 = A_1A_3$. Since the four points A, A_2, A_1, A_3 are on a circle, we know that $AA_2, AA_3 \equiv AY$ are reflections of the line AA_1. Since X is on the segment A_1A_2, the only possibility for $X \equiv Y$ is that A_1 and A_2 are both on the perpendicular bisector of BC. But this makes B_1 and C_1 also reflect on this line, which means $A_1B_1 = A_1C_1$. This contradicts the condition of the scalene triangle.

To sum up, we have two different points X, Y that have equal powers with respect to $\delta_A, \delta_B, \delta_C$. Thus, these circles have a common radical axis. And since X is inside δ_A (and also inside δ_B, δ_C), this radical axis has two intersections with the circle. Hence $\delta_A, \delta_B, \delta_C$ have two common points.

Proof 2. As shown in Figure 3.28, according to Proof 1, we know that A_1, B_1, C_1 are the circumcenters of $\triangle A_2BC, \triangle B_2AC, \triangle C_2AB$, respectively.

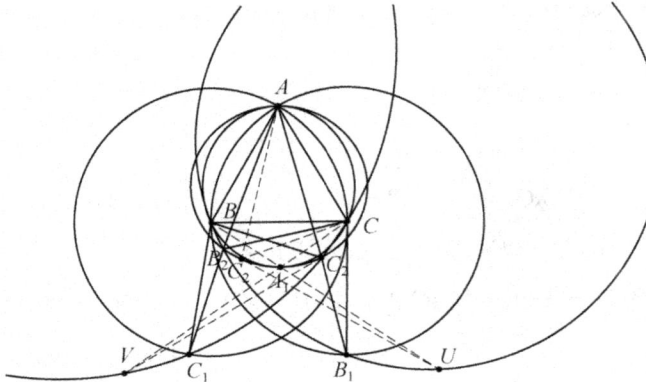

Figure 3.28

The concurrent point of three lines AA_1, BB_1, CC_1 is denoted as X. Then the powers of X with respect to $\odot AA_1A_2, \odot BB_1B_2, \odot CC_1C_2$ are equal.

And X is on the radical axis of $\odot BB_1B_2$, $\odot CC_1C_2$, and is inside $\odot BB_1B_2$, so the circle $\odot BB_1B_2$ intersects $\odot CC_1C_2$. Let their intersection points be P, Q.

Since PQ is the radical axis of $\odot BB_1B_2$ and $\odot CC_1C_2$, the three points P, X, Q are collinear, and

$$PX \cdot XQ = B_1X \cdot B_2X = A_1X \cdot A_2X,$$

so A_1, P, A_2, Q are concyclic. The following is to prove that $\odot AA_1A_2$ also passes through the two points P, Q. We just prove that the four points A, A_1, P, Q are concyclic.

Therefore, let A be the inversion center and perform an inversion with any inversion power. In the inversion figure, A_1 lies on the perpendicular bisector of BC and is the intersection of $\odot ABC_2$ and $\odot ACB_2$. The symmetric points of point A with respect to BC_2, CB_2 are C_1, B_1, respectively. The problem is transformed into proving that A_1, P, Q are collinear, that is, A_1 is on the radical axis of $\odot BB_1B_2$ and $\odot CC_1C_2$.

Let $\odot BB_1B_2, \odot CC_1C_2$ intersect lines BA_1, CA_1 again at U, V respectively. Then by $BA_1 = CA_1$, it is only necessary to prove that U, V are symmetric with respect to AA_1. Note that C is the circumcenter of $\triangle ABB_1$. Hence

$$\angle B_2UB = \angle B_2B_1B = \angle B_2B_1A - \angle BB_1A = \angle C_1AB_1 - 30°.$$

Similarly, $\angle C_2VC = \angle C_1AB_1 - 30°$, and thus $\angle B_2UB = \angle C_2VC$, so it is only necessary to prove that the symmetric point C_2' of C_2 with respect to AA_1 is on BU. And

$$\angle ABU = 150° - \angle BA_1A = 150° - \angle BC_2A = 60° + \angle C_1AB_1 = 90° + \angle B_2UB.$$

Therefore, AB is perpendicular to B_2U. Since

$$\angle AC_2'C = \angle AC_2B = 90° - \angle C_1AB_1 = \angle AB_2C,$$

we see that A, C, C_2', B_2 are concyclic. Thus,

$$\angle AB_2C_2' = 180° - \angle ACC_2' = 180° - \angle ABC_2 = 90° + \angle BAB_2.$$

This indicates that B_2C_2' is perpendicular to AB, from which C_2' is on B_2U. This completes the proof.

【Score Situation】 This particular problem saw the following distribution of scores among contestants: 6 contestants scored 7 points, 4 contestants scored 6 points, 1 contestant scored 5 points, 1 contestant scored 4 points, 4 contestants scored 3 points, 36 contestants scored 2 points, 11 contestants scored 1 point, and 555 contestants scored 0 point. The average score of this problem is 0.275, indicating that it was extremely difficult.

Among the top five teams in the team scores, the scores of this problem are as follows: the China team scored 30 points (with a total team score of 240 points), the United States team scored 17 points (with a total team score of 222 points), the South Korea team scored 16 points (with a total team score of 215 points), the

Romania team scored 16 points (with a total team score of 208 points), and the Canada team scored 7 points (with a total team score of 183 points).

The gold medal cutoff for this IMO was set at 32 points (with 54 contestants earning gold medals), the silver medal cutoff was 25 points (with 90 contestants earning silver medals), and the bronze medal cutoff was 18 points (with 170 contestants earning bronze medals).

In this IMO, a total of five contestants achieved a perfect score of 42 points.

3.2.2   *Quantity relation problems*

Problem 3.10 (IMO 41-1, proposed by Russia). Two circles Γ_1 and Γ_2 intersect at M and N. Let l be the common tangent to Γ_1 and Γ_2 so that M is closer to l than N is. Let l touch Γ_1 at A and Γ_2 at B. Let the line through M and parallel to l meet the circle Γ_1 again at C and the circle Γ_2 again at D. Lines CA and DB meet at E; lines AN and CD meet at P; lines BN and CD meet at Q. Prove that $EP = EQ$.

Proof.   As shown in Figure 3.29, connect to form NM and extend to intersect AB to R. Then connect to form EM, AM, and BM. Since

$$AR^2 = RM \cdot RN = RB^2,$$

so $AR = RB$, and $PQ \parallel AB$. Hence $PM = MQ$.

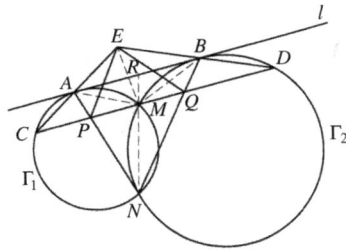

Figure 3.29

If we can prove that $EM \perp PQ$, then EM is the perpendicular bisector of PQ, so that $EP = EQ$ holds. Let us prove $EM \perp PQ$, which means that $EM \perp AB$.

Since $\angle EAB = \angle ECQ = \angle MAB$, similarly we can get $\angle ABE = \angle ABM$. Hence AB bisects EM vertically.

【Score Situation】 This particular problem saw the following distribution of scores among contestants: 220 contestants scored 7 points, 5 contestants scored 6 points,

4 contestants scored 5 points, 22 contestants scored 4 points, 37 contestants scored 3 points, 44 contestants scored 2 points, 11 contestants scored 1 point, and 118 contestants scored 0 point. The average score of this problem is 4.095, indicating that it was simple.

Among the top five teams in the team scores, the scores of this problem are as follows: the China team scored 42 points (with a total team score of 218 points), the Russia team scored 42 points (with a total team score of 215 points), the United States team scored 38 points (with a total team score of 184 points), the South Korea team scored 42 points (with a total team score of 172 points), the Vietnam team scored 42 points (with a total team score of 169 points), and the Bulgaria team scored 42 points (with a total team score of 169 points).

The gold medal cutoff for this IMO was set at 30 points (with 39 contestants earning gold medals), the silver medal cutoff was 21 points (with 71 contestants earning silver medals), and the bronze medal cutoff was 11 points (with 119 contestants earning bronze medals).

In this IMO, a total of four contestants achieved a perfect score of 42 points.

Problem 3.11 (IMO 50-2, proposed by Russia). Let ABC be a triangle with circumcenter O. Two points P and Q are interior points of the sides CA and AB, respectively. Let K, L, and M be the midpoints of the segments BP, CQ, and PQ, respectively, and let Γ be the circle passing through K, L, and M. Suppose that the line PQ is tangent to the circle Γ. Prove that $OP = OQ$.

Proof 1.   As shown in Figure 3.30, the necessary and sufficient condition for the line PQ to be tangent to the circle Γ is $\angle MLK = \angle QMK$.

Since $MK \parallel AB$, we have $\angle AQP = \angle QMK = \angle MLK$.

By a similar argument, $\angle APQ = \angle MKL$.

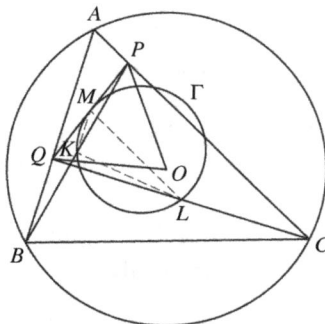

Figure 3.30

Thus, $\triangle APQ \backsim \triangle MKL$, so

$$\frac{AP}{AQ} = \frac{MK}{ML} = \frac{\frac{QB}{2}}{\frac{PC}{2}} = \frac{QB}{PC},$$

i.e., $AP \cdot PC = AQ \cdot QB$. The powers of points P, Q are equal with respect to the circumcircle of the triangle ABC. Hence $OP = OQ$.

Proof 2. Clearly, the line PQ touches the circle Γ at the point M. By the theorem of the tangent chord angle, $\angle QMK = \angle MLK$. Since the points K, M are the midpoints of the segments BP and PQ respectively, $KM \parallel BQ$, and we get $\angle QMK = \angle AQP$. Thus, $\angle MLK = \angle AQP$, and similarly $\angle MKL = \angle APQ$. As a result, $\triangle MKL$ and $\triangle APQ$ are similar, and thus

$$\frac{MK}{ML} = \frac{AP}{AQ}.$$

Since the points K, L, and M are the midpoints of the segments BP, CQ, and PQ respectively,

$$KM = \frac{1}{2}BQ, LM = \frac{1}{2}CP.$$

Substituting it into the previous equation, we get $\frac{BQ}{CP} = \frac{AP}{AQ}$, i.e.,

$$AP \cdot CP = AQ \cdot BQ.$$

By the power of a point,

$$OP^2 = OA^2 - AP \cdot CP = OA^2 - AQ \cdot BQ = OQ^2.$$

Therefore $OP = OQ$, as required.

【Score Situation】This particular problem saw the following distribution of scores among contestants: 214 contestants scored 7 points, 19 contestants scored 6 points, 15 contestants scored 5 points, 16 contestants scored 4 points, 51 contestants scored 3 points, 43 contestants scored 2 points, 106 contestants scored 1 point, and 101 contestants scored 0 point. The average score of this problem is 3.710, indicating that it was relatively straightforward.

Among the top five teams in the team scores, the scores of this problem are as follows: the China team scored 42 points (with a total team score of 221 points), the Japan team scored 42 points (with a total team score of 212 points), the Russia team scored 39 points (with a total team score of 203 points), the South Korea team scored 42 points (with a total team score of 188 points), and the North Korea team scored 35 points (with a total team score of 183 points).

The gold medal cutoff for this IMO was set at 32 points (with 49 contestants earning gold medals), the silver medal cutoff was 24 points (with 98 contestants earning silver medals), and the bronze medal cutoff was 14 points (with 135 contestants earning bronze medals).

In this IMO, only two contestants achieved a perfect score of 42 points, namely Dongyi Wei from China and Makoto Soejima from Japan.

Problem 3.12 (IMO 53-5, proposed by Czech Republic). Let ABC be a triangle with $\angle BCA = 90°$, and let D be the foot of the altitude from C. Let X be a point in the interior of the segment CD. Let K be the point on the segment AX such that $BK = BC$. Similarly, let L be a point on the segment BX such that $AL = AC$. Let M be the intersection point of AL and BK. Show that $MK = ML$.

Proof.   As shown in Figure 3.31, let C' be the symmetric point of C with respect to line AB, and ω_1, ω_2 be circles with centres A, B and radii AL, BK, respectively. Since $AC' = AC = AL$ and $BC' = BC = BK$, points C and C' are both on circles ω_1 and ω_2. Since $\angle BCA = 90°$, lines AC and BC are tangent to circles ω_2 and ω_1, respectively, at point C. Let K_1 be another intersection point of line AX and circle ω_2 different from K, and L_1 be another intersection point of line BX and circle ω_1 different from L.

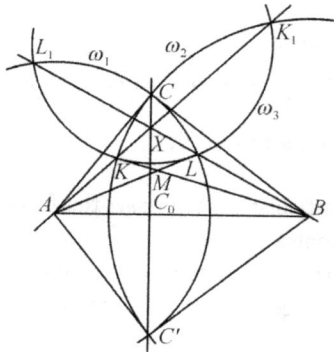

Figure 3.31

By the power of a point,

$$XK \cdot XK_1 = XC \cdot XC' = XL \cdot XL_1.$$

Thus, the four points K_1, L, K, L_1 are concyclic, and denote this circle as ω_3.

By the power of a point to circle ω_2,

$$AL^2 = AC^2 = AK \cdot AK_1.$$

This implies that the line AL is tangent to circle ω_3 at point L. By a similar argument, line BK is tangent to circle ω_3 at point K.

Therefore, MK and ML are two tangent lines from point M to circle ω_3, and thus $MK = ML$.

【Score Situation】 This particular problem saw the following distribution of scores among contestants: 86 contestants scored 7 points, 3 contestants scored 6 points, 4 contestants scored 5 points, 15 contestants scored 4 points, 45 contestants scored 3 points, 29 contestants scored 2 points, 17 contestants scored 1 point, and 348 contestants scored 0 point. The average score of this problem is 1.664, indicating that it was relatively challenging.

Among the top five teams in the team scores, the scores of this problem are as follows: the South Korea team scored 42 points (with a total team score of 209 points), the China team scored 38 points (with a total team score of 195 points), the United States team scored 23 points (with a total team score of 194 points), the Russia team scored 29 points (with a total team score of 177 points), the Thailand team scored 30 points (with a total team score of 159 points), and the Canada team scored 24 points (with a total team score of 159 points).

The gold medal cutoff for this IMO was set at 28 points (with 51 contestants earning gold medals), the silver medal cutoff was 21 points (with 88 contestants earning silver medals), and the bronze medal cutoff was 14 points (with 137 contestants earning bronze medals).

In this IMO, only one contestant achieved a perfect score of 42 points, namely Jeck Lim from Singapore.

3.3 Summary

In the first 64th IMOs, there are 12 problems on the power of a point, radical axis, and radical center as depicted in Figure 3.32, which can be broadly categorized into two types (Figure 3.32).

Problems 3.1–3.9 focus on "position structure problems;" among these nine problems, the one with the lowest average score is Problem 3.9 (IMO 64-6), proposed by the United States. Problems 3.10–3.12 deal with "quantity relation problems;" among these three problems, the one with the lowest average score is Problem 3.12 (IMO 53-5), proposed by Czech Republic.

These problems were proposed by eight countries and regions. Russia proposed four problems, while Ukraine contributed two problems.

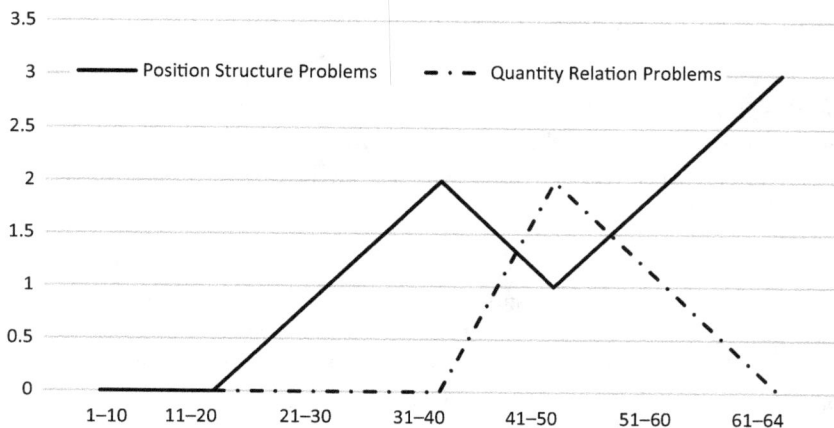

Figure 3.32 Numbers of Power of a Point, Radical Axis, and Radical Center Problems in the First 64 IMOs

From Table 3.2, it can be observed that in the first 64 IMOs, there were three problems with an average score of 0–1 point; three problems with average scores of 1–2 points; two problems with average scores of 3–4 points; four problems with average scores above 4 points. Overall, power of a point, radical axis, and radical center problems were relatively extreme, either quite simple, or quite difficult.

In the 24th to 64th IMOs, there were a total of 12 power of a point, radical axis, and radical center problems. Among these, three had an average score of 0–1 point; three had an average score of 1–2 points; two had an average score of 3–4 points; four had an average score above 4 points. Further analysis of the problem numbers of these 12 power of a point, radical axis, and radical center problems, as shown in Table 3.3, reveals that the frequency of these problems in each problem number was relatively balanced. The majority type of these problems were position structure problems, totaling nine problems, and quantity relation problems had only three problems.

From Table 3.2, it can be observed that the average score of the top five teams is generally more than 1.5 points higher than the average score of the problem. However, in simpler problems, the difference in average scores between the top five teams, 6th–15th teams and 16th–25th teams, is not significant, as seen in problems such as Problem 3.2 (IMO 36-1), Problem 3.4 (IMO 49-1), and Problem 3.5 (IMO 54-4).

Table 3.2 Score Details of Power of a Point, Radical Axis, and Radical Center Problems in the First 64 IMOs

Problem	3.1	3.2	3.3	3.4	3.5	3.6	3.7
Full points	7.000	7.000	7.000	7.000	7.000	7.000	7.000
Average score	1.847	5.056	1.811	4.979	5.442	0.653	0.372
Top five Mean	4.200	6.767	5.567	6.967	6.967	3.367	2.533
6th–15th mean	1.817	6.667	3.883	6.833	7	1.7	1
16th–25th mean	1.6	6.35	2.258	6.833	7	1.15	0.45
Problem number in IMO	26-5	36-1	40-5	49-1	54-4	56-3	62-3
Proposing country	Soviet Union	Bulgaria	Russia	Russia	Thailand	Ukraine	Ukraine

Problem	3.8	3.9	3.10	3.11	3.12
Full points	7.000	7.000	7.000	7.000	7.000
Average score	3.162	0.275	4.095	3.710	1.664
Top five mean	6.5	2.867	6.889	6.667	5.167
6th–15th mean	5.833	0.617	6.444	6.561	4.5
16th–25th mean	5.583	0.283	5.267	6.033	2.7
Problem number in IMO	64-2	64-6	41-1	50-2	53-5
Proposing country	Portugal	United States	Russia	Russia	Czech Republic

Note. Top five mean = Total score of the top five teams/Total number of contestants from the top five teams,
6th–15th mean = Total score of the 6th–15th teams/Total number of contestants from the 6th–15th teams,
16th–25th mean = Total score of the 16th–25th teams/Total number of contestants from the 16th–25th teams.

Table 3.3 Numbers of Power of a Point, Radical Axis, and Radical Center Problems in the 24th–64th IMOs

Power of a Point Radical Axis, and Radical Center Problems	Problem Number			Numbers of Problems in the First 64 IMOs
	1, 4	2, 5	3, 6	
Position structure problems	3	3	3	9
Quantity relation problems	1	2	0	3
Total	4	5	3	12

From Table 3.2, it can also be observed that after the 30th IMO, the average scores of the 6th–15th teams and 16th–25th teams also tend to be higher than the average score of the problem. This phenomenon is due to the smaller number of participating teams in early IMOs. It was not until the 30th IMO in 1989 that the number of participating teams exceeded 50. In addition, the appearance of extremely difficult geometry problems can have a depressing effect on the scores of many contestants, thus affecting the average score of the problem.

Chapter 4

Special Points and Special Lines in a Triangle

Triangles are most basic plane figures, and the theory of plane geometry about triangles is also most mature. There are many special points and lines in a triangle, and they have many special and beautiful properties. Regardless of the shape of a triangle, special points and lines with these special properties always exist.

There are many special points in a triangle, and we often encounter the "five centers" of a triangle (circumcenter, incenter, excenter, centroid, orthocenter), the Fermat point, Brocard point, Miquel point, Isogonic center, Gergonne point, and Nagel point. These special points have wonderful properties.

The special lines in a triangle are: the altitude of the triangle, median line of the triangle, interior angle bisector of the triangle, exterior angle bisector of the triangle, middle line of the triangle, etc.

Mastering and flexibly using the properties of special points and lines of triangles will provide us with ideas and methods for solving problems.

In the first 64th IMOs, there had been a total of eight problems related to special points and special lines in a triangle, accounting for 6.5% of all geometry problems. These problems are mainly categorized into two types: (1) position structure problems, totaling five problems; (2) quantity relation problems, totaling 3 problems. The statistical distribution of these two types of problems in the previous IMOs is presented in Table 4.1.

Table 4.1 Numbers of Special Points and Special Lines Problems in the First 64 IMOs

Content	Session							Total
	1–10	11–20	21–30	31–40	41–50	51–60	61–64	
Position structure problems	1	0	1	0	3	0	0	5
Quantity relation problems	1	1	0	0	1	0	0	3
Geometry problems	29	18	18	15	20	17	6	123
Percentage of special points and special lines problems among geometry problems	6.9%	5.6%	5.6%	0.0%	20.0%	0.0%	0.0%	6.5%

It can be seen that there are not many problems related to special points and special lines in a triangle, mainly appearing in the 41st to 50th IMOs, and there are no problems about special points and special lines in a triangle in the last ten IMOs. In fact, by using the properties of some special points and special lines, it can be relatively fast to verify whether a geometric figure meets a certain structure, and calculate the quantitative relations such as length, angle, area, and so on.

Therefore, this chapter will be divided into three parts. The first part introduces the properties and theorems related to the "five centers" of triangles, as well as some special lines, such as Euler's line and Simson's line, and through some examples to expound applications of these theorems.

The second part revolves around two types of problems: "position structure problems" and "quantity relation problems." These problems are presented in chronological order, and some problems include various solutions and generalizations.

It is important to note that for each problem, the solutions are followed by information on the scores, including the number of contestants in each score range, the average score, the scores of the top five teams. However, early IMOs often lacked information on contestant scores, so the number of contestants in each score range only represents the counted number of contestants, and some problems lack scores of the top five teams.

The third part provides a brief summary of this chapter.

4.1 Related Properties, Theorems, and Methods

4.1.1 *Special points in a triangle*

(1) *Circumcenter*

The perpendicular bisectors of the three sides of any triangle intersect at a point, which is called the center of the circumcircle of the triangle, referred to as the circumcenter.

With this circumcenter as the center, the circle can be constructed that passes through the three vertices of the triangle, and this circle is called the circumcircle of the triangle. The circumcircle of a triangle is unique.

Let the circumcenter of $\triangle ABC$ be O. Then

(i) As shown in Figure 4.1, if $\triangle ABC$ is an acute triangle, then the point O is inside the triangle and

$$\angle BOC = 2\angle BAC, \ \angle OBC = \angle OCB = 90° - \angle BAC;$$

$$\angle COA = 2\angle ABC, \ \angle OAC = \angle OCA = 90° - \angle ABC;$$

$$\angle AOB = 2\angle ACB, \ \angle OAB = \angle OBA = 90° - \angle ACB.$$

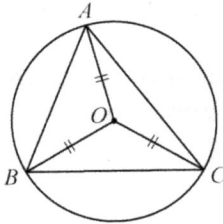

Figure 4.1

(ii) As shown in Figure 4.2, if $\triangle ABC$ is a right triangle, then the circumcenter is the midpoint of the hypotenuse.

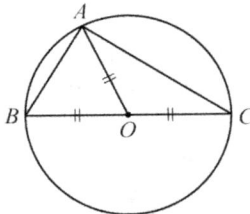

Figure 4.2

(iii) As shown in Figure 4.3, if $\triangle ABC$ is an obtuse triangle, then point O is outside the triangle.

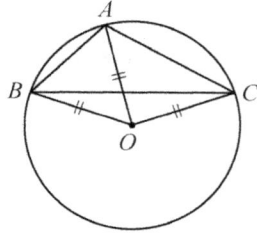

Figure 4.3

Without loss of generality, let $\angle BAC > 90°$. Then points O, A are on the different sides of BC, and

$$\angle BOC = 2(180° - \angle BAC), \quad \angle OBC = \angle OCB = \angle BAC - 90°;$$

$$\angle COA = 2\angle ABC, \quad \angle OAC = \angle OCA = 90° - \angle ABC;$$

$$\angle AOB = 2\angle ACB, \quad \angle OAB = \angle OBA = 90° - \angle ACB.$$

Denote the three interior angles of $\triangle ABC$ as A, B, C, their opposite sides as a, b, c, respectively, and the radius of the circumcircle as R. Then

$$\frac{a}{\sin A} = \frac{b}{\sin B} = \frac{c}{\sin C} = 2R, \quad S_{\triangle ABC} = \frac{abc}{4R}.$$

(2) *Incenter and excenter*

The angle bisectors of the three interior angles of any triangle intersect at a point, which is called the center of the inscribed circle of the triangle, referred to as the incenter.

With the incenter as the center, a circle can be constructed that is tangent to all three sides of the triangle, and this circle is called the inscribed circle of the triangle.

Let the incenter of $\triangle ABC$ be I. Then

$$\angle BIC = 90° + \frac{1}{2}\angle BAC, \quad \angle CIA = 90° + \frac{1}{2}\angle ABC,$$

$$\angle AIB = 90° + \frac{1}{2}\angle ACB.$$

As shown in Figure 4.4, in $\triangle ABC$, denote $a = BC$, $b = CA$, $c = AB$, and $p = \frac{a+b+c}{2}$ (the same below). Let the inscribed circle of $\triangle ABC$ be tangent to sides BC, CA, AB at points D, E, F, respectively. Then

$$AE = AF = \frac{-a+b+c}{2} = p - a,$$

$$BD = BF = \frac{a-b+c}{2} = p - b,$$

$$CD = CE = \frac{a+b-c}{2} = p - c.$$

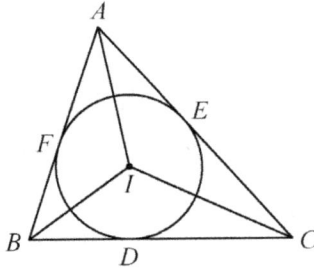

Figure 4.4

In a triangle, an interior angle bisector and the exterior angle bisectors of the other two interior angles intersect at one point, which is called the center of excircle of the triangle, referred to as the excenter. A triangle has three excenters.

With the excenter as the center, a circle can be constructed that is tangent to one side of the triangle and tangent to the extension lines of the other two sides of the triangle. This circle is called the excircle of the triangle. A triangle has three excircles.

As shown in Figure 4.5, let the excenter of $\triangle ABC$ inside $\angle BAC$ be I_a. Then

$$\angle BI_aC = 90° - \frac{1}{2}\angle BAC,$$

$$\angle AI_aB = \frac{1}{2}\angle ACB,$$

$$\angle AI_aC = \frac{1}{2}\angle ABC.$$

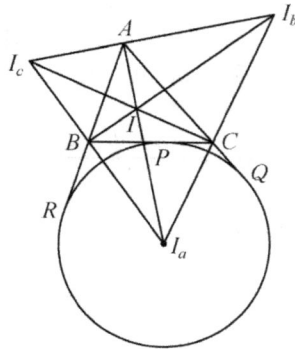

Figure 4.5

Theorem 4.1 (Euler Triangle Formula). Let the incenter and the circumcenter of $\triangle ABC$ be I, O, respectively. And let the radii of the circumcircle and inscribed circle of $\triangle ABC$ be R, r, respectively. Then

$$OI^2 = R^2 - 2Rr.$$

As shown in Figure 4.6, let the excircle $\odot I_a$ of $\triangle ABC$ inside $\angle BAC$ be tangent to side BC at point P, and tangent to the extension lines of AC, AB at points Q, R, respectively. Then

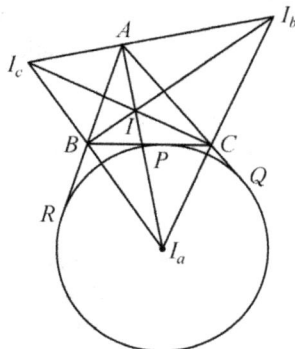

Figure 4.6

$$AQ = AR = \frac{a+b+c}{2} = p,$$

$$BP = BR = \frac{a+b-c}{2} = p - c,$$

$$CP = CQ = \frac{a-b+c}{2} = p - b.$$

It can be seen that the tangent point of the inscribed circle and the tangent point of the excircle on the same side of a triangle are symmetric with respect to the midpoint of this side.

An important property of the incenter and the excenter of the triangle (commonly known as the "the foot of chicken theorem"): As shown in Figure 4.7, let the incenter of $\triangle ABC$ be I, and the excenters of $\triangle ABC$ inside $\angle BAC$, $\angle ABC$, $\angle ACB$ be I_a, I_b, I_c, respectively. On the circumcircle of $\triangle ABC$, the midpoints of $\overset{\frown}{BC}$, $\overset{\frown}{BAC}$ are M, N respectively. Then

$$MB = MC = MI = MI_a.$$

$$NI_b = NI_c = NB = NC.$$

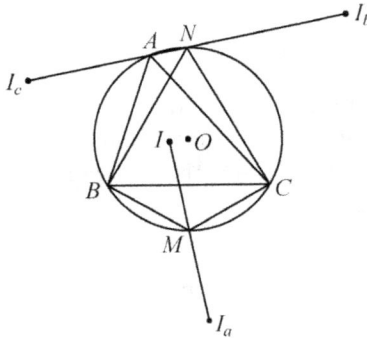

Figure 4.7

The radii of the inscribed circle and the excircle: Let the three interior angles of $\triangle ABC$ be A, B, C respectively. The radius of the inscribed circle is r. Let the radii of the excircles inside $\angle A$, $\angle B$, $\angle C$ be r_a, r_b, r_c, respectively. Then

$$r = \frac{S}{p}, \quad r_a = \frac{S}{p_a}, \quad r_b = \frac{S}{p_b}, \quad r_c = \frac{S}{p_c},$$

where S represents the area of $\triangle ABC$, $p_a = p - a$, $p_b = p - b$, and $p_c = p - c$.

(3) *Median point*

The three medians of a triangle intersect at a point, which is called the median point of triangle (see Figure 4.8).

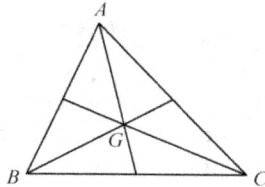

Figure 4.8

The median point is at a trisection point (away from the vertex) on any median of the triangle.

Three triangles composed of the median point and any two vertices of the triangle have the same area. The triangle is divided into six triangles by three medians, all of which have the same area.

The sum of squares of the distance between the median point and the three vertices of the triangle is the smallest.

(4) *Orthocenter*

The lines where the three altitudes of a triangle intersect at one point, which is called the orthocenter of the triangle.

An important property of the orthocenter of a triangle: The symmetry points of the orthocenter of the triangle about the three sides are on the circumcircle of this triangle.

Theorem 4.2 (Carnot's Theorem I). *Let the circumcenter and the orthocenter of* $\triangle ABC$ *be* O, H, *and the midpoints of sides* BC, CA, AB *be* M_a, M_b, M_c, *respectively (see Figures 4.9 and 4.10). Then*

$$\overrightarrow{AH} = 2\overrightarrow{OM_a}, \quad \overrightarrow{BH} = 2\overrightarrow{OM_b}, \quad \overrightarrow{CH} = 2\overrightarrow{OM_c}.$$

Theorem 4.3 (Carnot's Theorem II). *Let the distances between the circumcenter* O *of acute triangle* $\triangle ABC$ *and the three sides be* d_1, d_2, d_3, *respectively. Then*

$$d_1 + d_2 + d_3 = R + r,$$

where R, r *are the radii of the circumcircle and the inscribed circle of* $\triangle ABC$ *respectively. If* $\triangle ABC$ *is an obtuse triangle, then the sign before* d_1, d_2, d_3 *in the conclusion is appropriately changed.*

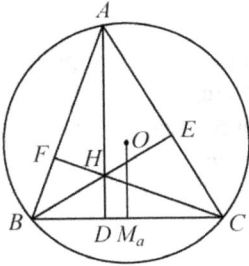

Figure 4.9

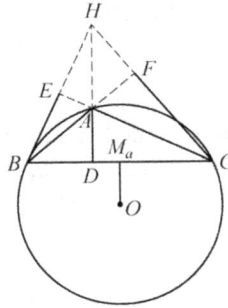

Figure 4.10

(5)   *The center of nine-point circle*

As shown in Figure 4.11, let the midpoints of sides BC, CA, AB of $\triangle ABC$ be M_a, M_b, M_c, respectively, and the feet of the altitudes on the three sides are H_a, H_b, H_c, respectively. Let the orthocenter be H, and the midpoints of AH, BH, CH be E_a, E_b, E_c, respectively. Then the nine points M_a, M_b, M_c, H_a, H_b, H_c, E_a, E_b, E_c are concyclic, which is called the nine-point circle of $\triangle ABC$. The center of the nine-point circle is the midpoint of segment OH.

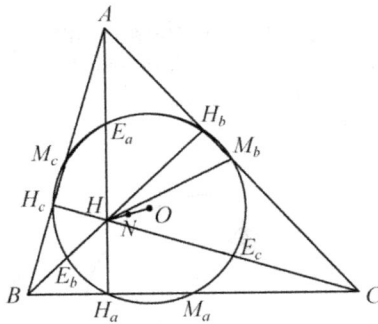

Figure 4.11

Theorem 4.4 (Feuerbach's Theorem). *The nine-point circle of a triangle is tangent to the inscribed circle and the three excircles.*

4.1.2 *Special lines in a triangle*

(1) *Euler line*

The circumcenter, orthocenter, and median point of a triangle are collinear, and this line is called the Euler line of the triangle.

As shown in Figure 4.12, let the circumcenter, orthocenter, and centroid of triangle ABC be O, H, G, respectively. Then

$$\overrightarrow{OG} = \frac{1}{3}\overrightarrow{OH},$$

that is, G is a trisection point of the segment OH.

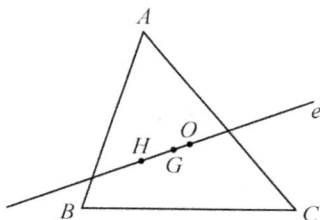

Figure 4.12

Obviously, the center of the nine-point circle is also on the Euler line.

(2) *Simson line*

Let the projections of any point P on the circumcircle of $\triangle ABC$ onto sides (or the lines) BC, CA, AB be D, E, F, respectively. Then the three points D, E, F are collinear, and the line is called the Simson line of point P with respect to $\triangle ABC$ (see Figure 4.13).

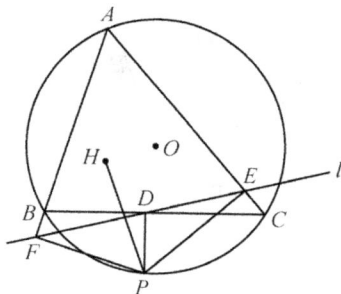

Figure 4.13

Theorem 4.5 (Steiner's Theorem). *Let the orthocenter of $\triangle ABC$ be H and the Simson line of any point P on the circumcircle be l. Then l bisects PH.*

The above two conclusions have an equivalent proposition: The symmetry points of any point on the circumcircle of a triangle about the three sides are collinear, and the line passes through the orthocenter of the triangle.

Example 4.1. As shown in Figure 4.14, $ABCD$ is a parallelogram and G is the center of gravity of $\triangle ABD$. Points P, Q are on the line BD such that $GP \perp PC$ and $GQ \perp QC$. Prove: AG bisects $\angle PAQ$.

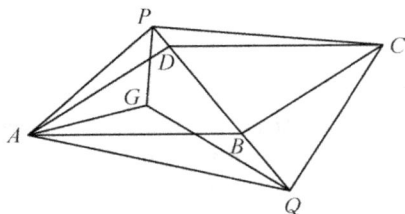

Figure 4.14

Proof. As shown in Figure 4.15, connect to form AC, which intersects BD at point M. By the property of a parallelogram, point M is the midpoint of AC, BD. So, point G is on the segment AC.

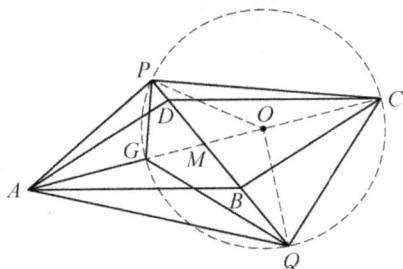

Figure 4.15

Since $\angle GPC = \angle GQC = 90°$, the four points P, G, Q, C are concyclic, and this circumcircle has GC as its diameter. By the intersecting chords theorem,

$$PM \cdot MQ = GM \cdot MC. \tag{1}$$

Take the midpoint O of GC. Noticing that $AG : GM : MC = 2 : 1 : 3$, we have $OC = \frac{1}{2}GC = AG$. Thus, G, O are symmetric with respect to point M. So,

$$GM \cdot MC = AM \cdot MO. \tag{2}$$

Combining (1) and (2), we obtain $PM \cdot MQ = AM \cdot MO$, so the points A, P, O, Q are concyclic.

And $OP = OQ = \frac{1}{2}GC$, from which $\angle PAO = \angle QAO$, that is AG bisects $\angle PAQ$.

Explanation. Since $\frac{AG}{GM} = \frac{AC}{CM} = 2$, the circle with diameter GC is the Apollonius circle, which contains points P, Q, thus $\frac{PA}{PM} = \frac{QA}{QM} = 2$. By the inverse of the angle bisector theorem, AG bisects $\angle PAQ$.

Example 4.2. As shown in Figure 4.16, in $AB = AC$ in $\triangle ABC$, and I is the incenter of $\triangle ABC$. Construct the circle Γ_1 with center A and radius AB, and construct the circle Γ_2 with center I and radius IB. A circle Γ_3 passing through points B, I intersects circles Γ_1 and Γ_2 at points P and Q, respectively (different from the point B). Let IP and BQ intersect at point R. Prove: $BR \perp CR$.

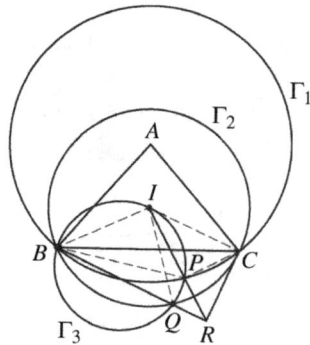

Figure 4.16

Proof. Connect to form IB, IC, IQ, PB, and PC.

Since point Q is on the circle Γ_2, then $IB = IQ$, so $\angle IBQ = \angle IQB$.

And the four points B, I, P, Q are concyclic, so $\angle IQB = \angle IPB$, and then $\angle IBQ = \angle IPB$. Thus $\triangle IBP \backsim \triangle IRB$. Consequently $\angle IRB = \angle IBP$ and

$$\frac{IB}{IR} = \frac{IP}{IB}.$$

Note that $AB = AC$, and I is the incenter of $\triangle ABC$, so $IB = IC$. Then

$$\frac{IC}{IR} = \frac{IP}{IC}.$$

Thus, $\triangle ICP \backsim \triangle IRC$, from which $\angle IRC = \angle ICP$.

And point P is on the arc BC of the circle Γ_1, so $\angle BPC = 180° - \frac{1}{2}\angle A$. Hence

$$\angle BRC = \angle IRB + \angle IRC = \angle IBP + \angle ICP$$
$$= 360° - \angle BIC - \angle BPC$$
$$= 360° - \left(90° + \frac{1}{2}\angle A\right) - \left(180° - \frac{1}{2}\angle A\right)$$
$$= 90°,$$

namely $BR \perp CR$.

Example 4.3. In a convex quadrilateral $ABCD$, the lines on the sides of AD and BC intersect at point P. Let I_1, I_2 be the centers of the incircles of $\triangle PAB$, $\triangle PDC$, respectively, O be the center of circumcircle of $\triangle PAB$, and H be the orthocenter of $\triangle PDC$.

Prove: The circumcircle of $\triangle AI_1 B$ is tangent to the circumcircle of $\triangle DHC$ if and only if the circumcircle of $\triangle AOB$ is tangent to that of $\triangle DI_2C$.

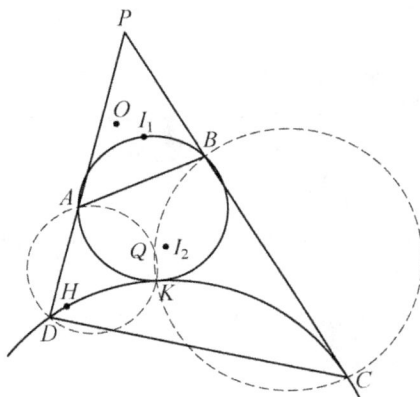

Figure 4.17

Proof. As shown in Figure 4.17, let the circumcircle of $\triangle AI_1B$ be tangent to the circumcircle of $\triangle DHC$ at point K, and point Q be another intersection point of the circumcircles of $\triangle AKD$, BKC. So

$$\angle DHC = \angle DKC = 180° - \angle P.$$

$$\angle PDK + \angle PCK = \angle DKC - \angle P = 180° - 2\angle P.$$

Since A, Q, K, D are concyclic,

$$\angle AQK = 180° - \angle PDK.$$

And since B, Q, K, C are concyclic,

$$\angle BQK = 180° - \angle PCK.$$

Thus

$$\angle AQB = 360° - \angle AQK - \angle BQK = \angle PDK + \angle PCK$$

$$= 180° - 2\angle P = 180° - \angle AOB,$$

so A, O, B, Q are concyclic. We also have $\angle AKD = \angle AQD$, $\angle BKC = \angle BQC$, and $\angle AQB = \angle DKC - \angle P$. Therefore,

$$\angle CQD = 360° - \angle AQB - \angle AQD - \angle BQC$$

$$= 360° - (\angle DKC - \angle P) - \angle AKD - \angle BKC$$

$$= \angle AKB + \angle P = 180° - \angle AI_1B + \angle P$$

$$= 180° - \left(90° + \frac{1}{2}\angle P\right) + \angle P = 90° + \frac{1}{2}\angle P,$$

while $\angle CI_2D = 90° + \frac{\angle P}{2}$, thus quadrilateral $CDQI_2$ is the inscribed quadrilateral.

The following shows that the circumcircle of $\triangle AOB$ is tangent to the circumcircle of $\triangle DI_2C$ at point Q. In fact, it is only necessary to prove that

$$\angle ABQ + \angle DCQ = \angle AQD.$$

Since the circumcircle of $\triangle AI_1B$ is tangent to the circumcircle of $\triangle DHC$ at point K, we know that $\angle ABK + \angle DCK = \angle AKD$, that is,

$$(\angle ABQ + \angle KBQ) + (\angle DCQ - \angle KCQ) = \angle AKD.$$

And $\angle KBQ = \angle KCQ$, $\angle AKD = \angle AQD$, so $\angle ABQ + \angle DCQ = \angle AQD$. Thus, the circumcircle of $\triangle AOB$ is tangent to the circumcircle of $\triangle DI_2C$ at Q.

For the other side of the problem, we assume that the circumcircle of $\triangle CI_2D$ is tangent to the circumcircle of $\triangle AOB$ at point Q, and point K is another intersection of the circumcircles of $\triangle AQD$, $\triangle BQC$. Similarly, we can prove that the circumcircle of $\triangle AI_1B$ is tangent to the circumcircle of $\triangle DHC$ at point K.

Example 4.4. As shown in Figure 4.18, there are two points P, Q on the side BC of $\triangle ABC$, and their distances from the midpoint of BC are equal. Cross points P, Q to construct two perpendicular lines of BC, which intersect AC, AB at points E, F, respectively. Let M be the intersection point of PF, EQ, and H_1, H_2 be the orthocenters of $\triangle BFP$, $\triangle CEQ$, respectively. Prove: $AM \perp H_1H_2$.

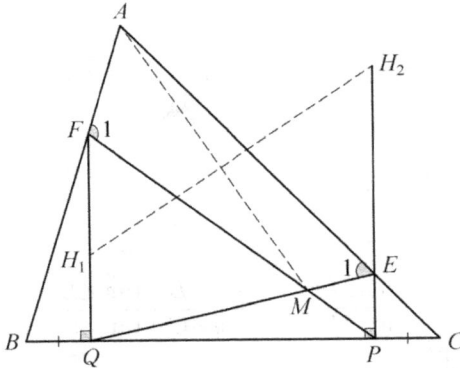

Figure 4.18

Proof. First, it will be proved that the line AM is independent of the positions of points P, Q. In fact, we just need to prove that $\frac{\sin \angle MAB}{\sin \angle MAC}$ is a definite value. In $\triangle AFM$, $\triangle AEM$, by the law of sines,

$$\frac{\sin \angle MAB}{\sin \angle MAC} = \frac{\sin \angle AFM}{\sin \angle AEM} \cdot \frac{FM}{EM}. \tag{1}$$

In the $\triangle FBP$, $\triangle CEQ$, we have

$$\sin \angle AFM = \frac{BP}{PF} \sin \angle B, \quad \sin \angle AEM = \frac{CQ}{EQ} \sin \angle C.$$

Since $BP = CQ$, so

$$\frac{\sin \angle AFM}{\sin \angle AEM} = \frac{\sin \angle B}{\sin \angle C} \cdot \frac{EQ}{FP}. \tag{2}$$

By (1) and (2),

$$\frac{\sin \angle MAB}{\sin \angle MAC} = \frac{\sin \angle B}{\sin \angle C} \cdot \frac{EQ}{FP} \cdot \frac{FM}{EM}. \tag{3}$$

Since $\triangle FMQ \backsim \triangle EMP$,

$$\frac{FM}{FP} = \frac{FQ}{FQ + EP},$$

$$\frac{EQ}{EM} = \frac{FQ + EP}{EP}.$$

Substituting into (3), we get

$$\frac{\sin \angle MAB}{\sin \angle MAC} = \frac{\sin \angle B}{\sin \angle C} \cdot \frac{FQ}{EP}. \tag{4}$$

On the other hand,

$$\tan \angle B = \frac{FQ}{BQ}, \quad \tan \angle C = \frac{EP}{CP}, \quad BQ = CP,$$

substituting into (4), we obtain

$$\frac{\sin \angle MAB}{\sin \angle MAC} = \frac{\sin \angle B}{\sin \angle C} \cdot \frac{\tan \angle B}{\tan \angle C},$$

which is a constant.

Let α be the angle between $H_1 H_2$ and BC. Therefore,

$$\tan \alpha = \frac{H_2 P - H_1 Q}{QP}. \tag{5}$$

Since H_1, H_2 are orthocenters of $\triangle BFP$, $\triangle CEQ$ respectively,

$$QF \cdot H_1 Q = BQ \cdot QP,$$

$$EP \cdot H_2 P = CP \cdot PQ,$$

that is

$$H_1 Q = \frac{BQ \cdot QP}{FQ},$$

$$H_2 P = \frac{CP \cdot PQ}{EP}.$$

Notice that $CP = BQ$, and so

$$H_2P - H_1Q = \frac{PQ \cdot BQ \cdot (FQ - EP)}{EP \cdot FQ}.$$

Substituting into (5), we see that

$$\tan \alpha = \frac{BQ \cdot (FQ - EP)}{EP \times FQ} = \frac{BQ}{EP} - \frac{BQ}{FQ}$$

$$= \frac{CP}{EP} - \frac{BQ}{FQ} = \cot C - \cot B.$$

Let θ be the angle between AM and BC. In the following we prove that

$$\tan \alpha \cdot \tan \theta = 1. \tag{6}$$

Let AM, BC intersect at point X. Then

$$\frac{BX}{CX} = \frac{\sin \angle MAB}{\sin \angle MAC} \cdot \frac{\sin \angle C}{\sin \angle B} = \frac{\tan \angle B}{\tan \angle C}.$$

Let AD be the altitude of $\triangle ABC$. Then

$$\frac{BX}{CX} = \frac{\tan \angle B}{\tan \angle C} = \frac{\dfrac{AD}{BD}}{\dfrac{AD}{CD}} = \frac{CD}{BD},$$

from which $BD = CX$. Thus

$$\tan \theta = \frac{AD}{DX} = \frac{AD}{CD - CX} = \frac{AD}{CD - BD}$$

$$= \frac{1}{\dfrac{CD}{AD} - \dfrac{BD}{AD}} = \frac{1}{\cot \angle C - \cot \angle B}.$$

Therefore, equation (6) holds, i.e., $AM \perp H_1H_2$.

4.2 Problems and Solutions

4.2.1 *Position structure problems*

Problem 4.1 (IMO 7-5, proposed by Romania). Consider $\triangle OAB$ with acute angle $\angle AOB$. Through a point $M \neq O$ the perpendiculars are drawn to OA and OB, the feet of which are P and Q respectively. The point of intersection of the altitudes of $\triangle OPQ$ is H. What is the locus of H if M is permitted to range over (a) the side AB, (b) the interior of $\triangle OAB$?

Solution. (a) First we prove a general lemma.

As shown in Figure 4.19, in any convex quadrilateral $ADCB$, points H and H' are on CD, points M, P, Q are respectively on AB, AD, BC, and $QH' \parallel BD \parallel MP$ and $PH \parallel AC \parallel MQ$. Then H and H' coincide, that is, the quadrilateral $MPHQ$ is an internal parallelogram.

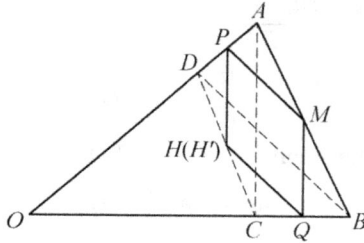

Figure 4.19

The proof is as follows. Since

$$\frac{DH}{DC} = \frac{DP}{DA} = \frac{BM}{BA} = \frac{BQ}{BC} = \frac{DH'}{DC},$$

so $DH = DH'$, then H and H' coincide. The lemma is proved.

We may as well suppose that the letters above apply to problem (a). By $AC \parallel MQ$ we know that $AC \perp OB$, and similarly $BD \perp OA$.

Hence H is the orthocenter of $\triangle OPQ$.

Because of $\frac{DH}{DC} = \frac{BM}{BA}$, when M moves from B to A, the point H moves from D to C. Therefore, the required locus of (a) is segment CD.

Note: When $\triangle OAB$ is a right or obtuse triangle, the conclusion still holds, but CD is not necessarily inside $\triangle OAB$. If you are interested, you can draw a diagram to see.

(b) Let M be a point inside $\triangle OAB$. A line parallel to AB through point M intersects the segments OA, OB at the inner points A', B', respectively. It is easy to know that if $A'C'$, $B'D'$ are perpendicular to OB, OA and intersect at C', D' respectively, then the points inside segment $A'B'$ correspond one-to-one with the points inside segment $C'D'$.

The interior points of $\triangle OAB$ can be regarded as the interior points of a set of segments $A'B'$ parallel to AB on OA and OB, so the corresponding points are the interior points of a set of segments $C'D'$ parallel to CD on OC and OD. Therefore, the inner point set of $\triangle OAB$ corresponds one-to-one with the inner point set of $\triangle OCD$, so the required locus of point H is the interior of $\triangle OCD$.

【Score Situation】 This particular problem saw the following distribution of scores among contestants: 26 contestants scored 7 points, 9 contestants scored 6 points, 3 contestants scored 5 points, 3 contestants scored 4 points, 4 contestants scored 3 points, 8 contestants scored 2 points, 12 contestants scored 1 point, and 15 contestants scored 0 point. The average score of this problem is 3.788, indicating that it was relatively straightforward.

Among the top five teams in the team scores, the scores of this problem are as follows: the Soviet Union team scored 42 points (with a total team score of 281 points), the Hungary team scored 45 points (with a total team score of 244 points), the Romania team scored 53 points (with a total team score of 222 points), the Poland team scored 40 points (with a total team score of 178 points), and the German Democratic Republic team scored 37 points (with a total team score of 175 points).

The gold medal cutoff for this IMO was set at 38 points (with 8 contestants earning gold medals), the silver medal cutoff was 30 points (with 12 contestants earning silver medals), and the bronze medal cutoff was 20 points (with 17 contestants earning bronze medals).

In this IMO, only two contestants achieved a perfect score of 40 points, namely László Lovász from Hungary and Pavel Bleher from the Soviet Union.

Problem 4.2 (IMO 22-5, proposed by the Soviet Union). Three congruent circles have a common point O and lie inside a given triangle. Each circle touches a pair of sides of the triangle. Prove that the incenter and the circumcenter of the triangle and the point O are collinear.

Proof 1. The three vertices of the triangle are represented by A, B, C respectively. The centers of each circle are represented by A', B', C' respectively, as shown in Figure 4.20. Since the radii of the three circles are equal, the corresponding sides of $\triangle ABC$ are parallel to the corresponding

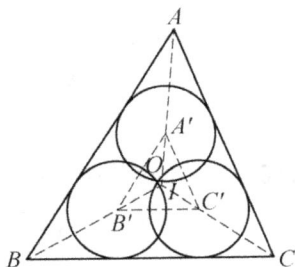

Figure 4.20

sides of $\triangle A'B'C'$. The segment AA' bisects $\angle BAC$, and after extension it bisects $\angle B'A'C'$; then BB' and CC' have similar properties.

Thus the incenter I of $\triangle ABC$ is also the incenter of $\triangle A'B'C'$, and is at the same time the homothetic center of these two triangles. An important property of homothety is that the corresponding points are collinear with the homothetic center. Since O is the circumcenter of $A'B'C'$, the circumcenter O' of $\triangle ABC$, the incenter I, and the point O must be collinear.

Proof 2. In the given $\triangle ABC$, the points K, L, M are the centers of the three circles, respectively, and AK, BL, CM are the internal angle bisectors of the triangle (see Figure 4.21). Since O is the intersection of the three circles, O is the circumcenter of $\triangle KLM$, and the angle bisectors AK, BL, CM intersect at the incenter I of $\triangle ABC$.

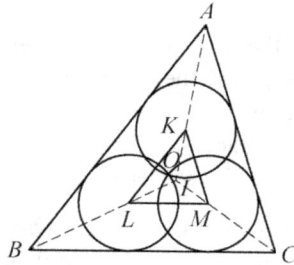

Figure 4.21

Since $\frac{IK}{IA} = \frac{IL}{IB} = \frac{IM}{IC}$, the triangle $\triangle KLM$ is the image of the homothetic transformation of $\triangle ABC$ with I as the homothetic center and $\lambda = \frac{IK}{IA}$ as the ratio.

Therefore, the circumcenter Ω of $\triangle ABC$ is the image of the homothetic transformation of point O with I as the homothetic center and $\frac{1}{\lambda}$ as the ratio. So I, O, and Ω are collinear.

【Score Situation】 This particular problem saw the following distribution of scores among contestants: 39 contestants scored 7 points, 5 contestants scored 6 points, 2 contestants scored 5 points, no contestant scored 4 points, 2 contestants scored 3 points, no contestant scored 2 points, 1 contestant scored 1 point, and 2 contestants scored 0 point. The average score of this problem is 6.275, indicating that it was simple.

Among the top five teams in the team scores, the scores of this problem are as follows: the United States team scored 47 points (with a total team score of 314 points), the Germany team scored 56 points (with a total team score of 312 points), the United Kingdom team scored 55 points (with a total team score of 301 points), the Austria team scored 48 points (with a total team score of 290 points), and the Bulgaria team scored 55 points (with a total team score of 287 points).

The gold medal cutoff for this IMO was set at 41 points (with 36 contestants earning gold medals), the silver medal cutoff was 34 points (with 37 contestants earning silver medals), and the bronze medal cutoff was 26 points (with 30 contestants earning bronze medals).

In this IMO, a total of 26 contestants achieved a perfect score of 42 points.

Problem 4.3 (IMO 41-6, proposed by Russia). Let AH_1, BH_2, and CH_3 be the altitudes of an acute-angled triangle ABC. The incircle of the triangle ABC touches the sides BC, CA, AB at T_1, T_2, T_3, respectively. Let the lines l_1, l_2, l_3 be the reflections of the lines H_2H_3, H_3H_1, H_1H_2 in the lines T_2T_3, T_3T_1, T_1T_2, respectively. Prove that l_1, l_2, l_3 determine a triangle whose vertices lie on the incircle of the triangle ABC.

Proof.   As shown in Figure 4.22, without loss of generality, let $\angle B \geq \angle C$ (denote $\angle A$, $\angle B$, $\angle C$ for $\angle BAC$, $\angle ABC$, $\angle ACB$, respectively). The proof is as follows: we do find a point on the incircle I (I is the incenter of $\triangle ABC$), where the symmetry of T_1T_3 is on H_1H_3 and the symmetry of T_1T_2 is on H_1H_2. In other words, this is equivalent to proving that the intersection of l_2 and l_3 is on $\odot I$, and the other two intersections are proved in the same way.

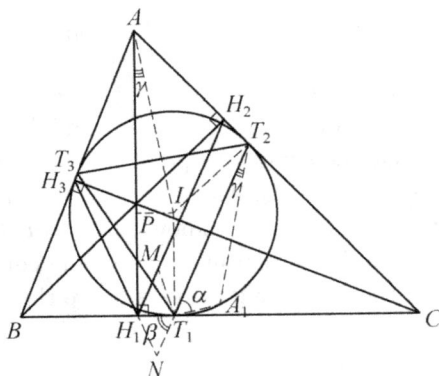

Figure 4.22

Using the analytical method, we first assume that the point A_1 has been found, so there must be points N and M on the lines H_3H_1 and H_2H_1, such that $NT_1 = MT_1 = T_1A_1$ and $\angle NT_1T_3 = \angle A_1T_1T_3$, $\angle MT_1T_2 = \angle A_1T_1T_2$.

Since $\angle MH_1T_1 = \angle NH_1T_1 = \angle H_3H_1B = \angle A$, $MT_1 = NT_1$, and H_1T_1 is the common side, so we conjecture that maybe $\triangle MH_1T_1 \cong \triangle NH_1T_1$. Then, let $\angle MT_1H_1 = \angle NT_1H_1 = \beta$ and $\angle T_2T_1A_1 = \alpha$ as shown in Figure 4.22. Since

$$\beta + 90° - \frac{\angle B}{2} = \angle T_3T_1N = \angle T_3T_1A_1 = \angle T_3T_1T_2 + \alpha$$

$$= 90° - \frac{\angle A}{2} + \alpha$$

$$= 90° - \frac{\angle A}{2} + \angle MT_1T_2$$

$$= 90° - \frac{\angle A}{2} + \angle H_1T_1T_2 - \beta$$

$$= 180° + \frac{\angle C - \angle A}{2} - \beta,$$

so

$$\beta = 90° - \frac{1}{2}\angle A, \quad \alpha = 90° - \frac{1}{2}\angle B,$$

and thus, $\triangle H_1T_1N$ is an isosceles triangle. Also

$$\angle T_3T_1A_1 = \angle T_3T_1T_2 + \alpha$$

$$= 90° - \frac{\angle A}{2} + 90° - \frac{\angle B}{2} = 90° + \frac{\angle C}{2}$$

$$= 180° - \left(90° - \frac{\angle C}{2}\right) = 180° - \angle T_1T_3T_2,$$

and consequently, $T_1A_1 \parallel T_3T_2$.

Thus, we make M, N, and A_1 as follows: extend H_3H_1 to N, such that $H_1N = H_1T_1$, and make the symmetry point M of N with respect to BC, which must lie on the line H_1H_2. Make an isosceles trapezoid $T_3T_1A_1T_2$ again. From the previous analysis we know that $\angle NT_1T_3 = \angle A_1T_1T_3$ and $\angle MT_1T_2 = \angle A_1T_1T_2$ are both guaranteed, and the equality $MT_1 = NT_1$ is also obvious. Then the only thing to do is prove that $NT_1 = A_1T_1$.

It is easy to know

$$\angle T_1T_2A_1 = \angle T_1T_3T_2 - \angle T_1T_2T_3$$

$$= 90° - \frac{1}{2}\angle C - \left(90° - \frac{1}{2}\angle B\right)$$

$$= \frac{1}{2}(\angle B - \angle C) = \frac{1}{2}\angle A - (90° - \angle B)$$

$$= \angle IAH_1 \triangleq \gamma.$$

Make $IP \perp AH_1$, and so P is the perpendicular foot. By $\triangle H_1 T_1 N \backsim \triangle AT_3 T_2$, it follows that

$$\frac{NT_1}{T_2 T_3} = \frac{H_1 T_1}{AT_2} = \frac{PI}{AT_2}$$

$$= \frac{AI \sin \gamma}{AI \sin \angle AIT_2} = \frac{\sin \gamma}{\sin \angle AT_2 T_3}$$

$$= \frac{\sin \gamma}{\sin \angle T_2 T_1 T_3} = \frac{A_1 T_1}{T_2 T_3},$$

and hence, $NT_1 = A_1 T_1$. Hence the proposition is proved.

【Score Situation】 This particular problem saw the following distribution of scores among contestants: 33 contestants scored 7 points, 2 contestants scored 6 points, 9 contestants scored 5 points, 11 contestants scored 4 points, 15 contestants scored 3 points, 51 contestants scored 2 points, 5 contestants scored 1 point, and 335 contestants scored 0 point. The average score of this problem is 1.050, indicating that it was relatively challenging.

Among the top five teams in the team scores, the scores of this problem are as follows: the China team scored 34 points (with a total team score of 218 points), the Russia team scored 32 points (with a total team score of 215 points), the United States team scored 18 points (with a total team score of 184 points), the South Korea team scored 21 points (with a total team score of 172 points), the Bulgaria team scored 32 points (with a total team score of 169 points), and the Vietnam team scored 23 points (with a total team score of 169 points).

The gold medal cutoff for this IMO was set at 30 points (with 39 contestants earning gold medals), the silver medal cutoff was 21 points (with 71 contestants earning silver medals), and the bronze medal cutoff was 11 points (with 119 contestants earning bronze medals).

In this IMO, a total of 4 contestants achieved a perfect score of 42 points.

Problem 4.4 (IMO 43-2, proposed by South Korea). A segment BC is a diameter of a circle with center O. A point A is on the circle with $\angle AOC > 60°$. A segment EF is a chord which is the perpendicular bisector of AO, and D is the midpoint of the minor arc AB. The line through O and parallel to AD meets AC at I. Show that I is the incenter of triangle CEF.

Proof. Obviously CA bisects $\angle ECF$. As shown in Figure 4.23, connect to form DO. Obviously,

$$\angle DOA = \frac{1}{2} \angle AOB = \angle OAC,$$

so $DO \parallel AC$.

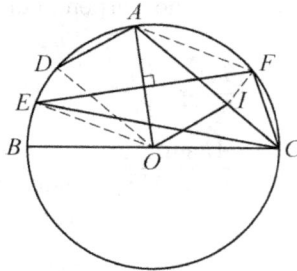

Figure 4.23

Further, $DA\|OI$ and quadrilateral $ADOI$ is a parallelogram, so $AI = OD = OE$. Since quadrilateral $AEOF$ is a rhombus, $OE = AF$. Then $AI = AF$, and thus

$$\angle IFE = \angle IFA - \angle EFA$$

$$= \angle AIF - \angle ECA$$

$$= \angle AIF - \angle ICF = \angle IFC.$$

Hence, IF bisects $\angle EFC$. Then I is the incenter of $\triangle CEF$.

【Score Situation】 This particular problem saw the following distribution of scores among contestants: 120 contestants scored 7 points, 129 contestants scored 6 points, 6 contestants scored 5 points, no contestant scored 4 points, 4 contestants scored 3 points, 1 contestant scored 2 points, 46 contestants scored 1 point, and 173 contestants scored 0 point. The average score of this problem is 3.557, indicating that it was relatively straightforward.

Among the top five teams in the team scores, the scores of this problem are as follows: the China team scored 41 points (with a total team score of 212 points), the Russia team scored 42 points (with a total team score of 204 points), the United States team scored 34 points (with a total team score of 171 points), the Bulgaria team scored 40 points (with a total team score of 167 points), and the Vietnam team scored 41 points (with a total team score of 166 points).

The gold medal cutoff for this IMO was set at 29 points (with 39 contestants earning gold medals), the silver medal cutoff was 23 points (with 73 contestants earning silver medals), and the bronze medal cutoff was 14 points (with 120 contestants earning bronze medals).

In this IMO, only three contestants achieved a perfect score of 42 points, namely Yunhao Fu and Botong Wang from China, and Andrei Khaliavine from Russia.

Problem 4.5 (IMO 46-1, proposed by Romania). Six points are chosen on the sides of an equilateral triangle ABC: A_1, A_2 on BC, B_1, B_2 on CA, and C_1, C_2 on AB, such that they are the vertices of a convex hexagon $A_1A_2B_1B_2C_1C_2$ with equal side lengths. Prove that the lines A_1B_2, B_1C_2, and C_1A_2 are concurrent.

Proof 1. As shown in Figure 4.24, choose a point P inside the equilateral triangle ABC such that $\triangle A_1A_2P$ is an equilateral triangle. By $A_1P \parallel C_2C_1$ and $A_1P = C_2C_1$, we see that $A_1PC_1C_2$ is a rhombus. Similarly, $A_2B_1B_2P$ is also a rhombus. Thus, $\triangle PB_2C_1$ is an equilateral triangle.

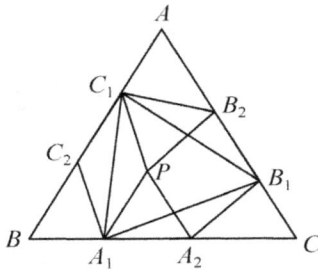

Figure 4.24

Denote $\angle A_1A_2B_1 = \alpha$, $\angle A_2B_1B_2 = \beta$, $\angle C_1C_2A_1 = \gamma$, so

$$\alpha + \beta = (\angle A_2B_1C + \angle C) + (\angle B_1A_2C + \angle C) = 240°.$$

And $\angle B_2PA_2 = \beta$, $\angle A_1PC_1 = \gamma$, so

$$\beta + \gamma = 360° - (\angle A_1PA_2 + \angle C_1PB_2) = 240°,$$

and thus $\alpha = \gamma$.

Similarly, $\angle B_1B_2C_1 = \alpha$.

The three triangles $\triangle A_1A_2B_1$, $\triangle B_1B_2C_1$, and $\triangle C_1C_2A_1$ are congruent. Hence, $\triangle A_1B_1C_1$ is an equilateral triangle, so A_1B_2, B_1C_2, C_1A_2 are vertical bisectors on the three sides B_1C_1, C_1A_1, A_1B_1 of the equilateral triangle $A_1B_1C_1$, respectively. Therefore, the lines are concurrent.

Proof 2. Assume $A_1A_2 = d$, and $AB = a$. Construct an equilateral triangle $A_0B_0C_0$ with side length $a - d$. Points A', B', C' are chosen on the sides of triangle $A_0B_0C_0$ such that $A'C_0 = A_2C$, $B'A_0 = B_2A$, and $C'B_0 = C_2B$ (see Figures 4.25 and 4.26).

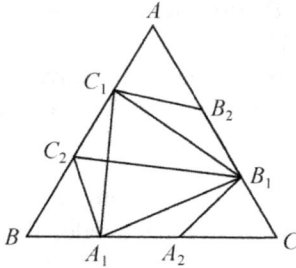
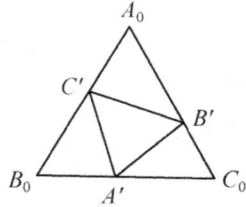

Figure 4.25   Figure 4.26

Therefore,

$$A'B_0 = a - d - A'C_0 = BC - A_1A_2 - A_2C = BA_1.$$

Similarly,

$$B'C_0 = B_1C, \quad C'A_0 = C_1A.$$

Since

$$\angle B_1CA_2 = \angle B'C_0A', \quad \angle B_2AC_1 = \angle B'A_0C', \quad \angle C_2BA_1 = \angle C'B_0A',$$

$$\triangle CB_1A_2 \cong \triangle C_0B'A', \quad \triangle AC_1B_2 \cong \triangle A_0C'B', \quad \triangle BA_1C_2 \cong \triangle B_0A'C',$$

which implies that $B'C' = C'A' = A'B' = d$ and triangle $A'B'C'$ is equilateral.

So

$$\angle A'C'B' = \angle C'B'A' = \angle A'B'C' = 60°,$$

$$\angle AB_2C_1 = \angle A_0B'C' = 180° - \angle C'B'A' - \angle A'B'C_0 = 120° - \angle A'B'C_0,$$

$$\angle C_1B_2B_1 = \angle B_1A_2A_1,$$

In view of $B_2C_1 = B_1B_2 = A_2B_1 = A_1A_2 = d$, triangles $C_1B_2B_1$ and $B_1A_2A_1$ are congruent, implying that $B_1C_1 = A_1B_1$. Together with $C_1C_2 = A_1C_2$, we show that C_2B_1 is the perpendicular bisector of A_1C_1, and C_2B_1 is the altitude on the side A_1C_1 of the triangle $A_1B_1C_1$.

Similarly, C_1A_2 and A_1B_2 are the altitudes on the sides A_1B_1 and B_1C_1 of the triangle $A_1B_1C_1$ respectively.

Therefore, the lines A_1B_2, B_1C_2, and C_1A_2 are concurrent.

【Score Situation】 This particular problem saw the following distribution of scores among contestants: 141 contestants scored 7 points, 5 contestants scored 6 points,

11 contestants scored 5 points, 5 contestants scored 4 points, 20 contestants scored 3 points, 65 contestants scored 2 points, 59 contestants scored 1 point, and 207 contestants scored 0 point. The average score of this problem is 2.614, indicating that it had a certain level of difficulty.

Among the top five teams in the team scores, the scores of this problem are as follows: the China team scored 32 points (with a total team score of 235 points), the United States team scored 27 points (with a total team score of 213 points), the Russia team scored 42 points (with a total team score of 212 points), the Iran team scored 36 points (with a total team score of 201 points), and the South Korea team scored 28 points (with a total team score of 200 points).

The gold medal cutoff for this IMO was set at 35 points (with 42 contestants earning gold medals), the silver medal cutoff was 23 points (with 79 contestants earning silver medals), and the bronze medal cutoff was 12 points (with 128 contestants earning bronze medals).

In this IMO, a total of 16 contestants achieved a perfect score of 42 points.

4.2.2 *Quantity relation problems*

Problem 4.6 (IMO 4-6, proposed by the German Democratic Republic). Consider an isosceles triangle. Let R be the radius of its circumscribed circle and r the radius of its inscribed circle. Prove that the distance d between the centres of these two circles is

$$d = \sqrt{R(R - 2r)}.$$

Proof 1. The conclusion of this problem holds for any triangle (not limited to isosceles triangle. It is Euler's formula).

As shown in Figure 4.27, let O, S be the circumcenter and incenter of $\triangle ABC$ respectively. Connect to form AS and extend it, intersecting $\odot O$ at D.

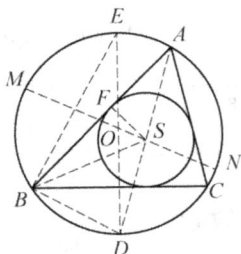

Figure 4.27

Make the diameter DE and connect to form BD, BE, BS, FS (F is the tangent point between the incircle and AB). Connect to form OS and extend it, intersecting $\odot O$ at points M, N.

Denote $OS = d$. So

$$d^2 = R^2 - (R^2 - d^2) = R^2 - (R + d)(R - d)$$
$$= R^2 - MS \cdot NS = R^2 - AS \cdot SD.$$

And then we just need to prove that

$$AS \cdot SD = 2Rr.$$

Obviously, $\triangle BED \backsim \triangle FAS$, so $BD \cdot AS = FS \cdot DE$.

Since

$$\angle SBD = \angle SBC + \angle CBD = \frac{1}{2}\angle ABC + \angle DAC = \frac{1}{2}(\angle ABC + \angle BAC),$$

and $\angle BSD = \angle ABS + \angle BAS = \frac{1}{2}(\angle ABC + \angle BAC)$, we have $BD = SD$.

Hence, $SD \cdot AS = FS \cdot DE = 2Rr$, so the proposition is proved and the conclusion holds.

Proof 2. If $AB = AC$ in triangle ABC, then the triangle is completely determined by $BC = 2a$ and $\angle BAC = 2x$. As shown in Figure 4.28, let $x \le 30°$. If $x > 30°$, we can have a similar discussion.

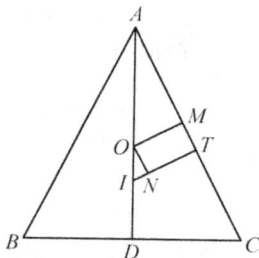

Figure 4.28

A direct calculation yields the following equalities:

$$AC = \frac{a}{\sin x}, \quad R = OA = \frac{a}{\sin 2x},$$

$$p = AC + DC = a + \frac{a}{\sin x},$$

$$S = \frac{1}{2}BC \cdot AD = \frac{a^2}{\tan x}.$$

By the last equality and $S = pr$, we get $r = \frac{a \cos x}{1 + \sin x}$, where $r = ID = IT$.

Thus,

$$R^2 - 2Rr = \frac{a^2 \cdot (1 - 3\sin x + 4\sin^3 x)}{(1 + \sin x)\sin^2 2x}$$

$$= \frac{a^2(1 - 4\sin x + 4\sin^3 x)}{\sin^2 2x}$$

$$= a^2 \left(\frac{1 - 2\sin x}{\sin 2x}\right)^2.$$

To calculate IO, consider the right triangle INO. Since $\angle ION = x$, so

$$IO = \frac{ON}{\cos x} = \frac{MT}{\cos x} = \frac{AT - AM}{\cos x}$$

$$= \left(p - 2a - \frac{AC}{2}\right)\frac{1}{\cos x} = \left(\frac{a}{\sin x} - a - \frac{a}{2\sin x}\right)\frac{1}{\cos x}$$

$$= \frac{a}{\cos x}\left(\frac{1}{2\sin x} - 1\right) = \frac{a(1 - 2\sin x)}{\sin 2x}.$$

【Score Situation】 This particular problem saw the following distribution of scores among contestants: 1 contestant scored 6 points, 1 contestant scored 5 points, 6 contestants scored 4 points, no contestant scored 3 points, 1 contestant scored 2 points, 1 contestant scored 1 point, and 7 contestants scored 0 point. The average score of this problem is 2.235, indicating that it had a certain level of difficulty.

Among the top five teams in the team scores, the Hungary team achieved a total score of 289 points, the Soviet Union team achieved a total score of 263 points, the Romania team achieved a total score of 257 points, the Czechoslovakia team achieved a total score of 212 points, and the Poland team achieved a total score of 212 points.

The gold medal cutoff for this IMO was set at 41 points (with 4 contestants earning gold medals), the silver medal cutoff was 34 points (with 12 contestants earning silver medals), and the bronze medal cutoff was 29 points (with 15 contestants earning bronze medals).

In this IMO, only one contestant achieved a perfect score of 46 points, namely Iosif Bernstein from the Soviet Union.

Problem 4.7 (IMO 12-1, proposed by Poland). Let M be a point on the side AB of ABC. Let r_1, r_2, and r be the radii of the inscribed circles of triangles AMC, BMC, and ABC. Let q_1, q_2, and q be the radii of the escribed circles of the same triangles that lie in the angle ACB. Prove that $\frac{r_1}{q_1} \cdot \frac{r_2}{q_2} = \frac{r}{q}$.

Proof.   As shown in Figure 4.29, let the center of the incircle and the center of the escribed circle of $\triangle AMC$ be O_1 and O_1', respectively, and let the center of the incircle and the center of the escribed circle of $\triangle CMB$ be O_2 and O_2', respectively.

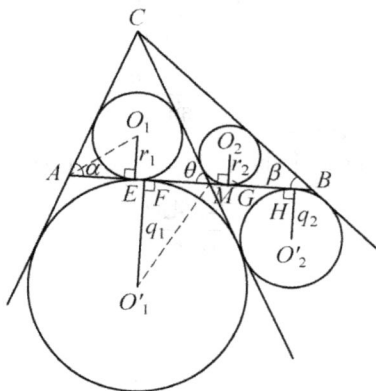

Figure 4.29

The projections of O_1, O_1', O_2, O_2' on the side of AB are E, F, G, H, respectively.

Denote $\angle CAB = \alpha$, $\angle CBA = \beta$, $\angle AMC = \theta$, and $\angle CMB = 180° - \theta$. Connect to form O_1A, MO_1'.

It is easy to know

$$AE = MF = \frac{1}{2}(AC + AM - MC).$$

While

$$AE = O_1E \cot \frac{\alpha}{2} = r_1 \cot \frac{\alpha}{2},$$

$$MF = O_1'F \cot \frac{180° - \theta}{2} = q_1 \tan \frac{\theta}{2},$$

from which

$$\frac{r_1}{q_1} = \tan \frac{\alpha}{2} \cdot \tan \frac{\theta}{2},$$

and similarly

$$\frac{r_2}{q_2} = \tan \frac{\beta}{2} \tan \frac{180° - \theta}{2} = \tan \frac{\beta}{2} \cot \frac{\theta}{2}, \quad \text{and} \quad \frac{r}{q} = \tan \frac{\alpha}{2} \tan \frac{\beta}{2}.$$

Hence

$$\frac{r_1}{q_1} \cdot \frac{r_2}{q_2} = \frac{r}{q}.$$

【Score Situation】 This particular problem saw the following distribution of scores among contestants: 23 contestants scored 5 points, no contestant scored 4 points, no contestant scored 3 points, no contestant scored 2 points, 7 contestants scored 1 point, and 5 contestants scored 0 point. The average score of this problem is 3.486, indicating that it was relatively straightforward.

Among the top five teams in the team scores, the Hungary team achieved a total score of 233 points, the German Democratic Republic team achieved a total score of 221 points, the Soviet Union team achieved a total score of 221 points, the Yugoslavia team achieved a total score of 209 points, and the Romania team achieved a total score of 208 points.

The gold medal cutoff for this IMO was set at 37 points (with 7 contestants earning gold medals), the silver medal cutoff was 30 points (with 11 contestants earning silver medals), and the bronze medal cutoff was 19 points (with 40 contestants earning bronze medals).

In this IMO, only three contestants achieved a perfect score of 40 points, namely Wolfgang Burmeister from the German Democratic Republic, Imre Ruzsa from Hungary and Andrei Hodulev from the Soviet Union.

Problem 4.8 (IMO 44-4, proposed by Finland). A quadrilateral $ABCD$ is cyclic. The feet of the perpendicular from D to the lines BC, CA, AB are P, Q, R respectively. Show that the angle bisectors of ABC and ADC meet on the line AC if and only if $PQ = QR$.

Proof 1. As shown in Figure 4.30, proving the conclusion is equivalent to proving that

$$\frac{AB}{BC} = \frac{AD}{CD}.$$

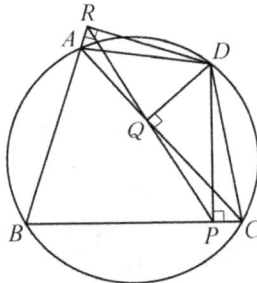

Figure 4.30

Since A, R, D, Q are concyclic, C, P, Q, D are concyclic, and the law of sines implies that

$$\frac{AB}{BC} = \frac{\sin \angle ACB}{\sin \angle BAC} = \frac{\sin \angle QDP}{\sin \angle RDQ} = \frac{\frac{QP}{CD}}{\frac{QR}{AD}} = \frac{AD}{CD} \cdot \frac{PQ}{QR}.$$

Hence, $\frac{AB}{BC} = \frac{AD}{CD} \Leftrightarrow PQ = QR$.

Note that the condition that A, B, C, D are concyclic is superfluous.

Proof 2. By Simson's theorem, P, Q, R are collinear. Moreover, since $\angle DPC$ and $\angle DQC$ are right angles, the points D, P, C, Q are concyclic and so

$$\angle DCA = \angle DPQ = \angle DPR.$$

Similarly, since D, Q, R, A are concyclic, $\angle DAC = \angle DRP$. Therefore $\triangle DCA \backsim \triangle DPR$.

Likewise, $\triangle DAB \backsim \triangle DQP$ and $\triangle DBC \backsim \triangle DRQ$. Then

$$\frac{DA}{DC} = \frac{DR}{DP} = \frac{DB \cdot \frac{QR}{BC}}{DB \cdot \frac{PQ}{BA}} = \frac{QR}{PQ} \cdot \frac{BA}{BC}.$$

Thus $PQ = QR$ if and only if $\frac{DA}{DC} = \frac{BA}{BC}$.

Now the bisectors of the angles ABC and ADC divide AC in the ratios of $\frac{BA}{BC}$ and $\frac{DA}{DC}$, respectively. This completes the proof.

【Score Situation】 This particular problem saw the following distribution of scores among contestants: 272 contestants scored 7 points, 2 contestants scored 6 points, 3 contestants scored 5 points, 1 contestant scored 4 points, 35 contestants scored 3 points, 23 contestants scored 2 points, 31 contestants scored 1 point, and 90 contestants scored 0 point. The average score of this problem is 4.632, indicating that it was simple.

Among the top five teams in the team scores, the scores of this problem are as follows: the Bulgaria team scored 42 points (with a total team score of 227 points), the China team scored 42 points (with a total team score of 211 points), the United States team scored 42 points (with a total team score of 188 points), the Vietnam team scored 42 points (with a total team score of 172 points), and the Russia team scored 42 points (with a total team score of 167 points).

The gold medal cutoff for this IMO was set at 29 points (with 37 contestants earning gold medals), the silver medal cutoff was 19 points (with 69 contestants earning silver medals), and the bronze medal cutoff was 13 points (with 104 contestants earning bronze medals).

In this IMO, only three contestants achieved a perfect score of 42 points, namely Bảo Lê Hùng Việt and Trọng Cảnh Nguyễn from Vietnam, and Yunhao Fu from China.

4.3 Summary

In the first 64 IMOs, there are eight problems on special points and special lines in a triangle as depicted in Figure 4.31, which can be broadly categorized into two types.

Problems 4.1–4.5 focus on "position structure problems;" among these five problems, the one with the lowest average score is Problem 4.3 (IMO 41-6), proposed by Russia. Problems 4.6–4.8 deal with "quantity relation problems;" among these three problems, the one with the lowest average score is Problem 4.6 (IMO 4-6), proposed by the German Democratic Republic (see Figure 4.31).

These problems were proposed by seven countries and regions. Romania proposed two problems.

From Table 4.2, it can be observed that in the first 64 IMOs, there was one problem with average scores of 1–2 points; two problems with average scores of 2–3 points; three problems with average scores of 3–4 points; two

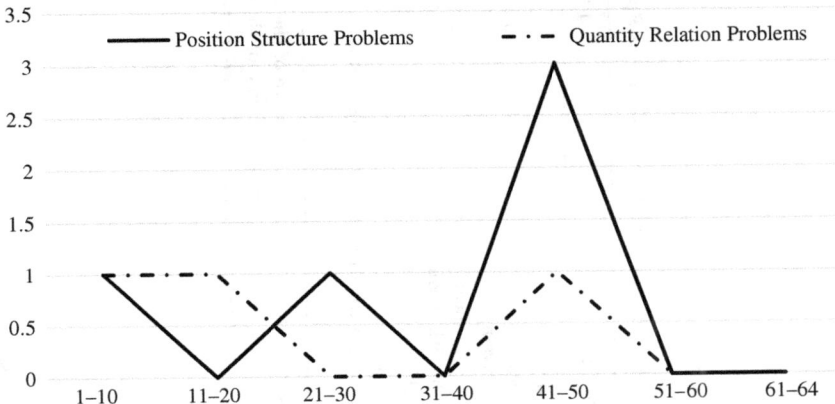

Figure 4.31 Numbers of Special Points and Special Lines Problems in the First 64 IMOs

IMO Problems, Theorems, and Methods: Geometry

Table 4.2 Score Details of Special Points and Special Lines Problems in the First 64 IMOs

Problem	4.1	4.2	4.3	4.4	4.5	4.6	4.7	4.8
Full points	7.000	7.000	7.000	7.000	7.000	6.000	5.000	7.000
Average score	3.788	6.275	1.050	3.557	2.614	2.235	3.486	4.632
Top five mean	5.425	6.525	4.444	6.600	5.500			7.000
6th–15th mean		5.740	2.5	5.75	4.717			6.533
16th–25th mean		4	1.1	5.367	4.4			6.153
Problem number in IMO	7-5	22-5	41-6	43-2	46-1	4-6	12-1	44-4
Proposing country	Romania	Soviet Union	Russia	South Korea	Romania	The German Democratic Republic	Poland	Finland

Note. Top five mean = Total score of the top five teams/Total number of contestants from the top five teams,

6th–15th mean = Total score of the 6th–15th teams/Total number of contestants from the 6th–15th teams,

16th–25th mean = Total score of the 16th–25th teams/Total number of contestants from the 16th–25th teams.

problems with average scores above 4 points. Overall, special points and special lines problems were relatively simple.

In the 24th to 64th IMOs, there were a total of 4 special points and special lines problems. Among these, one had an average score of 1–2 points; one had an average score of 2–3 points; one had an average score of 3–4 points; one had an average score above 4 points. Further analysis of the problem numbers of these 4 special points and special lines problems, as shown in Table 4.3, reveals that the frequency of these problems in each problem number was relatively balanced. The majority of these problems were of the type of position structure problems, with three in total, accounting for three quarters.

Table 4.3 Numbers of Special Points and Special Lines Problems in the 24th–64th IMOs

Special Points and Special Lines Problems	Problem Number			Numbers of Problems in the First 64 IMOs
	1, 4	2, 5	3, 6	
Position structure problems	1	1	1	5
Quantity relation problems	1	0	0	3
Total	2	1	1	8

From Table 4.2, it can be observed that, in most cases, the average score of the top five teams is typically about 1.5–3.5 points higher than the average score of the problem, except for individual problems. However, in the difficult problems, the difference in average scores between the top five teams, 6th–15th teams, and 16th–25th teams is relatively obvious. Especially, the average score of the 6th–15th teams is more than 1 point higher than the average score of 16th–25th teams, as seen in problems such as Problem 4.2 (IMO 22-5) and Problem 4.3 (IMO 41-6).

From Table 4.2, it can also be observed that after the 30th IMO, the average scores of the 6th–15th teams and 16th–25th teams also tend to be higher than the average score of the problem. This phenomenon is due to the smaller number of participating teams in early IMOs. It was not until the 30th IMO in 1989 that the number of participating teams exceeded 50. In addition, the appearance of geometry problems that are not extremely difficult facilitates the contestants score, thus raising the average score of the problems.

Chapter 5

Trigonometry, Areas, and Analytic Geometry

Trigonometric functions are an important tool in mathematics. Solving geometry problems by trigonometry is often called the trigonometric method. Using the trigonometric method can avoid adding some auxiliary lines, which provides a lot of convenience for us to solve geometry problems.

The trigonometric method often needs to use properties of trigonometric functions, such as monotonicity and boundedness, in addition to often use the law of sines, law of cosines, and so on.

Area is an important concept in plane geometry. When dealing with some geometry problems, the method that considers the area as the starting point of calculation or demonstration is called the area method. Area formulas can be used not only to calculate areas or solve area problems, but also to prove geometric propositions that are not obviously related to areas. For example, many proofs of the Pythagorean theorem are based on the idea of "calculating the area of the same figure in different ways." As a basic figure in plane geometry, the area of a triangle has many forms of expressions. In this way, we can not only derive area formulas of other figures, but also obtain some property theorems related to areas, such as equal area deformation, ratio theorem about common side, ratio theorem about common angle, etc., so that segment ratios and area ratios can transform each other.

The method of using analytic geometry to solve plane geometry problems is called the analytic method. The basic spirit of the analytic method is to translate the original problem into an algebraic problem through the coordinates of the figure, and then solve the problem through an algebraic calculation and demonstration.

Compared with the purely geometric method, the analytic method establishes the correlation between objects through more direct calculations, and reduces the difficulty of adding auxiliary lines. In general, when the coordinates of the figure are easy to achieve, the analytic method is expected (but not necessarily) to be better applied. Of course, the degree of difficulty of the algebraic problems formed after the same figure is coordinatized in different ways is usually different.

When establishing a rectangular coordinate system to solve plane geometry problems, the following items should be paid attention to:

(1) If the problem involves a right angle, consider using the two sides of this right angle as coordinate axes.

(2) If the problem involves the figure of a straight line, consider using the analytic method.

(3) If the problem involves a square relation, it is often relatively convenient to use the analytic method.

(4) If the problem involves a circle, it is not appropriate to use the analytic method, especially for problems with multiple circles, so as not to fall into a tedious calculation.

(5) In many cases, a comprehensive use of the analytic method, algebraic calculation, geometric reasoning, trigonometric function, and other methods may lead a better effect.

In the first 64th IMOs, there had been a total of 16 problems related to trigonometry, areas, and analytic geometry, accounting for 13.0% of all geometry problems. These problems are mainly categorized into three types: (1) existence problems, totaling three problems; (2) position structure problems, totaling six problems; (3) quantity relation problems, totaling seven problems. The statistical distribution of these three types of problems in the previous IMOs is presented in Table 5.1.

It can be seen that the problems related to trigonometry, areas, and analytic geometry are mainly concentrated in the first 30 IMOs, and rarely appear after the 30th IMO.

However, position structure problems and quantity relation problems of geometry are still the key content, especially in problems related to trigonometry, areas, and analytic geometry. It is extremely convenient to use algebraic methods to describe the relationship between the position

Table 5.1 Numbers of Trigonometry, Areas and Analytic Geometry Problems in the First 64 IMOs

Content	Session							Total
	1–10	11–20	21–30	31–40	41–50	51–60	61–64	
Existence problems	1	2	0	0	0	0	0	3
Position structure problems	2	2	2	0	0	0	0	6
Quantity relation problems	1	2	2	0	1	1	0	7
Geometry problems	29	18	18	15	20	17	6	123
Percentage of trigonometry, areas, and analytic geometry problems among geometry problems	13.8%	33.3%	22.2%	0.0%	5.0%	5.9%	0.0%	13.0%

structure and quantity relation, and the key is to find a suitable idea for calculation, and then the problem can often be simplified to verify that the left and right sides of an equation are equal.

Therefore, this chapter will be divided into three parts. The first part introduces common methods and theorems in trigonometry, such as the area theorem, ratio theorem about common side, etc., and through some examples to help understand the nature and application of these properties and theorems.

The second part revolves around three types of problems: "existence problems," "position structure problems," and "quantity relation problems." These problems are presented in chronological order, and some problems include various solutions and generalizations.

It is important to note that for each problem, the solutions are followed by information on the scores, including the number of contestants in each score range, the average score, the scores of the top five teams. However, early IMOs often lacked information on contestant scores, so the number of contestants in each score range only represents the counted number of contestants, and some problems lack scores of the top five teams.

The third part provides a brief summary of this chapter.

5.1 Related Properties, Theorems, and Methods

5.1.1 *Trigonometry*

In $\triangle ABC$, denote $A = \angle BAC, B = \angle ABC, C = \angle ACB, a = BC, b = CA$, and $c = AB$. Let R, r be respectively the radii of the circumcircle and the inscribed circle of $\triangle ABC$.

(1) *Area theorem*

$$S_{\triangle ABC} = \frac{1}{2}bc\sin A = \frac{1}{2}ac\sin B = \frac{1}{2}ab\sin C.$$

Note. In addition, the area of a triangle has the following formulas:

(i)

$$S_{\triangle ABC} = \frac{abc}{4R} = 2R^2 \sin A \sin B \sin C.$$

(ii)

$$S_{\triangle ABC} = \sqrt{p(p-a)(p-b)(p-c)} = \frac{1}{4}\sqrt{4a^2b^2 - (a^2 + b^2 - c^2)^2}.$$

(2) *Law of sines*

$$\frac{a}{\sin A} = \frac{b}{\sin B} = \frac{c}{\sin C} = 2R,$$

where R is the radius of the circumcircle of $\triangle ABC$.

(3) *Law of cosines*

$$a^2 = b^2 + c^2 - 2bc\cos A,$$
$$b^2 = a^2 + c^2 - 2ac\cos B,$$
$$c^2 = a^2 + b^2 - 2ab\cos C.$$

(4) *Stewart's theorem*

As shown in Figure 5.1, in $\triangle ABC$, a point P is on the side BC. Then

$$AP^2 = \frac{BP \cdot b^2 + CP \cdot c^2}{a} - BP \cdot CP.$$

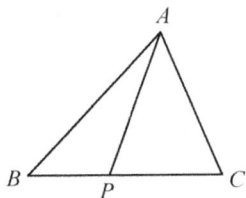

Figure 5.1

(5) *Conclusions related to a triangle*

(i) Radius of the inscribed circle $\odot I$ of $\triangle ABC$:

$$r = 4R \sin\frac{A}{2} \sin\frac{B}{2} \sin\frac{C}{2}.$$

And

$$AI = 4R\sin\frac{B}{2}\sin\frac{C}{2}, \quad BI = 4R\sin\frac{A}{2}\sin\frac{C}{2}, \quad CI = 4R\sin\frac{A}{2}\sin\frac{B}{2}.$$

(ii) Radius of the excircle $\odot I_a$ of $\triangle ABC$ inside $\angle BAC$:

$$r_a = 4R\sin\frac{A}{2}\cos\frac{B}{2}\cos\frac{C}{2}.$$

And

$$AI_a = 4R\cos\frac{B}{2}\cos\frac{C}{2}, \quad BI_a = 4R\sin\frac{A}{2}\cos\frac{C}{2},$$

$$CI_a = 4R\sin\frac{A}{2}\cos\frac{B}{2}.$$

(iii) Let the orthocenter of $\triangle ABC$ be H. Then

$$AH = 2R|\cos A| = a|\cot A|,$$

$$BH = 2R|\cos B| = b|\cot B|,$$

$$CH = 2R|\cos C| = c|\cot C|.$$

5.1.2 *Areas of triangles and area ratios*

(1) *Ratio theorem about a common side*

As shown in Figures 5.2–5.4, suppose $\triangle ABC, \triangle ABD$ have a common side AB, and the lines AB, CD intersect at point K. Then

$$\frac{S_{\triangle ABC}}{S_{\triangle ABD}} = \frac{CK}{DK}.$$

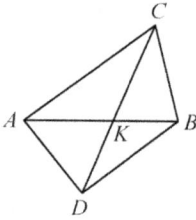

Figure 5.2

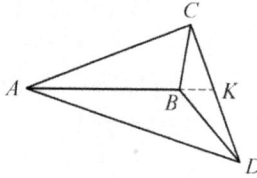

Figure 5.3

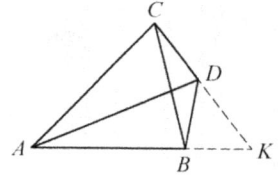

Figure 5.4

Note. The above formula can also be written in the directed form

$$\frac{\overline{S_{\triangle ABC}}}{\overline{S_{\triangle ABD}}} = \frac{\overline{CK}}{\overline{DK}}.$$

(2) *Ratio theorem about a common angle*

As shown in Figures 5.5–5.8, in $\triangle OAB, \triangle OCD$, the points O, A, C are collinear, and the points O, B, D are collinear. Then

$$\frac{S_{\triangle OAB}}{S_{\triangle OCD}} = \frac{OA \cdot OB}{OC \cdot OD}.$$

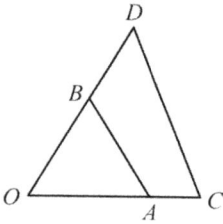

Figure 5.5

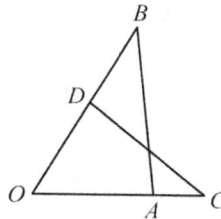

Figure 5.6

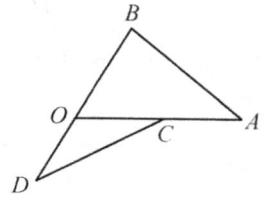

Figure 5.7

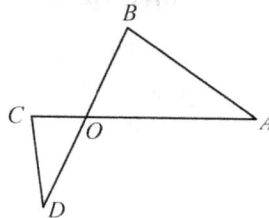

Figure 5.8

5.1.3 *Common trigonometric theorems*

(1) *Split angle theorem*

As shown in Figures 5.9 and 5.10, in $\triangle ABC$, given that P is a point on the line where side BC lies (not coincide with C), there is

$$\frac{BP}{CP} = \frac{AB}{AC} \cdot \frac{\sin \angle BAP}{\sin \angle CAP}.$$

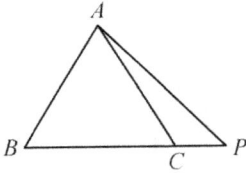

Figure 5.9

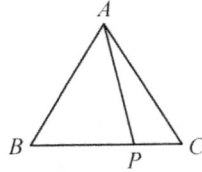

Figure 5.10

Note.

(i) The conclusion can also be written in the directed form:

$$\frac{\overline{BP}}{\overline{PC}} = \frac{AB}{AC} \cdot \frac{\sin \angle BAP}{\sin \angle PAC}.$$

(ii) This conclusion also has the following two variations:

$$\frac{\sin \angle BAP}{\sin \angle CAP} = \frac{BP}{CP} \cdot \frac{AC}{AB},$$

$$\frac{BP}{CP} = \frac{\sin \angle ACB}{\sin \angle ABC} \cdot \frac{\sin \angle BAP}{\sin \angle CAP}.$$

(2) *Spread angle theorem*

Given that P is a point inside $\angle BAC$, a sufficient and necessary condition for the points B, P, C to be collinear is

$$\frac{\sin \angle BAP}{AC} + \frac{\sin \angle CAP}{AB} = \frac{\sin \angle BAC}{AP}.$$

(3) *Three Chords theorem*

Given that P is a point inside $\angle BAC$, a sufficient and necessary condition for P to be on the circumcircle of $\triangle ABC$ is

$$AB \cdot \sin \angle CAP + AC \cdot \sin \angle BAP = AP \cdot \sin \angle BAC.$$

(4)   *The variation of the same method*

Given that $\alpha_1, \beta_1, \alpha_2, \beta_2 > 0$, suppose the following two conditions are satisfied:

(i) $\alpha_1 + \beta_1 = \alpha_2 + \beta_2 < \pi$;

(ii) $\dfrac{\sin \alpha_1}{\sin \beta_1} = \dfrac{\sin \alpha_2}{\sin \beta_2}$.

Then

$$\alpha_1 = \alpha_2, \quad \beta_1 = \beta_2.$$

The following are three conclusions that can be directly derived from the variation of the same method.

Conclusion 5.1. As shown in Figure 5.11, given that points P, Q are inside $\angle AOB(< \pi)$, a sufficient and necessary condition for points O, P, Q to be collinear is:

$$\frac{\sin \angle AOP}{\sin \angle BOP} = \frac{\sin \angle AOQ}{\sin \angle BOQ}.$$

Conclusion 5.2. As shown in Figure 5.12, given that points P, Q are inside $\angle AOB(< \pi)$, a sufficient and necessary condition for rays OP, OQ to be symmetric about the bisector of $\angle AOB$ is:

$$\frac{\sin \angle AOP}{\sin \angle BOP} = \frac{\sin \angle BOQ}{\sin \angle AOQ}.$$

Conclusion 5.3. As shown in Figure 5.13, given that segments AD, BC intersect at point O, and points P, Q are inside a pair of vertical angles $\angle AOB, \angle COD$ respectively, a sufficient and necessary condition for points P, O, Q to be collinear is:

$$\frac{\sin \angle AOP}{\sin \angle BOP} = \frac{\sin \angle DOQ}{\sin \angle COQ}.$$

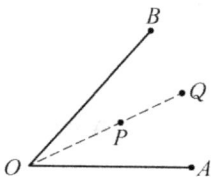

Figure 5.11

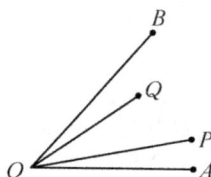

Figure 5.12

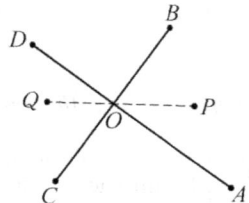

Figure 5.13

Example 5.1. As shown in Figure 5.14, the circumcenter and incenter of $\triangle ABC$ are O, I respectively. Crossing point A, construct the perpendicular line of side BC, and the perpendicular foot is H. Crossing point A, construct the perpendicular line of side OI, which intersects the side BC at point X. Then, crossing point X, construct two perpendicular lines of sides AC, AB, with the perpendicular feet E, F respectively. Prove: $AH = XE + XF$.

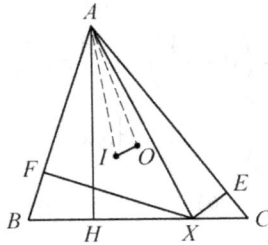

Figure 5.14

Proof. Denote the three internal angles of $\triangle ABC$ as A, B, C, respectively.

Let $\angle BAX = \theta$. Then by the definition of the trigonometric ratio, we only need to prove that

$$\sin \angle AXB = \sin \angle CAX + \sin \angle BAX,$$

that is

$$\sin(B + \theta) = \sin(A - \theta) + \sin \theta. \tag{1}$$

It is not difficult to get

$$\angle AOI = 90° + \angle OAX = 90° + \angle BAX - \angle BAO$$

$$= 90° + \theta - (90° - C) = C + \theta,$$

$$\angle AIO = 90° - \angle IAX = 90° - (\angle BAX - \angle BAI)$$

$$= 90° - \left(\theta - \frac{A}{2}\right),$$

and thus, by the law of sines,

$$\frac{AI}{AO} = \frac{\sin \angle AOI}{\sin \angle AIO} = \frac{\sin(C + \theta)}{\cos\left(\theta - \frac{A}{2}\right)}.$$

And by the properties of incenter and circumcenter,

$$\frac{AI}{AO} = 4\sin\frac{B}{2}\sin\frac{C}{2}.$$

Then

$$\frac{\sin(C+\theta)}{\cos\left(\theta - \frac{A}{2}\right)} = 4\sin\frac{B}{2}\sin\frac{C}{2}.$$

By simplifying and using trigonometric identities, we get

$$\sin(C+\theta) = 4\cos\left(\theta - \frac{A}{2}\right)\sin\frac{B}{2}\sin\frac{C}{2}$$

$$= 2\cos\left(\theta - \frac{A}{2}\right)\cdot\left[\cos\frac{B-C}{2} - \sin\frac{A}{2}\right]$$

$$= \sin(B+\theta) + \sin(\theta+C) - \sin\theta - \sin(A-\theta),$$

which is equivalent to equation (1), so the proposition is proved.

Example 5.2. As shown in Figure 5.15, let the incenter and orthocenter of an acute triangle $ABC(AB \neq AC)$ be I, H, respectively. Points B_1, C_1 are the midpoints of sides AC, AB, respectively. Ray B_1I intersects AB at point B_2, ray C_1I intersects the extension line of AC at point C_2, and side B_2C_2 intersects BC at point K. Let A_1 be the circumcenter of $\triangle BHC$. Prove: A sufficient and necessary condition for points A, I, A_1 to be collinear is that $\triangle BKB_2$ and $\triangle CKC_2$ have equal areas.

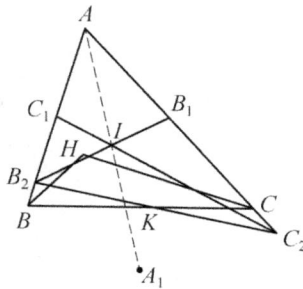

Figure 5.15

Proof. The following proves that both conditions of the conclusion are equivalent to $\angle BAC = \frac{\pi}{3}$.

Denote the three internal angles of $\triangle ABC$ as A, B, C, respectively.

First, we prove that the collinearity of three points A, I, A_1 is equivalent to $\angle BAC = \frac{\pi}{3}$.

If three points A, I, A_1 are collinear, then AA_1 bisects $\angle BAC$. Using the law of sines for $\triangle ABA_1$ and $\triangle ACA_1$, respectively, we get

$$\frac{AA_1}{\sin \angle ABA_1} = \frac{A_1 B}{\sin \angle BAA_1} = \frac{A_1 C}{\sin \angle CAA_1} = \frac{AA_1}{\sin \angle ACA_1},$$

and then $\sin \angle ABA_1 = \sin \angle ACA_1$.

If $\angle ABA_1 = \angle ACA_1$, then $\triangle ABA_1 \cong \triangle ACA_1$, so $AB = AC$, resulting in a contradiction.

Thus $\angle ABA_1 + \angle ACA_1 = \pi$, and so A, B, A_1, C are concyclic. Then

$$\pi = \angle BAC + \angle BA_1 C = A + 2(\pi - \angle BHC) = 3A.$$

That is $A = \frac{\pi}{3}$.

Conversely, if $A = \frac{\pi}{3}$, then A, B, A_1, C are concyclic. And since $A_1 B = A_1 C$, we see that AA_1 bisects $\angle BAC$, so the three points A, I, A_1 are collinear.

Again, it is proved that the areas of $\triangle BKB_2$ and $\triangle CKC_2$ are equal, which is equivalent to $\angle BAC = \frac{\pi}{3}$.

Using the spread angle theorem for $\triangle AB_1 B_2$ and AI, we get

$$\frac{\sin \frac{A}{2}}{AB_1} + \frac{\sin \frac{A}{2}}{AB_2} = \frac{\sin A}{AI} \Rightarrow \frac{2}{AC} + \frac{1}{AB_2} = \frac{2 \cos \frac{A}{2}}{AI}.$$

After simplification,

$$\frac{AC}{AB_2} = 2 \cos \frac{A}{2} \cdot \frac{AC}{AI} - 2 = 2 \cos \frac{A}{2} \cdot \frac{\sin \angle AIC}{\sin \angle ACI} - 2$$

$$= 2 \cos \frac{A}{2} \cdot \frac{\cos \frac{B}{2}}{\sin \frac{C}{2}} - 2 = 2 \cdot \frac{\cos \frac{A}{2} \cos \frac{B}{2} - \sin \frac{C}{2}}{\sin \frac{C}{2}}$$

$$= 2 \cdot \frac{\sin \frac{A}{2} \sin \frac{B}{2}}{\sin \frac{C}{2}}.$$

Similarly,

$$\frac{AB}{AC_2} = 2 \cdot \frac{\sin \frac{A}{2} \sin \frac{C}{2}}{\sin \frac{B}{2}}.$$

Then combing the ratio theorem about a common angle, we get

$$\frac{S_{\triangle ABC}}{S_{\triangle AB_2 C_2}} = \frac{AB \cdot AC}{AB_2 \cdot AC_2} = 4 \sin^2 \frac{A}{2}.$$

Therefore,

$$S_{\triangle BKB_2} = S_{\triangle CKC_2} \Leftrightarrow S_{\triangle ABC} = S_{\triangle AB_2C_2} \Leftrightarrow 4\sin^2\frac{A}{2} = 1 \Leftrightarrow A = \frac{\pi}{3}.$$

To sum up, the proposition holds.

Example 5.3. As shown in Figure 5.16, in a convex quadrilateral $ABCD$, point P satisfies $\angle APB = \angle DPC = 90°$. Construct the circumcircles O_1, O_2 of $\triangle APC$ and $\triangle BPD$, respectively. The connection line of the midpoints E, F of AB, CD passes through point P. The extension line of segment EF intersects with $\odot O_1, \odot O_2$ at points G, H respectively. If K, M are the midpoints of segments PG, PH, respectively. Prove: $KE = MF$.

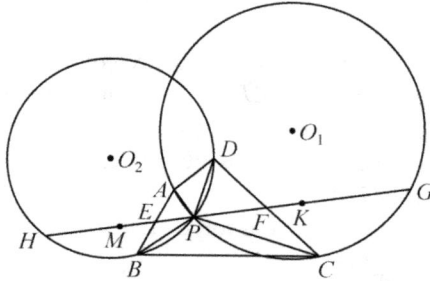

Figure 5.16

Proof. We only need to prove that $PE + \frac{1}{2}PG = PF + \frac{1}{2}PH$, which is equivalent to

$$AB + PG = CD + PH. \qquad (1)$$

Denote $AB = x, CD = y, \angle ABP = \alpha$, and $\angle DCP = \beta$.

For the chords PB, PH, PD in $\odot O_2$, using three chords theorem, we get

$$PB \cdot \sin\angle DPH + PD \cdot \sin\angle BPH = PH \cdot \sin\angle BPD. \qquad (2)$$

It is not difficult to get $PB = x\cos\alpha, PD = y\sin\beta, \angle BPH = \angle ABP = \alpha,$

$$\angle APD = \pi - \angle BPC = \angle BPE + \angle CPF = \alpha + \beta,$$

$$\angle BPD = \frac{\pi}{2} + \angle APD = \frac{\pi}{2} + \alpha + \beta,$$

$$\angle DPH = \angle APD + \angle APE = (\alpha + \beta) + \left(\frac{\pi}{2} - \alpha\right) = \frac{\pi}{2} + \beta.$$

Therefore, equality (2) becomes

$$(x\cos\alpha)\cdot\sin\left(\frac{\pi}{2}+\beta\right)+(y\sin\beta)\cdot\sin\alpha = PH\cdot\sin\left(\frac{\pi}{2}+\alpha+\beta\right),$$

that is

$$(x\cos\alpha)\cdot\cos\beta+(y\sin\beta)\cdot\sin\alpha = PH\cdot\cos(\alpha+\beta).$$

Then

$$PH = \frac{x\cos\alpha\cos\beta+y\sin\alpha\sin\beta}{\cos(\alpha+\beta)}.$$

Similarly

$$PG = \frac{y\cos\alpha\cos\beta+x\sin\alpha\sin\beta}{\cos(\alpha+\beta)}.$$

Thus,

$$PH - PG = \frac{(x-y)(\cos\alpha\cos\beta-\sin\alpha\sin\beta)}{\cos(\alpha+\beta)}$$

$$= x - y = AB - CD.$$

Hence equality (1) holds, and then the conclusion is proved.

Example 5.4. As shown in Figure 5.17, let $ABCD$ be an inscribed quadrilateral, whose circumcircle is Γ, where $AB = AD$, and let E be a point on the segment CD such that $BC = DE$. The extension line of AE intersects the circle Γ at point F. The chords AC and BF intersect at point M. Then let P be the symmetric point of C with respect to point M. Prove: PE and BF are parallel.

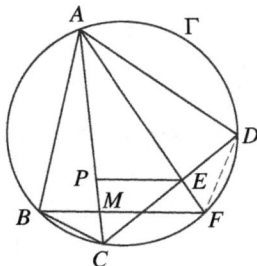

Figure 5.17

Proof. We only need to prove that $\frac{AM}{MP} = \frac{AF}{FE}$. Combined with the known conditions, it is necessary to prove that

$$\frac{AM}{MC} = \frac{AF}{FE}. \tag{1}$$

By the area ratio and the law of sines,

$$\frac{AM}{MC} = \frac{S_{\triangle ABM}}{S_{\triangle CBM}} = \frac{AB}{BC} \cdot \frac{\sin \angle ABM}{\sin \angle CBM}, \tag{2}$$

$$\frac{AF}{FE} = \frac{S_{\triangle ADF}}{S_{\triangle DEF}} = \frac{AD}{DE} \cdot \frac{\sin \angle ADF}{\sin \angle EDF}. \tag{3}$$

Note that $AB = AD$, $BC = DE$, $\angle ABM = \pi - \angle ADF$, and $\angle CBM = \angle EDF$, so the right sides of (2), (3) are equal, and then (1) holds. Hence $PE \parallel BF$.

Example 5.5. As shown in Figure 5.18, $\triangle ABC$ is inscribed in circle O. Point I is the incenter of $\triangle ABC$. The segments BI and AC intersect at point E, and the segments CI and AB intersect at point F. Let D be the midpoint of the arc BC of $\odot O$, and G be the symmetric point of O with respect to point D. Prove: $IG \perp EF$.

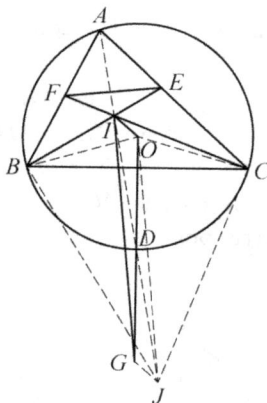

Figure 5.18

Proof. Let the excenter of $\triangle ABC$ within $\angle BAC$ be J. Then A, I, D, J are collinear, $JB \perp BE$, $JC \perp CF$, and D is the midpoint of IJ.

By the known conditions, D is the midpoint of OG. Therefore, the quadrilateral $IGJO$ is a parallelogram, and then $JO \parallel GI$.

So, we only need to prove that $JO \perp EF$.

Obviously, I, B, J, C are concyclic. Then

$$\angle BJO + \angle CJO = \angle BJC = 180° - \angle BIC = \angle IEF + \angle IFE,$$

that is

$$\angle BJO + \angle CJO = \angle IEF + \angle IFE. \tag{1}$$

And

$$\angle OBJ = \angle OBC + \angle JBC = (90° - \angle BAC) + \left(90° - \frac{1}{2}\angle ABC\right)$$

$$= \angle ACB + \frac{1}{2}\angle ABC = \angle AEI.$$

Similarly

$$\angle OCJ = \angle AFI.$$

Then

$$\frac{\sin \angle BJO}{\sin \angle CJO} = \frac{OB \cdot \sin \angle OBJ}{OC \cdot \sin \angle OCJ} = \frac{\sin \angle OBJ}{\sin \angle OCJ},$$

$$\frac{\sin \angle IEF}{\sin \angle IFE} = \frac{IF}{IE} = \frac{IF}{AI} \cdot \frac{AI}{IE} = \frac{\sin \angle IAF}{\sin \angle AFI} \cdot \frac{\sin \angle AEI}{\sin \angle IAE}$$

$$= \frac{\sin \angle AEI}{\sin \angle AFI}.$$

Thus

$$\frac{\sin \angle BJO}{\sin \angle CJO} = \frac{\sin \angle IEF}{\sin \angle IFE}. \tag{2}$$

Combining (1) and (2), we know

$$\angle BJO = \angle IEF, \quad \angle CJO = \angle IFE.$$

And $BI \perp BJ$. Therefore $JO \perp EF$, that is $GI \perp EF$.

5.2 Problems and Solutions

5.2.1 *Existence problems*

Problem 5.1 (IMO 1-4, proposed by Hungary). Construct a right triangle with given hypotenuse c such that the median drawn to the hypotenuse is the geometric mean of the two legs of the triangle.

Solution 1. Let the constructed right triangle be $\triangle ABC$, where AB is the hypotenuse and the length is c. As shown in Figure 5.19, denote $AC = b, BC = a$ respectively. And let the altitude on the side of AB be h. Then

$$ab = ch.$$

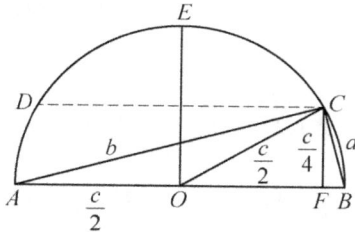

Figure 5.19

By the conditions, $ab = \left(\frac{c}{2}\right)^2$, so

$$h = \frac{c}{4}.$$

Therefore, a specific construction method is as follows:

First, construct a semicircle with AB as its diameter, and then construct a parallel line of AB, such that the distance between it and AB is $\frac{c}{4}$, and intersects the semicircle at two points C, D. Then $\triangle ABC$ is the desired triangle.

Solution 2. Denote the midpoint of AB as O, and let $\angle COB = \theta$. Then

$$S_{\triangle ABC} = \frac{1}{2} \cdot c \cdot \frac{c}{2} \cdot \sin\theta = \frac{c^2}{4}\sin\theta = \frac{1}{2}ab = \frac{c^2}{8}.$$

Thus $\sin\theta = \frac{1}{2}$, and $\theta = \frac{\pi}{6}$ or $\frac{5}{6}\pi$. Hence, it is not difficult to construct $\triangle ABC$.

【Score Situation】 This particular problem saw the following distribution of scores among contestants: 3 contestants scored 5 points, no contestant scored 4 points, no contestant scored 3 points, no contestant scored 2 points, 4 contestants scored 1 point, and 2 contestants scored 0 point. The average score of this problem is 2.111, indicating that it had a certain level of difficulty.

Among the top five teams in the team scores, the Romania team achieved a total score of 249 points, the Hungary team achieved a total score of 233 points, the Czechoslovakia team achieved a total score of 192 points, the Bulgaria team achieved a total score of 131 points, and the Poland team achieved a total score of 122 points.

The gold medal cutoff for this IMO was set at 37 points (with 3 contestants earning gold medals), the silver medal cutoff was 36 points (with 3 contestants earning silver medals), and the bronze medal cutoff was 33 points (with 5 contestants earning bronze medals).

In this IMO, only one contestant achieved a perfect score of 40 points, namely Bohuslav Diviš from Czechoslovakia.

Problem 5.2 (IMO 15-4, proposed by Yugoslavia). A soldier needs to check on the presence of mines in a region having the shape of an equilateral triangle. The radius of action of his detector is equal to half the altitude of the triangle. The soldier leaves from one vertex of the triangle. What path should he follow in order to travel the least possible distance and still accomplish his mission?.

Solution 1. Suppose that the soldier starts from vertex A and probes the equilateral triangle region. The altitude of $\triangle ABC$ is $2d$. The circle constructed with B as the center and d as the radius intersects AB, BC at M, N, respectively. The circle constructed with C as the center and d as the radius intersects AC, BC at P, Q, respectively, as shown in Figure 5.20.

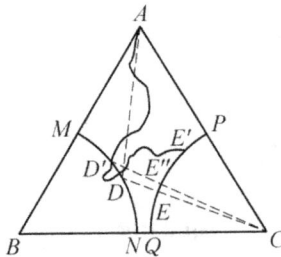

Figure 5.20

By the problem conditions, the soldier reaches $\widehat{MN}$ and $\widehat{PQ}$ at least once each. Might as well suppose he first arrived at a point D' of the $\widehat{MN}$, after arriving on a point E' of the $\widehat{PQ}$. Let D be the midpoint of $\widehat{MN}$. Join to form AD, and let CD, CD' intersect PQ at points E, E'', respectively.

It is easy to see that the length of the path S traveled by the soldier

$$S \geq AD' + D'E' \geq AD' + D'E'' \geq AD + DE.$$

In the above, the last part is because $AD + DC \leq AD' + D'C$. A brief explanation is as follows: construct the tangent l through point D parallel to AC. It intersects AB, BC at S, T respectively, and then $\angle ADS = \angle CDT$. By symmetry, D is a point on l that minimizes the sum of the distances to two points A, C. It is easy to see that D' and C are on different sides of l. Let CD' intersect l at point R. Then

$$AD + DC \leq AR + RC \leq AD' + D'R + RC = AD' + D'C.$$

Since the distance between point D and AC is d, the distance between any point on AD and AC is not greater than d. It is easy to know that the perpendicular line segment ($\leq d$) from the moving point on the fold line ADE to AC sweeps through the quadrilateral $ADEF(EF \perp AC, F$ is the foot of E on AC, inside the sector C). As for the region inside $\angle BAD$ and outside $\widehat{MN}$, and the region below DE and outside the two sectors, it is easy to prove that at any point G in the above two regions, point H must be found on AD or DE, satisfying $GH < d$. Thus ADE is the shortest path, and the other shortest path is to go first to $\widehat{PQ}$ and then to $\widehat{MN}$.

Notice that $\angle ABD = \angle CBD = 30°$, and by the law of cosines, the shortest path ADE has the length

$$2\sqrt{\left(\frac{4}{\sqrt{3}}d\right)^2 + d^2 - \frac{8}{\sqrt{3}}d^2 \cos 30° } - d$$

$$= \left(2\sqrt{\frac{7}{3}} - 1\right) d$$

$$= \frac{\sqrt{3}}{4}\left(2\sqrt{\frac{7}{3}} - 1\right) AB = \left(\frac{\sqrt{7}}{2} - \frac{\sqrt{3}}{4}\right) AB.$$

Solution 2. Let $\triangle ABC$ be an equilateral triangle with side length 2. Then $h = \sqrt{3}$.

Suppose the soldier starts from point A, and let Γ_B, Γ_C be two circular arcs with B, C as the centers, $\frac{\sqrt{3}}{2}$ as the radius, respectively, and lie in triangle ABC, as shown in Figure 5.21. In order to be able to check B, C, the path of the soldier must contain a point M on Γ_B and a point N on Γ_C. Such the shortest path is obtained from segments AM and MC, where M is a point on Γ_B, so that we only need to minimize the path AMC when M is a moving point on Γ_B.

Let M_0 be the intersection point of Γ_B and the altitude $BD(D \in AC)$, and we prove that the shortest path is AM_0C. In fact, if M_0 is on a line t

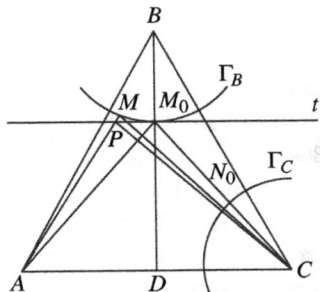

Figure 5.21

tangent to Γ_B at the point M_0, then AM_0C is the shortest path from A to C. For any $M \in \Gamma_B$, we prove that

$$AM + MC > AM_0 + M_0C.$$

Let AM and t intersect at point P. Then by the property of point M_0,

$$AM + MC = AP + PM + MC$$

$$> AP + PC > AM_0 + M_0C.$$

Let N_0 be the intersection point of M_0C and Γ_C, so the shortest path from point A to points B, C is AM_0N_0.

Now, we show that in this way, we can cover the triangle with circles of radius $\frac{\sqrt{3}}{2}$.

The circle centered at M_0 covers the point D. We just need to prove that $N_0D < \frac{\sqrt{3}}{2}$.

In triangle M_0DC, we can get the following segment lengths:

$$M_0C = \frac{\sqrt{7}}{2}, \quad M_0N_0 = \frac{\sqrt{7} - \sqrt{3}}{2}.$$

Applying Stewart's theorem to the segment DN_0, we get

$$DN_0^2 \cdot M_0C = DM_0^2 \cdot N_0C + DC^2 \cdot M_0N_0 - M_0N_0 \cdot N_0C \cdot M_0C.$$

After the calculation, we have

$$DN_0^2 = \frac{7}{4} - 2\frac{\sqrt{3}}{\sqrt{7}},$$

$$DN_0^2 < \frac{3}{4} \Leftrightarrow \frac{7}{4} - 2\frac{\sqrt{3}}{\sqrt{7}} < \frac{3}{4} \Leftrightarrow 1 < 2\frac{\sqrt{3}}{\sqrt{7}} \Leftrightarrow \sqrt{7} < 2\sqrt{3},$$

so, the conclusion is proved.

【Score Situation】 This particular problem saw the following distribution of scores among contestants: 38 contestants scored 6 points, 22 contestants scored 5 points, 20 contestants scored 4 points, 13 contestants scored 3 points, 9 contestants 2 points, 8 contestants scored 1 point, and 15 contestants scored 0 point. The average score of this problem is 3.864, indicating that it was relatively straightforward.

Among the top five teams in the team scores, the scores of this problem are as follows: the Soviet Union team scored 47 points (with a total team score of 254 points), the Hungary team scored 44 points (with a total team score of 215 points), the German Democratic Republic team scored 46 points (with a total team score of 188 points), the Poland team scored 37 points (with a total team score of 174 points), and the United Kingdom team scored 30 points (with a total team score of 164 points).

The gold medal cutoff for this IMO was set at 35 points (with 5 contestants earning gold medals), the silver medal cutoff was 27 points (with 15 contestants earning silver medals), and the bronze medal cutoff was 17 points (with 48 contestants earning bronze medals).

In this IMO, only one contestant achieved a perfect score of 40 points, namely Sergei Konyagin from the Soviet Union.

Problem 5.3 (IMO 16-2, proposed by Finland). In a triangle ABC, prove that there is a point D on side AB such that CD is the geometric mean of AD and DB if and only if $\sin A \sin B \le \sin^2 \frac{C}{2}$.

Proof 1.   As shown in Figure 5.22, construct the circumcircle of $\triangle ABC$ and extend CD to intersect the circle at point M. By the intersecting chords theorem, $CD \cdot MD = AD \cdot BD$.

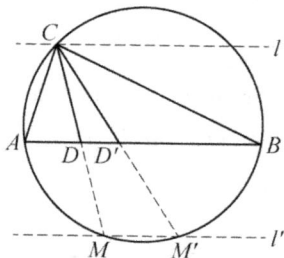

Figure 5.22

At the same time, if the condition $CD^2 = AD \cdot BD$ exists, then $CD = MD$, namely D is the midpoint of CM.

Construct lines l, l' through C, M that are parallel to AB. Then l, l' are symmetric with respect to AB. A sufficient and necessary condition for the existence of D is that l' has an intersection with the circumcircle.

Let P be the midpoint of $\overparen{MM'}$. Then the above sufficient and necessary condition becomes: the distance from point P to $AB \geq$ the distance from point C to AB, and this condition can also be converted to

$$S_{\triangle ABP} \geq S_{\triangle ABC}. \tag{1}$$

The radius of the circumcircle is denoted as R, let $\angle A = \alpha$, $\angle B = \beta$, and $\angle C = \gamma$. Then

$$S_{\triangle ABC} = \frac{1}{2} AC \cdot BC \cdot \sin C = 2R^2 \sin \alpha \sin \beta \sin \gamma.$$

And

$$\angle APB = 180° - \angle C = 180° - \gamma,$$

$$AP = BP = 2R \sin \frac{\gamma}{2}.$$

Thus

$$S_{\triangle ABP} = \frac{1}{2} AP \cdot BP \cdot \sin \angle APB$$

$$= 2R^2 \sin^2 \frac{\gamma}{2} \sin \gamma.$$

Therefore, (1) becomes

$$\sin \alpha \sin \beta \leq \sin^2 \frac{\gamma}{2}.$$

Note that the equality holds if and only if l' is tangent to the circumcircle, in which case there is only one point D on AB such that $CD^2 = AD \cdot BD$.

Proof 2. As shown in Figure 5.23, let $\angle ACD = \gamma_1$ and $\angle BCD = \gamma_2$.

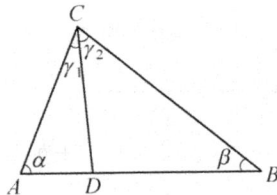

Figure 5.23

By the law of sines,

$$\frac{CD}{AD} = \frac{\sin \alpha}{\sin \gamma_1}, \quad \frac{CD}{BD} = \frac{\sin \beta}{\sin \gamma_2}.$$

Therefore, $CD^2 = AD \cdot BD$ is equivalent to

$$\sin \gamma_1 \sin \gamma_2 = \sin \alpha \sin \beta.$$

Denote $f(x) = \sin x \sin(\gamma - x), 0 < x < \gamma$.
By a trigonometric formula,

$$f(x) = \frac{1}{2} \cos(2x - \gamma) - \frac{1}{2} \cos \gamma \le \frac{1}{2}(1 - \cos \gamma) = \sin^2 \frac{\gamma}{2},$$

which is

$$\sin \alpha \sin \beta \le \sin^2 \frac{\gamma}{2}. \tag{2}$$

From $-\gamma < 2x - \gamma < \gamma$ and the continuity of the cosine function, the range of $f(x)$ is $\left(0, \sin^2 \frac{\gamma}{2}\right)$. As long as (2) is satisfied, there must exist $x \in (0, \gamma)$, (and since the cosine function is even, there are two solutions in general) such that

$$\sin x \sin(\gamma - x) = \sin \alpha \sin \beta.$$

Let $x = \gamma_1, \gamma - x = \gamma_2$. We get the position of point D.

When $\sin \alpha \sin \beta = \sin^2 \frac{\gamma}{2}$, there can only be $\cos(2x - \gamma) = 1$, that is, $x = \frac{\gamma}{2} = \gamma_1 = \gamma_2$. Then there is only one point D, and AD is the bisector of $\angle A$.

Proof 3. From some common formulas for triangles, the known condition can be converted to:

$$\sin A \sin B \le \sin^2 \frac{C}{2} \Leftrightarrow \frac{ab}{4R^2} \le \frac{(p-a)(p-b)}{ab}$$

$$\Leftrightarrow \frac{16S^2 \cdot ab}{4a^2b^2c^2} \le \frac{(p-a)(p-b)}{ab}$$

$$\Leftrightarrow \frac{4p(p-a)(p-b)(p-c)}{c^2} \le (p-a)(p-b)$$

$$\Leftrightarrow 4p(p-c) \le c^2$$

$$\Leftrightarrow 4 \cdot \frac{a+b+c}{2} \cdot \frac{a+b-c}{2} \le c^2$$

$$\Leftrightarrow (a+b)^2 \le 2c^2 \Leftrightarrow a+b \le \sqrt{2}c.$$

In the following, we convert the geometric condition. As shown in Figure 5.23, let D be a point on AB such that $AD = x, BD = c - x$, where $0 < x < c$. For $\triangle ABC$ and segment CD, by Stewart's theorem,

$$CD^2 \cdot AB = CA^2 \cdot BD + CB^2 \cdot AD - AD \cdot BD \cdot AB.$$

Using the known condition

$$CD^2 = x(c - x),$$

we see that there exists such a point D if and only if there exists x with $0 < x < c$, such that

$$f(x) = 2cx^2 - (2c^2 + b^2 - a^2)x + b^2c = 0.$$

This is a quadratic equation with the discriminant

$$\begin{aligned} \Delta &= (2c^2 + b^2 - a^2)^2 - 8b^2c^2 \\ &= (2c^2 + b^2 - a^2 + 2\sqrt{2}bc)(2c^2 + b^2 - a^2 - 2\sqrt{2}bc) \\ &= [(\sqrt{2}c + b)^2 - a^2][(\sqrt{2}c - b)^2 - a^2] \\ &= (b + \sqrt{2}c + a)(b + \sqrt{2}c - a)(\sqrt{2}c + a - b)(\sqrt{2}c - b - a), \end{aligned}$$

and the discriminant is positive if and only if

$$\sqrt{2}c > a + b.$$

Thus, the equation $f(x) = 0$ has two different real roots if and only if $a + b < \sqrt{2}c$.

By the inequality $a < \sqrt{2}c - b$, we get

$$x_1 + x_2 = \frac{2c^2 + b^2 - a^2}{2c} > 0,$$

that is, if the equation $f(x) = 0$ has two different real roots, then both real roots must be positive.

And since $f(0) = b^2c > 0$, $f(c) = a^2c > 0$, we need to prove that $0 < x < c$. It is simply to prove that

$$\frac{x_1 + x_2}{2} < c,$$

which is

$$2c^2 + a^2 > b^2.$$

By $b < \sqrt{2}c - a$, the above inequality clearly holds.

In summary, both known conditions are equivalent to $a + b \leq \sqrt{2}c$, so the two known conditions are equivalent.

【Score Situation】 This particular problem saw the following distribution of scores among contestants: 64 contestants scored 6 points, 7 contestants scored 5 points, 18 contestants scored 4 points, 4 contestants scored 3 points, 5 contestants scored 2 points, 14 contestants scored 1 point, and 28 contestants scored 0 point. The average score of this problem is 3.764, indicating that it was relatively straightforward.

Among the top five teams in the team scores, the scores of this problem are as follows: the Soviet Union team scored 46 points (with a total team score of 256 points), the United States team scored 36 points (with a total team score of 243 points), the Hungary team scored 44 points (with a total team score of 237 points), the German Democratic Republic team scored 39 points (with a total team score of 236 points), and the Yugoslavia team scored 38 points (with a total team score of 216 points).

The gold medal cutoff for this IMO was set at 38 points (with 10 contestants earning gold medals), the silver medal cutoff was 30 points (with 24 contestants earning silver medals), and the bronze medal cutoff was 23 points (with 37 contestants earning bronze medals).

In this IMO, a total of six contestants achieved a perfect score of 40 points.

5.2.2 *Position structure problems*

Problem 5.4 (IMO 1-5, proposed by Romania). An arbitrary point M is selected in the interior of a segment AB. Two squares $AMCD$ and $MBEF$ are constructed on the same side of AB, with the segments AM and BM as their respective bases. The circles circumscribed about these squares, with centres P and Q, intersect at M and also at another point N. Let N' denote the point of intersection of the straight lines AF, BC.

(a) Prove that the points N and N' coincide.
(b) Prove that the straight lines MN pass through a fixed point S independent of the choice of M.
(c) Find the locus of the midpoints of the segments PQ as M varies between A and B.

Solution 1. (a) As shown in Figure 5.24, connect to form AN, NF, BC, and CN.

It is easy to know that $\angle ANM = 45°, \angle MNF = 135°$, so

$$\angle ANM + \angle MNF = 45° + 135° = 180°,$$

and thus, the three points A, N, F are collinear.

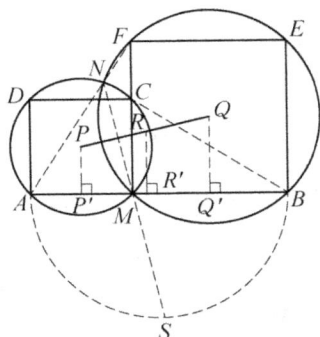

Figure 5.24

The segment AC is a diameter, so $\angle ANC = 90°$, and

$$\angle ANB = \angle ANM + \angle MNB = 45° + 45° = 90°,$$

so, the three points N, C, B are collinear.

(b) Construct a circle with AB as its diameter, and extend NM to intersect this circle at S (it is easy to know that N is on the circle). Since $\angle ANM = \angle MNB = 45°$, we see that S is the midpoint of the half circle $\overparen{ASB}$. Since $\overparen{ASB}$ is a fixed arc, S is necessarily a fixed point.

(c) Let the centers of the squares $AMCD, MBEF$ be P, Q, respectively, the midpoint of PQ be R, and the projections of P, Q, R on AB be P', Q', R' respectively.

It is easy to know

$$PP' = \frac{1}{2}AM,$$

$$QQ' = \frac{1}{2}MB,$$

so

$$RR' = \frac{1}{2}(PP' + QQ') = \frac{1}{4}AB.$$

This means that the distance from point R to AB is constant. When $M = A$, we have $AQ' = \frac{1}{2}AB$ and $AR' = \frac{1}{4}AB$. Then the locus of point R is a segment of length $\frac{AB}{2}$, parallel to AB and at a distance of $\frac{AB}{4}$ from AB, and the projection of its midpoint onto AB is the midpoint of AB.

Solution 2. We will omit the circumcircle of the square in the figure. Let AF and BC intersect at point N. We prove that point N lies on both circumcircles. For this, it is only necessary to show that AN is perpendicular to BN.

Denote $AB = a$, $AM = x$, $\angle FAM = \alpha$, and $\angle MBN = \beta$. Then $MB = MF = a - x$, $\tan\alpha = \frac{a-x}{x}$, and $\tan\beta = \frac{x}{a-x}$. Hence, $\tan\alpha \cdot \tan\beta = 1$, so $\alpha + \beta = \frac{\pi}{2}$. Then

$$\angle ANB = \frac{\pi}{2}.$$

Let S be the midpoint of segment AB and the perpendicular bisector of segment AB intersect MN at point T.

The angle formed by MN and AB is

$$\angle SMT = \angle BMN = \angle MAN + \angle ANM = \alpha + \frac{\pi}{4},$$

so

$$\tan\angle SMT = \tan\left(\alpha + \frac{\pi}{4}\right) = \frac{a}{2x-a}.$$

On the other hand,

$$\tan\angle SMT = \frac{ST}{SM} = \frac{ST}{AM - AS}$$

$$= \frac{ST}{x - \frac{a}{2}} = \frac{2ST}{2x-a},$$

and then we get $ST = \frac{a}{2}$. Therefore T is a fixed point (see Figure 5.25).

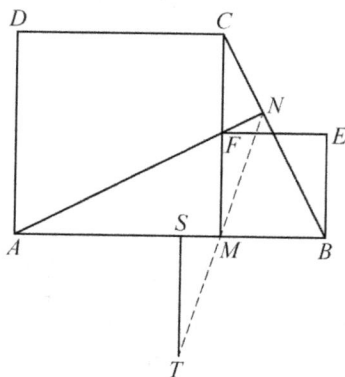

Figure 5.25

Problem (c) is not related to problems (a) and (b).

Let G be the midpoint of segment PQ, and P', Q', G' are the vertical projections of P, Q, G on the segment AB, respectively. Obviously, GG' is the midsegment of a trapezoid $PQQ'P'$, so $GG' = \frac{a}{4}$. Therefore, the locus of the midpoint of the segment PQ is a segment parallel to AB at a distance of $\frac{a}{4}$.

Solution 3. A Cartesian coordinate system is established with A as the origin and AB as the x-axis. Let $AB = a$, $AM = m(0 < m < a)$. Then the coordinates of points A, B, M, C, F are, respectively:

$$A(0,0), \quad B(a,0), \quad M(m,0), \quad C(m,m), \quad F(m, a-m).$$

The coordinates of the centers O_1, O_2 of the squares $AMCD$ and $MBEF$ are, respectively:

$$O_1\left(\frac{m}{2}, \frac{m}{2}\right), \quad O_2\left(\frac{a+m}{2}, \frac{a-m}{2}\right).$$

The equation for the circumcircle $\odot O_1$ of square $AMCD$ is

$$\left(x - \frac{m}{2}\right)^2 + \left(y - \frac{m}{2}\right)^2 = \left(\frac{\sqrt{2}}{2}m\right)^2,$$

which is

$$x^2 - mx + y^2 - my = 0. \tag{1}$$

The equation for the circumcircle $\odot O_2$ of square $MBEF$ is

$$\left(x - \frac{a+m}{2}\right)^2 + \left(y - \frac{a-m}{2}\right)^2 = \left[\frac{\sqrt{2}}{2}(a-m)\right]^2,$$

which is

$$x^2 - (a+m)x + y^2 - (a-m)y + am = 0. \tag{2}$$

The points M, N are the two intersection points of $\odot O_1$ and $\odot O_2$, and their coordinates are the solutions of the system of equations composed of equations (1) and (2). Therefore, the intersection of the two circles, in addition to point $M(m,0)$, is

$$N\left(\frac{am^2}{a^2 - 2am + 2m^2}, \frac{am(a-m)}{a^2 - 2am + 2m^2}\right).$$

(a) The equation of the line AF passing through $A(0,0)$ and $F(m, a-m)$ is

$$(a-m)x - my = 0. \tag{3}$$

The equation of the line BC passing through $B(a,0)$ and $C(m,m)$ is

$$mx + (a-m)y - am = 0. \tag{4}$$

It is not difficult to verify that the coordinates of point N satisfy both (3) and (4), that is, the line AF intersects BC at point N.

(b) The equation of the line MN passing through $M(m,0)$ and $N\left(\dfrac{am^2}{a^2-2am+2m^2}, \dfrac{am(a-m)}{a^2-2am+2m^2}\right)$ is

$$a(x+y) = m(2y+a). \tag{5}$$

Obviously, for any value of m, the numbers $x = \frac{a}{2}$, $y = -\frac{a}{2}$ satisfy the equation (5), that is, no matter where M is on the segment AB, point $\left(\frac{a}{2}, -\frac{a}{2}\right)$ is always on the line MN, so the line MN always passes through the same point $\left(\frac{a}{2}, -\frac{a}{2}\right)$.

(c) Since the coordinates of the centers O_1, O_2 of the two squares are, respectively,

$$O_1\left(\frac{m}{2}, \frac{m}{2}\right), \quad O_2\left(\frac{a+m}{2}, \frac{a-m}{2}\right),$$

so, the coordinates of the midpoint $S(x', y')$ of segment O_1O_2 are

$$x' = \frac{1}{2}\left(\frac{m}{2} + \frac{a+m}{2}\right) = \frac{a+2m}{4}, \quad y' = \frac{1}{2}\left(\frac{m}{2} + \frac{a-m}{2}\right) = \frac{a}{4},$$

that is, the coordinates of the midpoint are $S\left(\frac{a+2m}{4}, \frac{a}{4}\right)$.

It follows that for any m value, the ordinate of the midpoint S of O_1O_2 is always a constant value of $\frac{a}{4}$, that is, point S is always on the line $y = \frac{a}{4}$. Since point M is on segment AB, that is $0 < m < a$, in this case $\frac{a}{4} < \frac{a+2m}{4} < \frac{3a}{4}$, that is $\frac{a}{4} < x' < \frac{3a}{4}$.

When M moves continuously from point A to point B, the value of m increases continuously from 0 to a, and the value of $x' = \frac{a+2m}{4}$ increases continuously from $\frac{a}{4}$ to $\frac{3a}{4}$. It follows that when point M is moving on segment AB, the locus of the midpoint of the connecting line between the centers of the two squares is a segment parallel to AB, and their two endpoints are $G\left(\frac{a}{4}, \frac{a}{4}\right), K\left(\frac{3a}{4}, \frac{a}{4}\right)$.

【Score Situation】 This particular problem saw the following distribution of scores among contestants: 1 contestant scored 8 points, no contestant scored 7 points, no contestant scored 6 points, no contestant scored 5 points, no contestant scored 4 points, no contestant scored 3 points, 2 contestants scored 2 points, 1 contestant scored 1 point, and 5 contestants scored 0 point. The average score of this problem is 1.444, indicating that it was relatively challenging.

Among the top five teams in the team scores, the Romania team achieved a total score of 249 points, the Hungary team achieved a total score of 233 points, the Czechoslovakia team achieved a total score of 192 points, the Bulgaria team achieved a total score of 131 points, and the Poland team achieved a total score of 122 points.

The gold medal cutoff for this IMO was set at 37 points (with 3 contestants earning gold medals), the silver medal cutoff was 36 points (with 3 contestants earning silver medals), and the bronze medal cutoff was 33 points (with 5 contestants earning bronze medals).

In this IMO, only one contestant achieved a perfect score of 40 points, namely Bohuslav Diviš from Czechoslovakia.

Problem 5.5 (IMO 8-2, proposed by Hungary). Let a, b, c be the lengths of the sides of a triangle, and α, β, γ, respectively, the angles opposite these sides.

Prove that if $a+b = \tan \frac{\gamma}{2}(a \tan \alpha + b \tan \beta)$, then the triangle is isosceles.

Proof. The equality given in the question is equivalent to

$$\sin \alpha + \sin \beta = \frac{\cos \frac{\alpha+\beta}{2}}{\sin \frac{\alpha+\beta}{2}} \left(\frac{\sin^2 \alpha}{\cos \alpha} + \frac{\sin^2 \beta}{\cos \beta} \right).$$

After simplification,

$$\sin \frac{\alpha + \beta}{2} \cos \alpha \cos \beta (\sin \alpha + \sin \beta) = \cos \frac{\alpha + \beta}{2} \left(\sin^2 \alpha \cos \beta + \sin^2 \beta \cos \alpha \right),$$

$$\sin \alpha \cos \beta \left(\sin \frac{\alpha + \beta}{2} \cos \alpha - \cos \frac{\alpha + \beta}{2} \sin \alpha \right)$$

$$= \sin \beta \cos \alpha \left(\cos \frac{\alpha + \beta}{2} \sin \beta - \sin \frac{\alpha + \beta}{2} \cos \beta \right),$$

$$\sin \alpha \cos \beta \sin \frac{\beta - \alpha}{2} = \sin \beta \cos \alpha \sin \frac{\beta - \alpha}{2}.$$

Thus, we get $\sin \frac{\beta-\alpha}{2} = 0$ or $\sin(\alpha - \beta) = 0$, and then $\alpha = \beta$.

【Score Situation】This particular problem saw the following distribution of scores among contestants: 20 contestants scored 7 points, 3 contestants scored 6 points, no contestant scored 5 points, 1 contestant scored 4 points, no contestant scored 3 points, no contestant scored 2 points, 1 contestant scored 1 point, and 2 contestants scored 0 point. The average score of this problem is 6.037, indicating that it was simple.

Among the top five teams in the team scores, the Soviet Union team achieved a total score of 293 points, the Hungary team achieved a total score of 281 points, the German Democratic Republic team achieved a total score of 280 points, the Poland team achieved a total score of 269 points, and the Romania team achieved a total score of 257 points.

The gold medal cutoff for this IMO was set at 39 points (with 13 contestants earning gold medals), the silver medal cutoff was 34 points (with 15 contestants earning silver medals), and the bronze medal cutoff was 31 points (with 11 contestants earning bronze medals).

In this IMO, a total of 11 contestants achieved a perfect score of 40 points.

Problem 5.6 (IMO 11-4, proposed by the Netherlands). A semicircular arc γ is drawn on AB as diameter. C is a point on γ other than A, B, and D is the foot of the perpendicular from C to AB. We consider three circles $\gamma_1, \gamma_2, \gamma_3$, all tangent to the line AB. Of these, γ_1 is inscribed in $\triangle ABC$, while γ_2 and γ_3 are both tangent to CD and to γ, one on each side of CD. Prove that γ_1, γ_2, and γ_3 have a second tangent in common.

Proof 1. As shown in Figure 5.26, let O be the midpoint of AB and D is on AO.

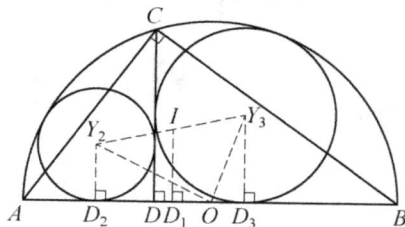

Figure 5.26

Let the centers of $\gamma_1, \gamma_2, \gamma_3$ be Y_1, Y_2, Y_3 respectively. The projection of centers Y_2, Y_3 on AB are D_2, D_3, and the radii of $\odot Y_2$ and $\odot Y_3$ are r_2, r_3, respectively. Then $AB = 2r$.

So, by $Y_2D_2^2 + D_2O^2 = Y_2O^2$, we get
$$r_2^2 + (r_2 + DO)^2 = (r - r_2)^2.$$

This is
$$r_2^2 + 2(DO + r)r_2 = r^2 - DO^2$$
$$= (r + DO)(r - DO),$$

or
$$r_2^2 + 2r_2BD = AD \cdot BD.$$

Let AB, BC, CA be c, a, b, respectively, and by similarity,
$$BD = \frac{a^2}{c}, AD = \frac{b^2}{c}.$$

Then since $BC = BD_2, BC^2 = CD^2 + BD^2,$

$$\left(r_2 + \frac{a^2}{c}\right)^2 = \frac{a^2 b^2}{c^2} + \frac{a^4}{c^2} = \frac{a^2}{c^2}(a^2 + b^2) = a^2,$$

and so,

$$r_2 = a - \frac{a^2}{c}.$$

Similarly,

$$r_3 = b - \frac{b^2}{c}.$$

Thus

$$r_2 + r_3 = a + b - \frac{a^2 + b^2}{c} = a + b - c = 2(p - c),$$

where

$$p = \frac{1}{2}(a + b + c).$$

Find the midpoint I of $Y_2 Y_3$ and construct $I D_1 \perp AB$, where D_1 is the foot. Then

$$I D_1 = \frac{1}{2}(r_2 + r_3) = p - c,$$

and it is easy to know that this is the length of the radius of the incircle of $\triangle ABC$. And

$$D_1 B = D_1 D_3 + D_3 B = D_1 D_3 + BD - DD_3$$
$$= \frac{1}{2}(r_2 + r_3) + \frac{a^2}{c} - r_3 = \frac{1}{2}(r_2 - r_3) + \frac{a^2}{c}$$
$$= \frac{1}{2}\left(a - \frac{a^2}{c} - b + \frac{b^2}{c}\right) + \frac{a^2}{c}$$
$$= \frac{1}{2}(a - b) + \frac{1}{2}\frac{a^2 + b^2}{c}$$
$$= \frac{1}{2}(c + a - b) = p - b.$$

Thus, D_1 is one of the points of tangency of the incircle of $\triangle ABC$. Also we know that $I D_1$ is exactly the length of the radius of the incircle. So, I is Y_1, which means Y_1 is exactly the midpoint of $Y_2 Y_3$, and the three circles are symmetric with respect to line $Y_2 Y_3$. Hence, they are tangent to AB, and there must be a second common tangent line, which is symmetric to AB with respect to the line $Y_2 Y_3$.

Proof 2.   We create the following Cartesian coordinate system: Let the midpoint O of AB be the origin and OB be the x-axis, the radius of γ be 1, and the coordinates of point C be $C(a, b)$, that is, $OD = a, CD = b$, and $a^2 + b^2 = 1$.

Let O_1, O_2, O_3 be the centers of $\gamma_1, \gamma_2, \gamma_3$. A sufficient and necessary condition for circles $\gamma_1, \gamma_2, \gamma_3$ to have a second common tangent line is that the three points O_1, O_2, O_3 are collinear. In this case, the required tangent line is the reflection of line AB with respect to line O_1O_3. Let r_1, r_2, r_3 be the radii of the circles $\gamma_1, \gamma_2, \gamma_3$ respectively, T_2, T_3 be the points of tangency of γ with γ_2, γ_3 respectively, and S_1, S_2, S_3 be the projections of O_1, O_2, O_3 on AB respectively, as shown in Figure 5.27.

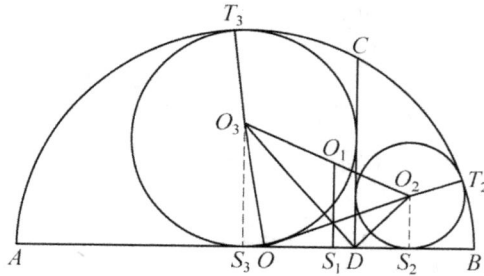

Figure 5.27

The coordinates of the center O_2 are $x_2 = a + r_2, y_2 = r_2$, where r_2 can be obtained from the condition $OO_2 + O_2T_2 = OT_2 = 1$. This is,

$$\sqrt{(a + r_2)^2 + r_2^2} + r_2 = 1,$$

so
$$r_2 = -(a + 1) + \sqrt{2(1 + a)}.$$

The coordinates of the center O_3 are $x_3 = a - r_3, y_3 = r_3$, where r_3 can be obtained from the condition $OO_3 + O_3T_3 = OT_3 = 1$. This is,

$$\sqrt{(a - r_3)^2 + r_3^2} + r_3 = 1,$$

so
$$r_3 = -(1 - a) + \sqrt{2(1 - a)}.$$

The coordinates of the center O_1 are $x_1 = 1 - BS_1, y_1 = r_1$.

Let S, p be the area and semiperimeter of triangle ABC, respectively. Then $S = b = \sqrt{1 - a^2}$. By the power of a point, we get the length of

AC, BC:

$$AC^2 = AD \cdot AB, \quad BC^2 = BD \cdot AB,$$

from which

$$AC = \sqrt{2(1+a)}, \quad BC = \sqrt{2(1-a)}.$$

Thus

$$p = 1 + \sqrt{\frac{1+a}{2}} + \sqrt{\frac{1-a}{2}}.$$

By the formula $r_1 = \frac{S}{p}$, we get

$$y_1 = r_1 = \frac{\sqrt{1-a^2}}{1 + \sqrt{\frac{1+a}{2}} + \sqrt{\frac{1-a}{2}}}$$

$$= -1 + \sqrt{\frac{1+a}{2}} + \sqrt{\frac{1-a}{2}},$$

$$BS_1 = p - AC = 1 + \sqrt{\frac{1-a}{2}} - \sqrt{\frac{1+a}{2}}.$$

Then

$$x_1 = \sqrt{\frac{1+a}{2}} - \sqrt{\frac{1-a}{2}}.$$

Therefore, we can verify the decision condition of collinearity of points. We can further prove that $x_1 = \frac{x_2 + x_3}{2}, y_1 = \frac{y_2 + y_3}{2}$. So, O_1 is the midpoint of segment $O_2 O_3$. In fact,

$$\frac{x_2 + x_3}{2} = a + \frac{r_2 - r_3}{2}$$

$$= a + \frac{-(1+a) + \sqrt{2(1+a)} + (1-a) - \sqrt{2(1-a)}}{2}$$

$$= a - a + \sqrt{\frac{1+a}{2}} - \sqrt{\frac{1-a}{2}} = x_1,$$

$$\frac{y_2 + y_3}{2} = \frac{r_2 + r_3}{2}$$

$$= \frac{-(1+a) + \sqrt{2(1+a)} - (1-a) + \sqrt{2(1-a)}}{2}$$

$$= -1 + \sqrt{\frac{1+a}{2}} + \sqrt{\frac{1-a}{2}} = y_1.$$

【Score Situation】 This particular problem saw the following distribution of scores among contestants: 20 contestants scored 6 points, 5 contestants scored 5 points, 6 contestants scored 4 points, 9 contestants scored 3 points, 15 contestants scored 2 points, 28 contestants scored 1 point, and 29 contestants scored 0 point. The average score of this problem is 2.268, indicating that it had a certain level of difficulty.

Among the top five teams in the team scores, the scores of this problem are as follows: the Hungary team scored 34 points (with a total team score of 247 points), the German Democratic Republic team scored 35 points (with a total team score of 240 points), the Soviet Union team scored 30 points (with a total team score of 231 points), the Romania team scored 28 points (with a total team score of 219 points), and the United Kingdom team scored 16 points (with a total team score of 193 points).

The gold medal cutoff for this IMO was set at 40 points (with 3 contestants earning gold medals), the silver medal cutoff was 30 points (with 20 contestants earning silver medals), and the bronze medal cutoff was 24 points (with 21 contestants earning bronze medals).

In this IMO, only three contestants achieved a perfect score of 40 points, namely Tibor Fiala from Hungary, Vladimir Drinfeld from the Soviet Union and Simon Phillips Norton from the United Kingdom.

Problem 5.7 (IMO 19-1, proposed by the Netherlands). Equilateral triangles ABK, BCL, CDM, DAN are constructed inside a square $ABCD$. Prove that the midpoints of the four segments KL, LM, MN, NK and the midpoints of the eight segments $AK, BK, BL, CL, CM, DM, DN, AN$ are the twelve vertices of a regular dodecagon.

Proof.   As shown in Figure 5.28, with the center O of the square $ABCD$ as the origin and half of the side length of the square as the length unit of the Cartesian coordinate system, the coordinates of the four vertices of the square are $A(-1,-1)$, $B(1,-1)$, $C(1,1)$, $D(-1,1)$.

It is easy to know that the coordinates of the four points K, L, M, N are

$$K(0, \sqrt{3}-1), \quad L(1-\sqrt{3}, 0), \quad M(0, 1-\sqrt{3}), \quad N(\sqrt{3}-1, 0).$$

The midpoints of the segments CL, NK, CM are

$$P_1\left(1-\frac{\sqrt{3}}{2}, \frac{1}{2}\right), \quad P_2\left(\frac{\sqrt{3}-1}{2}, \frac{\sqrt{3}-1}{2}\right), \quad P_3\left(\frac{1}{2}, 1-\frac{\sqrt{3}}{2}\right).$$

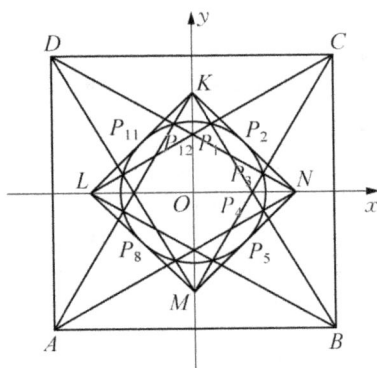

Figure 5.28

Using the distance formula, we get

$$|OP_1| = |OP_2| = |OP_3| = \sqrt{2 - \sqrt{3}}.$$
$$|P_1P_2| = |P_2P_3| = 2 - \sqrt{3}.$$

By symmetry, we know that the midpoints of these 12 segments, that is, $P_1, P_2, \ldots, P_{11}, P_{12}$ in the figure, all satisfy $|OP_i| = \sqrt{2 - \sqrt{3}}$, $i = 1, 2, \ldots, 12$, $|P_iP_{i+1}| = 2 - \sqrt{3}$ ($i = 1, 2, \ldots, 12$, $P_{13} = P_1$).

Thus, these 12 points are the vertices of a regular dodecagon.

Note. To take full advantage of symmetry, it is only necessary to prove that

$$P_2P_3 = P_3P_4, \angle P_2P_3P_4 = 150°.$$

【Score Situation】 This particular problem saw the following distribution of scores among contestants: 26 contestants scored 6 points, 3 contestants scored 5 points, 4 contestants scored 4 points, 2 contestants scored 3 points, no contestant scored 2 points, no contestant scored 1 point, and 2 contestants scored 0 point. The average score of this problem is 5.216, indicating that it was simple.

Among the top five teams in the team scores, the United States team achieved a total score of 202 points, the Soviet Union team achieved a total score of 192 points, the Hungary team achieved a total score of 190 points, the United Kingdom team achieved a total score of 190 points, and the Netherlands team achieved a total score of 185 points.

The gold medal cutoff for this IMO was set at 34 points (with 13 contestants earning gold medals), the silver medal cutoff was 24 points (with 29 contestants

earning silver medals), and the bronze medal cutoff was 17 points (with 35 contestants earning bronze medals).

In this IMO, a total of five contestants achieved a perfect score of 40 points.

Problem 5.8 (IMO 27-2, proposed by China). A triangle A_1, A_2, A_3 and a point P_0 are given in a plane. We define $A_s = A_{s-3}$ for all $s \geq 4$. We construct a sequence of points $P_1, P_2, P_3, \ldots$, such that P_{k+1} is the image of P_k under the rotation with center A_{k+1} through angle $120°$ clockwise (for $k = 0, 1, 2, \ldots$). Prove that if $P_{1986} = P_0$, then the triangle A_1, A_2, A_3 is equilateral.

Proof.   Let $u = e^{\frac{i\pi}{3}} = \cos\frac{\pi}{3} + i\sin\frac{\pi}{3} = \frac{1}{2} + \frac{\sqrt{3}i}{2}$. All the capital letters denote complex numbers in the plane. According to the problem,

$$P_n - A_n = (A_n - P_{n-1})u, \quad n = 1, 2, 3, \ldots.$$

Rewrite the recurrence relation as

$$P_n = (1 + u)A_n - uP_{n-1}.$$

Then,

$$P_3 = (1 + u)A_3 - uP_2$$

$$= (1 + u)A_3 - u(1 + u)A_2 + u^2 P_1$$

$$= (1 + u)(A_3 - uA_2 + u^2 A_1) - u^3 P_0$$

$$= w + P_0.$$

Here, $u^3 = -1$ and $w = (1 + u)(A_3 - uA_2 + u^2 A_1)$ are irrelevant to P_0. Similarly, $P_6 = P_3 + w, \ldots, P_{1986} = 662w + P_0 = P_0$, and thus $w = 0$.

Since $1 + u \neq 0$, so $A_3 - uA_2 + u^2 A_1 = 0$ or $A_3 - uA_2 + (u - 1)A_1 = 0$ (because $u^2 - u + 1 = 0$), that is, $A_3 - A_1 = u(A_2 - A_1)$. This shows that vector $\overrightarrow{A_1 A_3}$ is obtained by rotating vector $\overrightarrow{A_1 A_2}$ by $\frac{\pi}{3}$ counterclockwise. Hence A_1, A_2, and A_3 form the three vertices of an equilateral triangle.

【Score Situation】 This particular problem saw the following distribution of scores among contestants: 108 contestants scored 7 points, 6 contestants scored 6 points, 3 contestants scored 5 points, 2 contestants scored 4 points, 2 contestants scored 3 points, 9 contestants scored 2 points, 12 contestants scored 1 point, and 68 contestants scored 0 point. The average score of this problem is 4.052, indicating that it was simple.

Among the top five teams in the team scores, the scores of this problem are as follows: the Soviet Union team scored 42 points (with a total team score of

203 points), the United States team scored 42 points (with a total team score of 203 points), the Germany team scored 42 points (with a total team score of 196 points), the China team scored 42 points (with a total team score of 177 points), and the German Democratic Republic team scored 42 points (with a total team score of 172 points).

The gold medal cutoff for this IMO was set at 34 points (with 18 contestants earning gold medals), the silver medal cutoff was 26 points (with 41 contestants earning silver medals), and the bronze medal cutoff was 17 points (with 48 contestants earning bronze medals).

In this IMO, only three contestants achieved a perfect score of 42 points, namely Vladimir Roganov and Stanislav Smirnov from the Soviet Union, and Géza Kós from Hungary.

Problem 5.9 (IMO 29-1, proposed by Luxembourg). Consider two coplanar circles of radii R and r $(R > r)$ with the same center. Let P be a fixed point on the smaller circle and B a variable point on the larger circle. The line BP meets the larger circle again at C. The perpendicular l to BP at P meets the smaller circle again at A (If l is tangent to the circle at P, then $A = P$).

(a) Find the set of values of $BC^2 + CA^2 + AB^2$.
(b) Find the locus of the midpoint of AB.

Solution 1. (a) As shown in Figure 5.29, let the centers of the two circles be O. By constructing $OQ \perp BC$ through O, it is easy to know that

$$BP = \frac{BC}{2} + PQ,$$

$$CP = \frac{BC}{2} - PQ$$

(without loss of generality, let $BP \geq CP$).

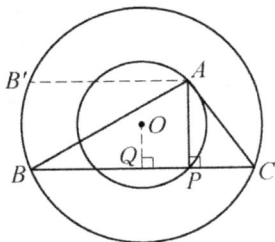

Figure 5.29

So, by the Pythagorean theorem and $AP = 2OQ$,

$$AB^2 + BC^2 + CA^2 = BP^2 + CP^2 + 2AP^2 + BC^2$$

$$= \frac{3}{2}BC^2 + 2AP^2 + 2PQ^2$$

$$= \frac{3}{2}(BC^2 + AP^2) + \frac{1}{2}AP^2 + 2PQ^2$$

$$= 6(BQ^2 + OQ^2) + 2(PQ^2 + OQ^2)$$

$$= 6R^2 + 2r^2.$$

This is a constant, and it is irrelevant to the position of point B.

(b) Construct rectangle $BPAB'$. It is easy to know that B' is on the larger circle, as shown in the figure.

So, the midpoint of AB is the midpoint of $B'P$. When B moves one round on the larger circle, B' also moves one round on the larger circle. It is easy to know that the locus is a circle with the midpoint of OP as the center and $\frac{R}{2}$ as the radius.

Solution 2. Let the centers of the two circles be O and let $t = \angle OPA$. The diameter DE passes through point P, and points M, N are the midpoints of PA and BC, respectively.

(a) The required sum is

$$S = BC^2 + AC^2 + AB^2$$

$$= (BP + PC)^2 + PC^2 + PA^2 + PA^2 + PB^2$$

$$= 2(PA^2 + PB^2 + PC^2 + PB \cdot PC).$$

Also considering

$$PA = 2r\cos t, \quad PB = PN + NB, \quad PC = CN - PN = NB - PN,$$

we obtain

$$PB^2 + PC^2 = 2(PN^2 + BN^2).$$

In the right triangles PNO and ONB, the segments PN and BN can be expressed as

$$PN = r\sin t, \quad BN^2 = R^2 - r^2 \cos^2 t,$$

and the product $PB \cdot PC$ can be expressed as

$$PB \cdot PC = PD \cdot PE = (R - r)(R + r) = R^2 - r^2,$$

so

$$S = 2(4r^2 \cos^2 t + 2r^2 \sin^2 t + 2R^2 - 2r^2 \cos^2 t + R^2 - r^2)$$
$$= 2(2r^2 \cos^2 t + 2r^2 \sin^2 t + 3R^2 - r^2)$$
$$= 2(3R^2 + r^2),$$

which is a constant.

(b) The line through point A and parallel to BC intersects the larger circle at G, H (see Figure 5.30). This line and BC are symmetric with respect to OM, so the quadrilateral $BPAG$ is a rectangle, and the required locus is the intersection of its diagonals. So, the problem is to find the locus of the midpoint of the segment PG, where point G varies on the larger circle.

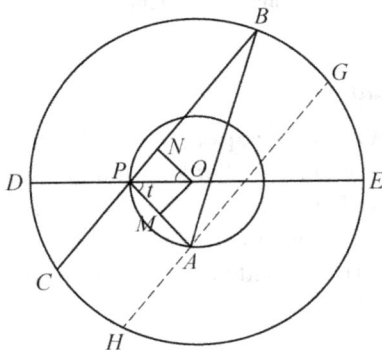

Figure 5.30

A Cartesian coordinate system is constructed with DE as the x-axis and O as the origin. Then we have $P(-r, 0)$.

Let the coordinates of the midpoint Q of AB be (x, y). Then the coordinates of G are $(2x + r, 2y)$. Since G is a moving point on the larger circle,

$$(2x + r)^2 + (2y)^2 = R^2,$$

which is

$$\left(x + \frac{r}{2}\right)^2 + y^2 = \frac{R^2}{4}.$$

【Score Situation】 This particular problem saw the following distribution of scores among contestants: 98 contestants scored 7 points, 8 contestants scored 6 points, 14 contestants scored 5 points, 22 contestants scored 4 points, 24 contestants scored 3 points, 19 contestants scored 2 points, 56 contestants scored 1 point, and 27 contestants scored 0 point. The average score of this problem is 3.948, indicating that it was relatively straightforward.

Among the top five teams in the team scores, the scores of this problem are as follows: the Soviet Union team scored 42 points (with a total team score of 217 points), the China team scored 42 points (with a total team score of 201 points),the Romania team scored 42 points (with a total team score of 201 points), the Germany team scored 42 points (with a total team score of 174 points), and the Vietnam team scored 39 points (with a total team score of 166 points).

The gold medal cutoff for this IMO was set at 32 points (with 17 contestants earning gold medals), the silver medal cutoff was 23 points (with 48 contestants earning silver medals), and the bronze medal cutoff was 14 points (with 65 contestants earning bronze medals).

In this IMO, a total of five contestants achieved a perfect score of 42 points.

5.2.3   *Quantity relation problems*

Problem 5.10 (IMO 2-3, proposed by Romania). In a given right triangle ABC, the hypotenuse BC of length a is divided into n equal parts (n is an odd integer). Let α be an acute angle subtending from A the segment of the partition, which contains the midpoint of the hypotenuse. Let h be the length of the altitude to the hypotenuse of the triangle.

Prove: $\tan \alpha = \frac{4nh}{(n^2-1)a}$.

Proof 1. Let the middle part of the hypotenuse be PQ. It is easy to know that the midpoint O of BC is also the midpoint of PQ, as shown in Figure 5.31.

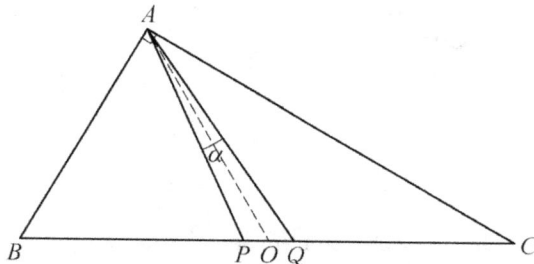

Figure 5.31

By the median length formula,

$$AP^2 + AQ^2 = \frac{1}{2}PQ^2 + 2AO^2 = \left(\frac{1}{2n^2} + \frac{1}{2}\right)a^2.$$

And by the law of cosines,

$$AP \cdot AQ \cdot \cos\alpha = \frac{1}{2}(AP^2 + AQ^2 - PQ^2) = \frac{1}{4}\left(1 - \frac{1}{n^2}\right)a^2. \qquad (1)$$

While by the area formula,

$$\frac{1}{2}AP \cdot AQ \cdot \sin\alpha = S_{\triangle APQ} = \frac{1}{n}S_{\triangle ABC} = \frac{1}{2n}ah. \qquad (2)$$

$(2) \div (1)$ gives

$$\tan\alpha = \frac{4ah}{n(1 - \frac{1}{n^2})a^2} = \frac{4nh}{(n^2 - 1)a}.$$

Proof 2. Let H be the foot of the altitude through vertex A (see Figure 5.32). Let KL be a segment containing the midpoint O of BC. For convenience, we denote $BH = x$ and $\angle HAK = \beta$. So x can be given by the following equality

$$h^2 = x(a - x).$$

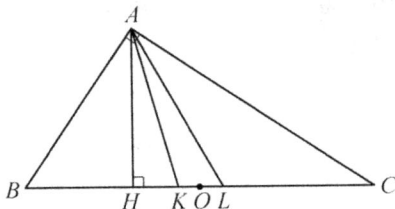

Figure 5.32

The value of $\tan\alpha$ can be given by the triangles HAK and HAL:

$$\tan\alpha = \frac{\tan(\alpha + \beta) - \tan\beta}{1 + \tan(\alpha + \beta)\tan\beta} = \frac{\frac{LH}{h} - \frac{KH}{h}}{1 + \frac{LH}{h} \cdot \frac{KH}{h}} = \frac{h \cdot LK}{h^2 + LH \cdot KH}.$$

The segments LH, KH, LK can be expressed as

$$LK = \frac{a}{n}, LH = \frac{n+1}{2n}a - x, KH = \frac{n-1}{2n}a - x,$$

and substituting and calculating, we get the required equality.

【Score Situation】 This particular problem saw the following distribution of scores among contestants: 2 contestants scored 6 points, no contestant scored 5 points, no contestant scored 4 points, no contestant scored 3 points, no contestant scored 2 points, no contestant scored 1 point, and 8 contestants scored 0 point. The average score of this problem is 1.200, indicating that it was relatively challenging.

Among the top five teams in the team scores, the Czechoslovakia team achieved a total score of 257 points, the Hungary team achieved a total score of 248 points, the Romania team achieved a total score of 248 points, the Bulgaria team achieved a total score of 175 points, and the German Democratic Republic team achieved a total score of 38 points.

The gold medal cutoff for this IMO was set at 40 points (with 4 contestants earning gold medals), the silver medal cutoff was 37 points (with 4 contestants earning silver medals), and the bronze medal cutoff was 33 points (with 4 contestants earning bronze medals).

In this IMO, no contestant achieved a perfect score of 44 points.

Problem 5.11 (IMO 15-1, proposed by Czechoslovakia). Point O lies on line l; $\overrightarrow{OP_1}$, $\overrightarrow{OP_2}, \ldots, \overrightarrow{OP_n}$ are unit vectors such that points $P_1, P_2, \ldots, P_n$ all lie in a plane containing l and on one side of l. Prove that if n is odd, $|\overrightarrow{OP_1} + \overrightarrow{OP_2} + \cdots + \overrightarrow{OP_n}| \geq 1$. Here $|\overrightarrow{OM}|$ denotes the length of vector $\overrightarrow{OM}$.

Proof.   It is easy to know that if the angle between vectors $\overrightarrow{u}$ and $\overrightarrow{v}$ does not exceed $\frac{\pi}{2}$, then

$$|\overrightarrow{u} + \overrightarrow{v}| \geq \max\{|\overrightarrow{u}|, |\overrightarrow{v}|\}.$$

Let $n = 2k + 1$ be odd and use mathematical induction for the non-negative integer k below.

Without loss of generality, let P_i be arranged counterclockwise along the semicircle of center O, with their subscripts increasing in turn. Denote $\overrightarrow{OP_i}$ $(1 \leq i \leq n)$ by $\overrightarrow{u_i}$ and $\sum_{i=1}^{n} \overrightarrow{u_i}$ by $\overrightarrow{u}$, as shown in Figure 5.33.

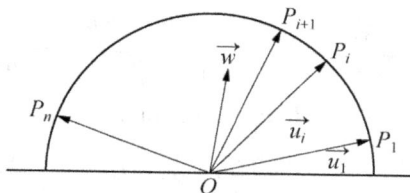

Figure 5.33

When $k = 0$, we have $n = 1$, in which case $|\vec{u}| = |\vec{u_1}| = 1$, and the proposition holds.

Suppose the proposition holds for non-negative integers less than a positive integer k. Denote

$$\vec{v} = \sum_{i=2}^{2k} \vec{u_i}.$$

Then by the induction hypothesis, $|\vec{v}| \geq 1$, and $\vec{v}$ is between $\vec{u_2}$ and $\vec{u_{2k}}$, that is, inside $\angle P_{2k+1}OP_1$.

Let $\vec{w} = \vec{u_1} + \vec{u_{2k+1}}$. Then

$$\vec{u} = \vec{v} + \vec{w} = \sum_{i=1}^{2k+1} \vec{u_i}.$$

When $\vec{w} = 0$, we see that $|\vec{u}| = |\vec{v}| \geq 1$.

When $\vec{w} \neq 0$, the vector $\vec{w}$ bisects the angle $\angle P_{2k+1}OP_1$, so the angle between $\vec{w}$ and $\vec{u_1}$, and the angle between $\vec{w}$ and $\vec{u_{2k+1}}$, are less than $\frac{\pi}{2}$, so the angle between $\vec{w}$ and $\vec{v}$ is less than $\frac{\pi}{2}$. Thus

$$|\vec{u}| = |\vec{v} + \vec{w}| \geq |\vec{v}| \geq 1,$$

that is, the proposition also holds for k.

Consequently, the proposition holds for all positive odd integers n.

【Score Situation】 This particular problem saw the following distribution of scores among contestants: 37 contestants scored 6 points, 6 contestants scored 5 points, 3 contestants scored 4 points, 5 contestants scored 3 points, 14 contestants scored 2 points, 15 contestants scored 1 point, and 45 contestants scored 0 point. The average score of this problem is 2.576, indicating that it had a certain level of difficulty.

Among the top five teams in the team scores, the scores of this problem are as follows: the Soviet Union team scored 35 points (with a total team score of 254 points), the Hungary team scored 34 points (with a total team score of 215 points), the German Democratic Republic team scored 39 points (with a total team score of 188 points), the Poland team scored 35 points (with a total team score of 174 points), and the United Kingdom team scored 17 points (with a total team score of 164 points).

The gold medal cutoff for this IMO was set at 35 points (with 5 contestants earning gold medals), the silver medal cutoff was 27 points (with 15 contestants earning silver medals), and the bronze medal cutoff was 17 points (with 48 contestants earning bronze medals).

In this IMO, only one contestant achieved a perfect score of 40 points, namely Sergei Konyagin from the Soviet Union.

Problem 5.12 (IMO 17-3, proposed by the Netherlands). On the sides of an arbitrary triangle ABC, triangles ABR, BCP, and CAQ are constructed externally with $\angle CBP = \angle CAQ = 45°, \angle BCP = \angle ACQ = 30°$, and $\angle ABR = \angle BAR = 15°$. Prove that $\angle QRP = 90°$ and $QR = RP$.

Proof 1. As shown in Figure 5.34, we first discuss the case where A, C are on the different sides of QR, and C, B are on the different sides of PR. If they are on the same side, then the same argument can be used.

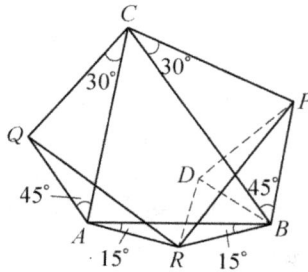

Figure 5.34

Since

$$\angle QRA + \angle AQR = 120° - \angle CAB,$$

$$\angle PRB + \angle BPR = 120° - \angle CBA,$$

so

$$\angle QRA + \angle AQR + \angle PRB + \angle BPR$$

$$= 240° - \angle CAB - \angle CBA$$

$$= 60° + \angle ACB.$$

At least one of $\min\{\angle QRA, \angle PRB\} < 60°$ and $\min\{\angle AQR, \angle BPR\} < \angle ACB$ is valid. For convenience, let

$$\angle PRB < 60° \quad \text{or} \quad \angle BPR < \angle ACB. \tag{1}$$

Construct an equilateral triangle $\triangle DRB$ with RB as an edge inward. Connect to form PD. Then it is easy to know

$$\angle PBD = \angle PBR - 60° = \angle CBA.$$

And

$$\frac{PB}{BC} = \frac{\sin 30°}{\sin 75°} = \frac{1}{2\cos 15°},$$

while

$$\frac{BD}{AB} = \frac{BR}{AB} = \frac{1}{2\cos 15°}.$$

Thus

$$\frac{PB}{BC} = \frac{BD}{AB},$$

so $\triangle BPD \backsim \triangle BCA$.

By (1), $\angle DRB = 60°$ and $\angle DPB = \angle ACB$, so B, D are on the different sides of RP. From

$$\angle PDR = 60° + \angle PDB$$
$$= 60° + \angle CAB$$
$$= \angle QAR,$$

$DR = AR$.

Since $\triangle CQA \backsim \triangle CPB$,

$$\frac{DP}{AC} = \frac{BP}{BC} = \frac{AQ}{AC},$$

and then $DP = AQ$.

So, $\triangle QAR \cong \triangle PDR(\text{SAS})$, from which $QR = RP$.

In addition,

$$\angle QRP = \angle QRD + \angle DRP$$
$$= \angle QRD + \angle ARQ$$
$$= \angle DRA$$
$$= 150° - 60°$$
$$= 90°.$$

Proof 2. This problem can be solved by the complex number method.

As shown in Figure 5.35, construct a complex plane. Let $A = -1$, $B = 1$, and R, P, Q, C all denote complex numbers.

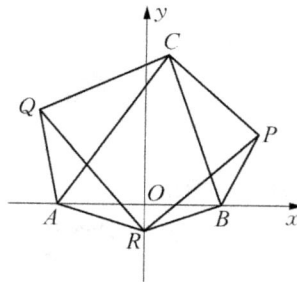

Figure 5.35

Clearly, $R = -i\tan 15°$.
And

$$\frac{P-B}{C-B} = \frac{|BP|}{|BC|}(\cos 45° - i\sin 45°);$$

this is

$$\frac{P-1}{C-1} = \frac{\sin 30°}{\sin 75°}(\cos 45° - i\sin 45°).$$

Consequently,

$$P = \frac{C-1}{4\cos 15°}(\sqrt{2} - \sqrt{2}i) + 1$$

Similarly,

$$Q = \frac{C+1}{4\cos 15°}(\sqrt{2} + \sqrt{2}i) - 1,$$

$$Q - iP = \frac{\sqrt{2} + \sqrt{2}i}{2\cos 15°} - 1 - i$$

$$= (1+i)\left(\frac{\sqrt{2}}{2\cos 15°} - 1\right)$$

$$= -(1+i)\tan 15°$$

$$= (1-i)R.$$

This is

$$Q - R = i(P - R).$$

Thus, $\triangle QRP$ is an isosceles right triangle.

Proof 3. Let segment RS be obtained by rotating segment RB counter-clockwise $90°$ around the point R (see Figure 5.36). Then

$$RA = RS, \angle SRA = 150° - 90° = 60°,$$

so $\triangle RSA$ is an equilateral triangle, from which $AS = AR$.

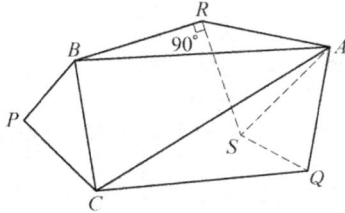

Figure 5.36

It is easy to prove that $\angle SAQ = \angle BAC$. Using the law of sines for $\triangle AQC$, we get

$$\frac{AQ}{\sin 30°} = \frac{AC}{\sin 105°},$$

and thus

$$AQ = AC \cdot \frac{\sin 30°}{\sin 105°} = AC \cdot \frac{2 \sin 15° \cos 15°}{\cos 15°} = 2AC \sin 15°. \qquad (1)$$

Using the law of sines for $\triangle ARB$, we obtain

$$\frac{AR}{\sin 15°} = \frac{AB}{\sin 150°},$$

from which

$$AR = AB \cdot \frac{\sin 15°}{\sin 150°} = 2AB \sin 15°. \qquad (2)$$

By (1) and (2),

$$\frac{AQ}{AS} = \frac{AC}{AB},$$

so, $\triangle CAB$ and $\triangle QAS$ are similar. Consequently

$$\angle ABC = \angle ASQ, \frac{AB}{AS} = \frac{BC}{SQ}.$$

Then

$$SQ = \frac{AS}{AB} \cdot BC = \frac{AQ}{AC} \cdot BC = 2BC \sin 15°.$$

Using the law of sines for $\triangle BPC$, we find $BP = 2BC \sin 15°$, so $SQ = BP$. While

$$\angle RBP = \angle ABC + 60° = \angle RSQ,$$

we see that $\triangle RBP$ and $\triangle RSQ$ are congruent. This proves that $\angle PRQ = 90°$, $PR = RQ$.

【Score Situation】 This particular problem saw the following distribution of scores among contestants: 58 contestants scored 7 points, 3 contestants scored 6 points, 7 contestants scored 5 points, 4 contestants scored 4 points, 7 contestants scored 3 points, 4 contestants scored 2 points, 14 contestants scored 1 point, and 31 contestants scored 0 point. The average score of this problem is 4.047, indicating that it was simple.

Among the top five teams in the team scores, the scores of this problem are as follows: the Hungary team scored 56 points (with a total team score of 258 points), the German Democratic Republic team scored 49 points (with a total team score of 249 points), the United States team scored 37 points (with a total team score of 247 points), the Soviet Union team scored 43 points (with a total team score of 246 points), and the United Kingdom team scored 24 points (with a total team score of 241 points).

The gold medal cutoff for this IMO was set at 38 points (with 8 contestants earning gold medals), the silver medal cutoff was 32 points (with 25 contestants earning silver medals), and the bronze medal cutoff was 23 points (with 36 contestants earning bronze medals).

In this IMO, a total of six contestants achieved a perfect score of 40 points.

Problem 5.13 (IMO 26-1, proposed by the United Kingdom). A circle has center on the side AB of the cyclic quadrilateral $ABCD$. The other three sides are tangent to the circle. Prove that $AD + BC = AB$.

Proof 1.   As shown in Figure 5.37, let the center of the circle on AB be O. The circle $\odot O$ is tangent to AD, DC, CB at points E, F, G, respectively. Connect to form OE, OF, and OG. It is easy to know $OE = OF = OG$.

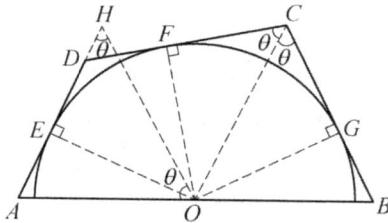

Figure 5.37

Let $\angle DCB = 2\theta$. Then $\angle FCO = \theta$. Rotate $\triangle OFC$ around point O to get $\triangle OEH$. Point H is on line AD. Since A, B, C, D are concyclic, $\angle A = 180° - 2\theta$. Then

$$\angle HOA = 180° - \angle A - \theta = \theta = \angle AHO,$$

and thus

$$AO = AH = AE + FC = AE + CG.$$

Similarly, we can prove

$$BO = BG + ED.$$

Adding the two equalities above, we get

$$AB = AD + BC.$$

Proof 2. Without loss of generality, let $\angle B \leq \angle A$. Connect to form OD, OC. Since

$$\angle AOD = 180° - \angle A - \angle ADO$$

$$= 90° + \frac{\angle B}{2} - \angle A$$

$$\leq 90° - \frac{\angle B}{2}$$

$$= \angle ADO.$$

So, $AD \leq AO$. As shown in Figure 5.38, take a point E in the segment AO such that $AD = AE$. Obviously, the next step is to prove that $BE = BC$. Connect to form DE, CE.

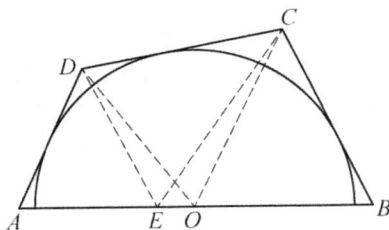

Figure 5.38

Then we see that

$$\angle AED = 90° - \frac{\angle A}{2} = \frac{\angle DCB}{2} = \angle DCO.$$

Thus, D, E, O, C are concyclic. On the other hand,

$$\angle CEB = \angle ODC = \frac{\angle ADC}{2} = 90° - \frac{\angle B}{2},$$

$$\angle ECB = 180° - \angle CEB - \angle B = 90° - \frac{\angle B}{2} = \angle CEB,$$

so $\qquad BE = BC.$

Note. If the points E and O coincide, $\angle CEB$ becomes a "tangent chord angle," and the conclusion still holds.

Proof 3. The circle is tangent to BC, CD, DA respectively at points E, F, G with radius R. Then $OE = OF = OG = R$. Denote

$$\alpha = \angle BAD, \quad \beta = \angle ABC, \quad \gamma = \angle OCB = \angle OCD, \quad \delta = \angle ODC = \angle ODA.$$

Since $ABCD$ is the cyclic quadrilateral, so $\alpha + 2\gamma = \beta + 2\delta = \pi$. In the right triangle,

$$GA = \frac{R}{\tan \alpha}, \quad BE = \frac{R}{\tan \beta},$$

$$CE = \frac{R}{\tan \gamma}, \quad DG = \frac{R}{\tan \delta},$$

$$AO = \frac{R}{\sin \alpha}, \quad BO = \frac{R}{\sin \beta}.$$

For the problem, we need to prove the following equality:

$$\frac{1}{\tan \alpha} + \frac{1}{\tan \beta} + \frac{1}{\tan \gamma} + \frac{1}{\tan \delta} = \frac{1}{\sin \alpha} + \frac{1}{\sin \beta}.$$

Based on symmetry, we rewrite the above equality as follows:

$$\left(\frac{1}{\tan \alpha} - \frac{1}{\sin \alpha} + \frac{1}{\tan \gamma} \right) + \left(\frac{1}{\tan \beta} - \frac{1}{\sin \beta} + \frac{1}{\tan \delta} \right) = 0.$$

The two parts above with respect to the angles have similar properties. So, to prove the original problem we need to prove that every value in the brackets is equal to zero. After a simple calculation we can get

$$\frac{1}{\tan \alpha} - \frac{1}{\sin \alpha} + \frac{1}{\tan \gamma} = \frac{-\cos 2\gamma}{\sin 2\gamma} - \frac{1}{\sin 2\gamma} + \frac{\cos \gamma}{\sin \gamma}$$

$$= \frac{-\cos 2\gamma - 1 + 2\cos^2 \gamma}{\sin 2\gamma} = 0.$$

In addition, if $\alpha = \frac{\pi}{2}$, then the quadrilateral $ABCD$ is a rectangle, and the conclusion obviously holds.

【Score Situation】 This particular problem saw the following distribution of scores among contestants: 103 contestants scored 7 points, 2 contestants scored 6 points, 2 contestants scored 5 points, 4 contestants scored 4 points, 9 contestants scored 3 points, 11 contestants scored 2 points, 46 contestants scored 1 point, and 32 contestants scored 0 point. The average score of this problem is 4.086, indicating that it was simple.

Among the top five teams in the team scores, the scores of this problem are as follows: the Romania team scored 42 points (with a total team score of 201 points), the United States team scored 28 points (with a total team score of 180 points), the Hungary team scored 42 points (with a total team score of 168 points), the Bulgaria team scored 42 points (with a total team score of 165 points), and the Vietnam team scored 42 points (with a total team score of 144 points).

The gold medal cutoff for this IMO was set at 34 points (with 14 contestants earning gold medals), the silver medal cutoff was 22 points (with 35 contestants earning silver medals), and the bronze medal cutoff was 15 points (with 52 contestants earning bronze medals).

In this IMO, only two contestants achieved a perfect score of 42 points, namely G éza Kós from Hungary and Daniel Tătaru from Romania.

Problem 5.14 (IMO 28-2, proposed by the Soviet Union). In an acute-angled triangle ABC, the interior bisector of the angle A intersects BC at L and intersects the circumcircle of ABC again at N. From point L the perpendiculars are drawn to AB and AC, the feet of these perpendiculars being K and M, respectively. Prove that the quadrilateral $AKNM$ and the triangle ABC have an equal area.

Proof 1. As shown in Figure 5.39, construct the altitude AH of the triangle $\triangle ABC$. Then the five points A, K, H, L, M are concyclic. Connect to form KH, HM, HN, BN, and NC. Then

$$\angle KHB = \angle BAN = \angle NAC = \angle NBC.$$

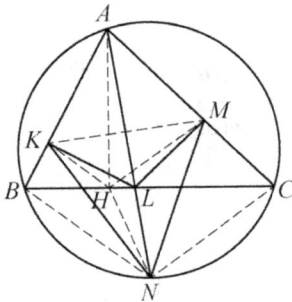

Figure 5.39

Thus $KH \parallel BN$, so $S_{\triangle KHN} = S_{\triangle BKH}$. Similarly, $S_{\triangle MHN} = S_{\triangle HMC}$. Hence, $S_{\triangle ABC} = S_{AKNM}$.

Proof 2.   Following the figure in Proof 1, we connect to form KM. It is easy to know that AN bisects KM vertically. By the fact that A, K, L, M are concyclic and the law of sines,

$$S_{AKNM} = \frac{1}{2}KM \cdot AN = \frac{1}{2}AL \cdot AN \cdot \sin \angle BAC,$$

while $S_{\triangle ABC} = \frac{1}{2}AB \cdot AC \cdot \sin \angle BAC$.

To prove that the two areas are equal, we just prove

$$AL \cdot AN = AB \cdot AC.$$

The proof is as follows. Since

$$\angle BAL = \angle NAC,$$

$$\angle ABL = \angle ANC,$$

so $$\triangle ABL \sim \triangle ANC.$$

Then

$$\frac{AB}{AL} = \frac{AN}{AC},$$

which is $$AL \cdot AN = AB \cdot AC.$$

Proof 3.   Following the figure in Proof 1, since AN is the interior bisector of $\angle BAC$, we see that $LK \perp AB$ at K and $LM \perp AC$ at M. So $\triangle AKL$ and $\triangle AML$ are congruent, and then $AK = AM$.

Thus, $\triangle NKA$ and $\triangle NMA$ are congruent. Therefore, it is only necessary to prove that the area of $\triangle AKN$ is half the area of $\triangle ABC$. In general,

$$S_{\triangle ABC} = \frac{1}{2}bc \sin A.$$

By the internal angle bisector's length formula,

$$AL = \frac{2bc}{b+c} \cos \frac{A}{2}.$$

Thus

$$AK = AL \cos \frac{A}{2} = \frac{2bc}{b+c} \cos^2 \frac{A}{2}.$$

In $\triangle ABN$, by the law of sines,

$$\frac{AN}{\sin(B + \frac{A}{2})} = 2R = \frac{a}{\sin A}.$$

From the triangle area formula,

$$S_{\triangle AKN} = \frac{1}{2} AK \cdot AN \sin \frac{A}{2}$$

$$= \frac{abc}{2(b+c)} \cos \frac{A}{2} \sin \left(B + \frac{A}{2}\right).$$

Therefore, it is only necessary to prove

$$\frac{a}{b+c} = \frac{\sin A}{2 \cos \frac{A}{2} \sin \left(B + \frac{A}{2}\right)}.$$

By the law of sines, that is to prove

$$\sin B + \sin C = 2 \cos \frac{A}{2} \sin \left(B + \frac{A}{2}\right).$$

A calculation gives

$$\sin B + \sin C = 2 \sin \frac{B+C}{2} \cos \frac{C-B}{2}$$

$$= 2 \sin \frac{\pi - A}{2} \cos \frac{C-B}{2}$$

$$= 2 \cos \frac{A}{2} \cos \frac{C-B}{2},$$

$$\sin \left(B + \frac{A}{2}\right) = \sin \frac{2B+A}{2}$$

$$= \sin \frac{\pi - (C-B)}{2} = \cos \frac{C-B}{2},$$

therefore, the proposition is proved.

Note. The condition "$\triangle ABC$ is an acute triangle" is not necessary.

【Score Situation】 This particular problem saw the following distribution of scores among contestants: 141 contestants scored 7 points, 4 contestants scored 6 points, 4 contestants scored 5 points, 7 contestants scored 4 points, 11 contestants scored 3 points, 7 contestants scored 2 points, 9 contestants scored 1 point, and 54 contestants scored 0 point. The average score of this problem is 4.705, indicating that it was simple.

Among the top five teams in the team scores, the scores of this problem are as follows: the Romania team scored 42 points (with a total team score of 250 points), the Germany team scored 42 points (with a total team score of 248 points), the Soviet Union team scored 42 points (with a total team score of 235 points), the German Democratic Republic team scored 42 points (with a total team score of

231 points), and the United States team scored 42 points (with a total team score of 220 points).

The gold medal cutoff for this IMO was set at 42 points (with 22 contestants earning gold medals), the silver medal cutoff was 32 points (with 42 contestants earning silver medals), and the bronze medal cutoff was 18 points (with 56 contestants earning bronze medals).

In this IMO, a total of 22 contestants achieved a perfect score of 42 points.

Problem 5.15 (IMO 48-4, proposed by Czech Republic). In triangle ABC the bisector of angle BCA intersects the circumcircle again at R, the perpendicular bisector of BC at P, and the perpendicular bisector of AC at Q. The midpoint of BC is K and the midpoint of AC is L. Prove that the triangles RPK and RQL have the same area.

Proof. If $AC = BC$, then $\triangle ABC$ is an isosceles triangle, and $\triangle RQL$ and $\triangle RPK$ are symmetric with respect to the angular bisector CR. The conclusion obviously holds.

If $AC \neq BC$, without loss of generality, let $AC < BC$. Denote the circumcenter of $\triangle ABC$ by O, and note that the right triangles CLQ and CKP are similar. Thus

$$\angle CPK = \angle CQL \quad \text{and} \quad \frac{QL}{PK} = \frac{CQ}{CP}. \tag{1}$$

As shown in Figure 5.40, let l be the perpendicular bisector of the chord CR. Then l passes through the circumcenter O.

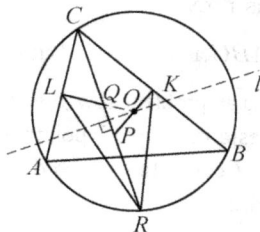

Figure 5.40

Since $\triangle OPQ$ is an isosceles triangle, the points P and Q are symmetric on CR with respect to l, so

$$RP = CQ \quad \text{and} \quad RQ = CP. \tag{2}$$

Thus, by (1) and (2),

$$\frac{S_{\triangle RQL}}{S_{\triangle RPK}} = \frac{\frac{1}{2}RQ \cdot QL \cdot \sin \angle RQL}{\frac{1}{2}RP \cdot PK \cdot \sin \angle RPK}$$

$$= \frac{RQ}{RP} \cdot \frac{QL}{PK} = \frac{CP}{CQ} \cdot \frac{CQ}{CP} = 1.$$

Hence, $S_{\triangle RQL} = S_{\triangle RPK}$.

【Score Situation】 This particular problem saw the following distribution of scores among contestants: 363 contestants scored 7 points, 51 contestants scored 6 points, 4 contestants scored 5 points, 3 contestants scored 4 points, 9 contestants scored 3 points, 9 contestants scored 2 points, 30 contestants scored 1 point, and 51 contestants scored 0 point. The average score of this problem is 5.681, indicating that it was simple.

Among the top five teams in the team scores, the scores of this problem are as follows: the Russia team scored 42 points (with a total team score of 184 points), the China team scored 41 points (with a total team score of 181 points), the South Korea team scored 42 points (with a total team score of 168 points), the Vietnam team scored 41 points (with a total team score of 168 points), and the United States team scored 42 points (with a total team score of 155 points).

The gold medal cutoff for this IMO was set at 29 points (with 39 contestants earning gold medals), the silver medal cutoff was 21 points (with 83 contestants earning silver medals), and the bronze medal cutoff was 14 points (with 131 contestants earning bronze medals).

In this IMO, no contestant achieved a perfect score of 42 points.

Problem 5.16 (IMO 59-6, proposed by Poland). A convex quadrilateral $ABCD$ satisfies $AB \cdot CD = BC \cdot DA$. Point X lies inside $ABCD$ so that $\angle XAB = \angle XCD$ and $\angle XBC = \angle XDA$.

Prove that $\angle BXA + \angle DXC = 180°$.

Proof. First notice that we only need to prove that

$$\frac{XB}{XD} = \frac{AB}{CD} \tag{1}$$

and

$$\frac{XA}{XC} = \frac{DA}{BC}. \tag{2}$$

This is because from (1) and the law of sines,

$$\frac{\sin \angle AXB}{\sin \angle XAB} = \frac{AB}{XB} = \frac{CD}{XD} = \frac{\sin \angle CXD}{\sin \angle XCD},$$

and then by the condition $\angle XAB = \angle XCD$, we get $\sin \angle AXB = \sin \angle CXD$.

Similarly, from (2),

$$\sin \angle DXA = \sin \angle BXC.$$

If either $\angle AXB + \angle CXD = 180°$ or $\angle DXA + \angle BXC = 180°$, then the conclusion holds.

If $\angle AXB = \angle CXD, \angle DXA = \angle BXC$, then X is the intersection of AC, BD, and by the given condition, $ABCD$ is a parallelogram. Again by $AB \cdot CD = BC \cdot DA$, we know that $ABCD$ is a rhombus, in which case $AC \perp BD$, and the conclusion still holds.

The following proves that (1) and (2) hold.

Take X as the center and 1 as the radius for an inversion transformation. The inversion images of A, B, C, D are represented by A', B', C', D' respectively, as shown in Figure 5.41.

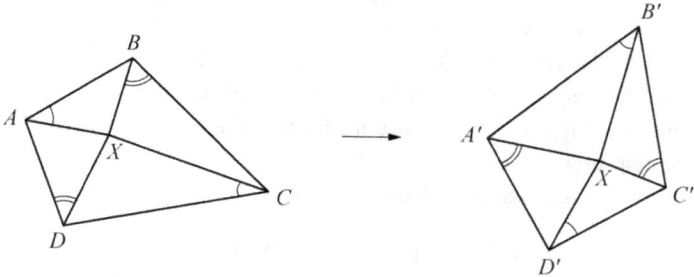

Figure 5.41

Since

$$XA \cdot XA' = XB \cdot XB' = XC \cdot XC' = XD \cdot XD',$$

the triangles XAB and $XB'A'$ are similar, and the triangles XBC and $XC'B'$ are similar. Then

$$\angle XB'A' = \angle XAB = \angle XCD, \quad \angle XCB = \angle XB'C'.$$

Hence

$$\angle BCD = \angle BCX + \angle XCD = \angle XB'C' + \angle A'B'X = \angle A'B'C'.$$

Similarly,

$$\angle CDA = \angle B'C'D', \quad \angle DAB = \angle C'D'A', \quad \angle ABC = \angle D'A'B'.$$

Therefore, the corresponding internal angles of the quadrilaterals $ABCD$ and $D'A'B'C'$ are equal. And using similarity, we obtain

$$\frac{A'B'}{AB} = \frac{XB'}{XA} = \frac{1}{XA \cdot XB}.$$

Hence

$$A'B' = \frac{AB}{XA \cdot XB}.$$

For $B'C', C'D', D'A'$, there also exist similar equalities:

$$\begin{aligned} A'B' \cdot C'D' &= \frac{AB}{XA \cdot XB} \cdot \frac{CD}{XC \cdot XD} \\ &= \frac{BC}{XB \cdot XC} \cdot \frac{DA}{XD \cdot XA} \\ &= B'C' \cdot D'A'. \end{aligned}$$

Thus, $ABCD$ and $D'A'B'C'$ have the same internal angles, and the same property of equal products of opposite sides, and next we prove that they are similar quadrilaterals.

Lemma. *Let $XYZT$ and $X'Y'Z'T'$ be two convex quadrilaterals with equal corresponding internal angles, and*

$$XY \cdot ZT = YZ \cdot TX, \quad X'Y' \cdot Z'T' = Y'Z' \cdot T'X'.$$

Then the two quadrilaterals are similar.

Proof of the Lemma. As shown in Figure 5.42, make a quadrilateral XYZ_1T_1 similar to $X'Y'Z'T'$, with T_1 and Z_1 on the rays XT and YZ, respectively. If $XYZT$ and $X'Y'Z'T'$ are not similar, then $T_1 \neq T$ and $Z_1 \neq Z$. And since the internal angles are the same, $T_1Z_1 \| TZ$.

Without loss of generality, let point T_1 be inside segment XT. Let segments XZ, T_1Z_1 intersect at point U. So

$$\frac{T_1X}{T_1Z_1} < \frac{T_1X}{T_1U} = \frac{TX}{ZT} = \frac{XY}{YZ} < \frac{XY}{YZ_1}.$$

Then, $T_1X \cdot YZ_1 < T_1Z_1 \cdot XY$, a contradiction.

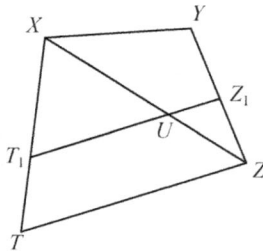

Figure 5.42

Returning to the original problem, we have proved that $ABCD$ and $D'A'B'C'$ are similar. Then

$$\frac{BC}{AB} = \frac{A'B'}{D'A'} = \frac{AB}{XA \cdot XB} \cdot \frac{XD \cdot XA}{DA} = \frac{AB}{AD} \cdot \frac{XD}{XB},$$

and consequently,

$$\frac{XB}{XD} = \frac{AB^2}{BC \cdot AD} = \frac{AB^2}{AB \cdot CD} = \frac{AB}{CD}.$$

Thus, we complete the proof of (1) and similarly, we can prove (2).

【Score Situation】 This particular problem saw the following distribution of scores among contestants: 18 contestants scored 7 points, 5 contestants scored 6 points, 2 contestants scored 5 points, 5 contestants scored 4 points, 11 contestants scored 3 points, 26 contestants scored 2 points, 108 contestants scored 1 point, and 419 contestants scored 0 point. The average score of this problem is 0.638, indicating that it was extremely difficult.

Among the top five teams in the team scores, the scores of this problem are as follows: the United States team scored 31 points (with a total team score of 212 points), the Russia team scored 23 points (with a total team score of 201 points), the China team scored 19 points (with a total team score of 199 points), the Ukraine team scored 14 points (with a total team score of 186 points), and the Thailand team scored 16 points (with a total team score of 183 points).

The gold medal cutoff for this IMO was set at 31 points (with 48 contestants earning gold medals), the silver medal cutoff was 25 points (with 98 contestants earning silver medals), and the bronze medal cutoff was 16 points (with 143 contestants earning bronze medals).

In this IMO, only two contestants achieved a perfect score of 42 points, namely Agnijo Banerjee from the United Kingdom and James Lin from the United States.

5.3 Summary

In the first 64th IMOs, there are 16 problems on trigonometric, areas, and analytic geometry, as depicted in Figure 5.43 which can be broadly categorized into three types.

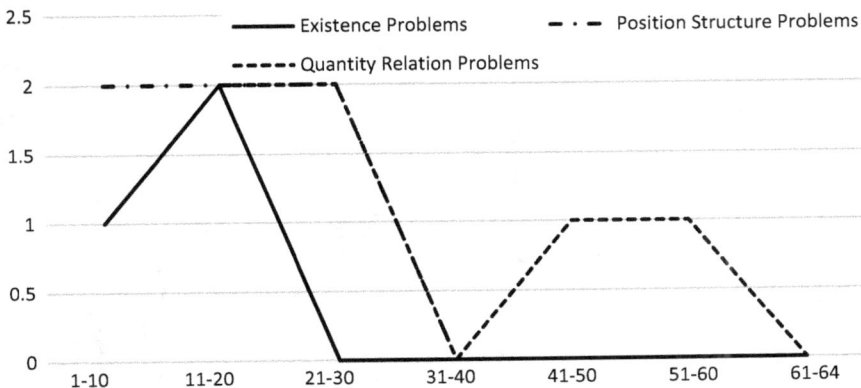

Figure 5.43 Numbers of Trigonometry, Areas and Analytic Geometry Problems in the First 64 IMOs.

Problems 5.1–5.3 focus on "existence problems;" among these three problems, the one with the lowest average score is Problem 5.1 (IMO 1-4), proposed by Hungary. Problems 5.4–5.9 deal with "position structure problems;" among these six problems, the one with the lowest average score is Problem 5.4 (IMO 1-5), proposed by Romania. Problems 5.10–5.16 are about "quantity relation problems;" among these seven problems, the one with the lowest average score is Problem 5.16 (IMO 59-6), proposed by Poland.

These problems were proposed by 12 countries and regions. The Netherlands proposed three problems, while Romania and Hungary each contributed two problems.

From Table 5.2, it can be observed that in the first 64 IMOs, there was one problem with an average score of 0–1 point; two problems with average scores of 1–2 points; three problems with average scores of 2–3 points; three problems with average scores of 3–4 points; seven problems with average scores above 4 points. Overall, trigonometric, areas, and analytic geometry problems were relatively simple.

In the 24th to 64th IMOs, there were a total of six trigonometry, areas, and analytic geometry problems. Among these, one had an average score

Table 5.2 Score Details of Trigonometry, Areas and Analytic Geometry Problems in the First 64 IMOs.

Problem	5.1	5.2	5.3	5.4	5.5	5.6
Full points	5.000	6.000	6.000	8.000	7.000	6.000
Average score	2.111	3.864	3.764	1.444	6.037	2.268
Top five mean		5.100	5.075			3.575
6th–15th mean		3.388	3.857			
16th–25th mean						
Problem number in the IMO	1-4	15-4	16-2	1-5	8-2	11-4
Proposing country	Hungary	Yugoslavia	Finland	Romania	Hungary	Netherlands

Problem	5.7	5.8	5.9	5.10	5.11	5.12
Full points	6.000	7.000	7.000	6.000	6.000	7.000
Average score	5.216	4.052	3.948	1.200	2.576	4.047
Top five mean		7.000	6.900		4.000	5.225
6th–15th mean		5.767	5.288		1.913	4.228
16th–25th mean		4.121	3.741			
Problem number in the IMO	19-1	27-2	29-1	2-3	15-1	17-3
Proposing country	Netherlands	China	Luxembourg	Romania	Czechoslovakia	Netherlands

Problem	5.13	5.14	5.15	5.16
Full points	7.000	7.000	7.000	7.000
Average score	4.086	4.705	5.681	0.638
Top five mean	6.533	7.000	6.933	3.433
6th–15th mean	5.1	5.9	6.967	1.65
16th–25th mean	3.767	5.2	6.8	0.8
Problem number in the IMO	26-1	28-2	48-4	59-6
Proposing country	United Kingdom	Soviet Union	Czech Republic	Poland

Note. Top five mean = Total score of the top five teams ÷ Total number of contestants from the top five teams,

6th–15th mean = Total score of the 6th–15th teams ÷ Total number of contestants from the 6th–15th teams,

16th–25th mean = Total score of the 16th–25th teams ÷ Total number of contestants from the 16th–25th teams.

of 0–1 point; one had an average score of 3–4 points; four had an average score above 4 points. Further analysis of the problem numbers of these six trigonometric, areas, and analytic geometry problems, as shown in Table 5.3, reveals that these problems frequently appeared as the 1st/4th problem. The majority of these problems were of the type of quantity relation problems, with four in total, accounting for two thirds.

Table 5.3 Numbers of Trigonometry, Areas, and Analytic Geometry Problems in the 24th–64th IMOs.

Trigonometry, areas, and analytic geometry problem	Problem number			Number of problems in the first 64 IMOs
	1, 4	2, 5	3, 6	
Existence problems	0	0	0	3
Position structure problems	1	1	0	6
Quantity relation problems	2	1	1	7
Total	3	2	1	16

From Table 5.2, it can be observed that, in most cases, the average score of the top five teams is typically about 1–3 points higher than the average score of the problem. Moreover, except for individual problems, the difference in average scores between the top five teams and 6th–15th teams is relatively obvious. Concretely, the average score of the top five teams is more than 1 point higher than the average score of 6th–15th teams.

From Table 5.2, it can also be observed that the average scores of the 6th–15th teams and 16th–25th teams also tend to be higher than the average score of the problem. This phenomenon is due to the smaller number of participating teams in early IMOs. It was not until the 30th IMO in 1989 that the number of participating teams exceeded 50. In addition, the appearance of geometry problems that are not extremely difficult facilitates the contestants score, thus raising the average score of the problem.

Chapter 6

Solid Geometry

Solid geometry is the traditional name for the geometry of the three-dimensional Euclidean space. It mainly studies straight lines and planes in the space, cylinders, cones, spheres, prisms, pyramids, polyhedra, etc. Solid geometry plays an important role in cultivating students' spatial imagination ability, logical thinking ability, and drawing ability.

Lines and planes are basic concepts of solid geometry. Mastering the basic knowledge and skills for lines and planes is a basis of studying solid geometry.

Converting solid geometry problems into plane problems is also an important method to solve solid geometry problems.

In the first 64th IMOs, there had been a total of 17 problems related to solid geometry, accounting for 13.8% of all geometry problems. These problems are mainly categorized into three types: (1) existence problems, totaling five problems; (2) position structure problems, totaling six problems; (3) quantity relation problems, totaling six problems. The statistical distribution of these three types of problems in the previous IMOs is presented in Table 6.1.

It can be seen that the problems related to solid geometry only appeared in the first 20 IMOs, and accounted for a large proportion, the number of more than 30% of the geometry problems at that time. In particular, in position structure problems, a total of five problems are to find the locus of a point, which is quite similar to analytic geometry problems.

In fact, many of the methods used in analytic geometry are also applicable to solid geometry problems, such as coordinates and vectors. The difficulty is that the structure of three-dimensional graphics is not easy to

Table 6.1 Numbers of Solid Geometry Problems in the First 64 IMOs

Content	Session							Total
	1–10	11–20	21–30	31–40	41–50	51–60	61–64	
Existence problems	0	5	0	0	0	0	0	5
Position structure problems	5	1	0	0	0	0	0	6
Quantity relation problems	5	1	0	0	0	0	0	6
Geometry problems	29	18	18	15	20	17	6	123
Percentage of solid geometry problems among geometry problems	34.5%	38.9%	0.0%	0.0%	0.0%	0.0%	0.0%	13.8%

be represented by geometric construction, so it is a test of students' spatial imagination ability.

Therefore, this chapter will be divided into three parts. The first part introduces relationships between the position of lines and planes, the calculation of space angles, volumes of polyhedral and revolutions, and other knowledge, through some examples to expound the relevant calculation methods.

The second part revolves around three types of problems: "existence problems," "position structure problems," and "quantity relation problems." These problems are presented in chronological order, and some problems include various solutions and generalizations.

It is important to note that for each problem, the solutions are followed by information on the scores, including the number of contestants in each score range, the average score, the scores of the top five teams. However, early IMOs often lacked information on contestant scores, so the number of contestants in each score range only represents the counted number of contestants, and some problems lack scores of the top five teams.

The third part provides a brief summary of this chapter.

6.1 Related Properties, Theorems, and Methods

6.1.1 *Position relation of lines and planes*

(1) *Determination of skew lines*

Proof by contradiction is often used.

(2) *Determinations of parallel lines*

- **Axiom.** If two lines are parallel to a third line respectively, then they are parallel to each other.
- **Property of Lines Parallel to Planes.** If a line is parallel to a plane, and another plane contains the line and intersects the plane, then the line is parallel to the intersection line of the two planes.
- **Property of Parallel Planes.** If two parallel planes intersect with a third plane , then their intersection lines are parallel.
- **Property of Lines Perpendicular to a Plane.** If two lines are perpendicular to a plane, then the two lines are parallel.

(3) *Determinations of perpendicular lines*

- **Converted to Line Perpendicular to Plane**
- **Three Perpendicular Lines Theorem and Inverse Theorem**

(4) *Determinations and property of lines parallel to planes*

(i) Determinations

- **Determination Theorem.** If a line is parallel to a line contained in a plane, then the first line is parallel to the plane.
- **Property of Parallel Planes.** If two planes are parallel, then any line in one plane is parallel to the other plane.

(ii) Property

- **Property of Lines Parallel to Planes.** If a line is parallel to a plane, and another plane contains the line and intersects the plane, then the line is parallel to the intersection line of the two planes. When encountering a line parallel to a plane, it is often necessary to make an auxiliary plane that contains the given line and intersects the given plane in order to use the property of lines parallel to planes.

(5) *Determinations and properties of lines perpendicular to planes*

(i) Determinations

- **Determination Theorem I.** If a line is perpendicular to two intersecting lines in a plane, then the line is perpendicular to the plane.
- **Determination Theorem II.** If one of the two parallel lines is perpendicular to a plane, then the other line is also perpendicular to the plane.

(ii) Properties

- **Property I.** If a line is perpendicular to a plane, then the line is perpendicular to all lines in that plane.
- **Property II.** If two lines are perpendicular to the same plane, then the two lines are parallel.

(6) *Determination and property of parallel planes*

(i) Determination

- **Determination Theorem.** If two intersecting lines in a plane are parallel to another plane, then the two planes are parallel.

(ii) Property

- **Property of Parallel Planes.** If two parallel planes intersect a third plane, then their intersection lines are parallel.

(7) *Determinations and property of perpendicular planes*

(i) Determinations

- **Determination Theorem.** If a plane contains a line that is perpendicular to another plane, then the two planes are perpendicular to each other.
- **Definition Method.** The dihedral angle formed by two intersecting planes is a right dihedral angle.

(ii) Property

- **Property of Perpendicular Planes.** If two planes are perpendicular, then the line perpendicular to their intersection in one plane is perpendicular to the other plane.

6.1.2 Angles and distance in space

(1) *Angle between two skew lines*

(i) Scope: $\left(0, \frac{\pi}{2}\right]$.
(ii) Solution: The key to calculating the angle between two skew lines is translation (midpoint translation, vertex translation, and complement method: the space figure is filled into a familiar or complete geometry, such as cubes, parallelepipeds, cuboids, etc., so that it is easy to find a relationship of the skew lines), converted into the angle of the intersection of two lines in the plane.

(2) *Angle between a line and a plane*

(i) Definition: The acute angle formed by an oblique line and its projection in the plane is called the angle between a line and a plane.

(ii) Scope: $\left[0, \frac{\pi}{2}\right]$.

(iii) Solution: make the projection of the line on the plane.

(iv) Characteristic of the angle between a line and a plane: This angle is the smallest of the angles between the oblique line and all the lines in the plane.

(3) *Dihedral angles*

  (i) Scope: $[0, \pi]$.

  (ii) Solution:

  - **Definition Method.** Directly take a point (special point) on the edge of the dihedral angle, make the perpendicular lines of the edge in the two half planes respectively, and obtain the dihedral angle. When using the definition method, it is necessary to carefully observe the characteristics of the figure.
  - **Three Perpendicular Lines Method.** Through a point in one plane to make a line perpendicular to another plane, by the three perpendicular lines theorem or its reverse theorem, the dihedral angle is obtained.
  - **Perpendicular Plane Method.** Through a point, make a plane perpendicular to the edge, and then the intersection lines of the perpendicular plane and two half planes forms the dihedral angle.
  - **Area Projection Method.** Using the area projection formula $S_{\text{proj}} = S_{\text{orig}} \cdot \cos\theta$, where θ is the dihedral angle. For a class of dihedral angles that do not show an edge, it is necessary to extend the two half planes such that they intersect to form an edge, and then use the above method (especially considering the area projection method).

(4) *Distance between two skew lines*

The distance between two skew lines is often calculated in one of three ways:

  - **Direct Method.** Find a segment perpendicular to two skew lines.
  - **Conversion Method I.** Convert to the distance between the line and the plane, that is, construct a plane that contains one of the lines and is parallel to the other line.
  - **Conversion Method II.** Convert to the distance between the parallel planes, that is, construct two parallel planes that contain two skew lines respectively.

(5) *Distance from a point to a plane*

- **Perpendicular Plane Method.** Use the property of perpendicular planes to make a perpendicular line, and the key is to determine the plane through the given point and perpendicular to the given plane.
- **Volumetric Method.** Convert to the height of the triangular pyramid.
- **Equivalent Transfer Method**

(6) *Distance between a line and a plane*

For a line parallel to a plane, the distance from any point on the line to the plane is the same, which is converted to the distance from a point to a plane.

(7) *Distance between two parallel planes*

Convert to the distance from a point to a plane.

(8) *Spherical distance*

The length of an inferior arc of a great circle passing through two points on the sphere.

The steps to calculate the spherical distance between two points A, B on the sphere are as follows: first calculate the length of segment AB, then calculate the radian of the spherical central angle $\angle AOB$, and finally use the arc length formula to calculate the length of the inferior arc $\overset{\frown}{AB}$.

6.1.3 *Polyhedra and solids of revolution*

(1) *Lateral surface area (sum of each lateral area)*

 (i) Prism: Lateral surface area $S =$ perimeter of the vertical cross section (this section is perpendicular to each lateral edge) × lateral edge length.

　　In particular, the lateral surface area of a right prism $S =$ base perimeter × lateral edge length.

 (ii) Regular pyramid: The lateral surface area of a regular pyramid $S = \frac{1}{2}$ × base perimeter × slant height.

(iii) Cone: The lateral surface area of a cone $S = \pi$ × base radius × slant height.

(2) *Volume*

 (i) Column: Volume = base area × height = area of vertical cross section × lateral edge length.

In particular, the volume of a right prism = base area × lateral edge length. The volume of a triangular prism = $V = \frac{1}{2}Sd$ (where S is the area of one lateral surface of the triangular prism, and d is the distance between the edge parallel to the lateral surface).

(ii) Cone: Volume = $\frac{1}{3}$ × base area × height.

In particular, a common method for calculating the volume of a polyhedron is the decrement and supplement method (decrement or supplement into a polyhedron whose volume is easy to calculate).

(iii) Volume and surface area of a sphere: $V = \frac{4}{3}\pi R^3$, $S = 4\pi R^2$.

Area of a spherical crown: $S = 2\pi Rh$.

Example 6.1. As shown in Figure 6.1, E, F, G, H are points on the edges AB, AD, BC, CD of a space quadrilateral $ABCD$, and the lines EF, GH intersect at point M. Prove:

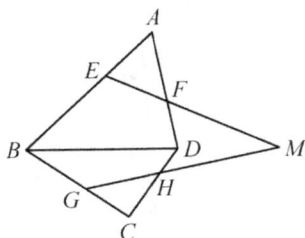

Figure 6.1

(a) B, D, M are collinear;

(b) $\frac{AE}{EB} \cdot \frac{BG}{GC} \cdot \frac{CH}{HD} \cdot \frac{DF}{FA} = 1.$

Proof. (a) Since $M \in EF$ and $EF \subseteq$ plane ABD, so $M \in$ plane ABD. Similarly, $M \in$ plane BCD. Then $M \in$ plane $ABD \cap$ plane BCD, that is $M \in BD$, so the conclusion holds.

(b) In $\triangle ABD$, the segment EFM is the transversal line. By Menelaus's theorem,

$$\frac{AE}{EB} \cdot \frac{BM}{MD} \cdot \frac{DF}{FA} = 1. \tag{1}$$

In $\triangle BCD$, the segment GHM is the transversal line. By Menelaus's theorem,

$$\frac{BG}{GC} \cdot \frac{CH}{HD} \cdot \frac{DM}{MB} = 1. \tag{2}$$

Multiplying (1) and (2), we get the conclusion.

Example 6.2. As shown in Figure 6.2, let O be the center of the base triangle ABC of a regular triangular pyramid $P - ABC$. The dynamic plane through O intersects PC at point S, and intersects the extension lines of PA, PB at points Q, R, respectively. Prove: $\frac{1}{PQ} + \frac{1}{PR} + \frac{1}{PS}$ is a definite value.

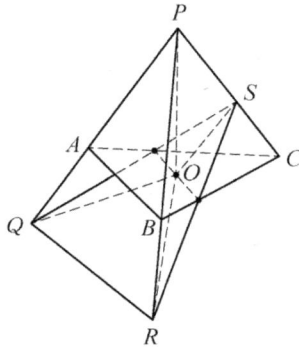

Figure 6.2

Proof.   The distances from point O to planes PAB, PBC, PCA are equal. Let $\angle APB = \alpha$. Then

$$V_{PQRS} = V_{O-PQR} + V_{O-PRS} + V_{O-PSQ}$$

$$= \frac{1}{6} d(PQ \cdot PR + PR \cdot PS + PS \cdot PQ) \sin \alpha,$$

where d is the distance from O to each plane.
And

$$V_{PQRS} = \frac{1}{3} S_{\triangle PRQ} \cdot h = \frac{1}{6} PQ \cdot PR \cdot PS \sin \alpha \sin \theta,$$

where h is the distance from S to plane PQR and θ is the angle between PS and plane PQR.

Therefore,

$$d(PQ \cdot PR + PR \cdot PS + PS \cdot PQ) = PQ \cdot PR \cdot PS \sin \theta.$$

Then $\frac{1}{PQ} + \frac{1}{PR} + \frac{1}{PS} = \frac{\sin \theta}{d}$ is a definite value.

Example 6.3. The edges SA, SB, SC of a triangular pyramid $S - ABC$ have points A_1, B_1, C_1, respectively, and satisfy

$$SA \cdot SA_1 = SB \cdot SB_1 = SC \cdot SC_1.$$

Prove: points A, B, C, A_1, B_1, C_1 are on the same sphere.

Proof. Through the four non-coplanar points A, B, C, and A_1, the sphere δ can be constructed, which intersects SB and SC at points B_1', C_1', respectively. The points A_1, A, C, C_1' are on the cross-section circle of the sphere δ truncated by the plane SAC, and the points A_1, A, B, B_1' are on the cross-section circle of the sphere δ truncated by the plane SAB.

By the intersecting secants theorem,

$$SA \cdot SA_1 = SB \cdot SB_1', \quad SA \cdot SA_1 = SC \cdot SC_1'.$$

So, from the known conditions,

$$SB_1' = \frac{SA \cdot SA_1}{SB} = SB_1,$$

$$SC_1' = \frac{SA \cdot SA_1}{SC} = SC_1,$$

that is, point B_1' coincides with B_1 and point C_1' coincides with C_1, so points A, B, C, A_1, B_1, C_1 are on the same sphere.

Example 6.4. As shown in Figure 6.3, in a triangular pyramid $S - ABC$, the edges SA, SB, SC are perpendicular in pairs. Suppose M is the median point of $\triangle ABC$ and D is the midpoint of AB. Construct the line DP parallel to SC.

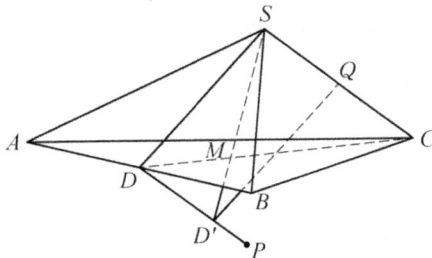

Figure 6.3

(a) Prove: DP intersects SM.

(b) Let the intersection point of DP and SM be D', and prove that D' is the center of the circumscribed sphere of the triangular pyramid $S - ABC$.

Proof.   (a) Since $DP \parallel SC$, we see that DP, CS are coplanar. So $DC \subseteq$ plane DPC.

Since $M \in DC$, we have $M \in$ plane DPC and $SM \in$ plane DPC.

Thus in the plane DPC, the segment SM intersects SC, so the line SM intersects DP.

(b) By the known conditions, the edges SA, SB, SC are perpendicular in pairs, so $SC \perp$ plane SAB and $SC \perp SD$.

Since $SC \parallel DP$, we know that $\triangle DD'M \backsim \triangle CSM$.

Since M is the median point of $\triangle ABC$, so $DM : MC = 1 : 2$, and then $DD' : SC = 1 : 2$.

Take the midpoint Q of SC and connect to form $D'Q$, and then $SQ = DD'$. Combining $DP \parallel SC$ and $SC \perp SD$, we know that the quadrilateral $DD'QS$ is a rectangular.

Thus $D'Q \perp SC$, and by the median and the altitude coincide theorem, $D'C = D'S$.

Similarly, $D'A = D'B = D'C = D'S$. Thus, the spherical D' is constructed with D' as the center and $D'S$ as the radius. Then A, B, C are all on this sphere, that is, D' is the center of the circumscribed sphere of the triangular pyramid $S - ABC$.

6.2   Problems and Solutions

6.2.1   *Existence problems*

Problem 6.1 (IMO 11-3, proposed by Poland). For each value of $k = 1, 2, 3, 4, 5$, find a necessary and sufficient condition on a number $a > 0$ for the existence of a tetrahedron with k edges of length a, and the remaining $(6 - k)$ edges of length 1.

Solution. There are five situations to discuss:

(1) When $k = 1$, without loss of generality, let tetrahedron $ABCD$ satisfy $AB = BC = CA = BD = CD = 1$, and $AD = a$. Let the midpoint of BC be E. Then $AE = DE = \frac{\sqrt{3}}{2}$. And let $AE = DE = \frac{\sqrt{3}}{2}$, $\angle AED = 0$. Then it is easy to know that $AD = \sqrt{3} \sin \frac{\theta}{2}$.

Since $0 < \theta < 180°$, we know that

$$0 < a < \sqrt{3}.$$

(2) When $k = 2$, it must be divided into two cases.

Case 1. The two edges of length a have a common vertex, denoted D, and the other three vertices, A, B, C, form an equilateral triangle with side length 1. Suppose $DA = 1$ and $BD = CD = a$. Let the angle between DA and the plane ABC be θ, and the midpoint of BC be E, so it is easy to know that

$$a = \sqrt{BE^2 + (AE + AD\cos\theta)^2 + (AD\sin\theta)^2}$$

$$= \sqrt{\frac{1}{4} + \left(\frac{\sqrt{3}}{2} + \cos\theta\right)^2 + \sin^2\theta}$$

$$= \sqrt{2 + \sqrt{3}\cos\theta},$$

where $0 < \theta < 180°$. Thus

$$\sqrt{2 - \sqrt{3}} < a < \sqrt{2 + \sqrt{3}}.$$

Case 2. The two edges of length a are opposite edges, and without loss of generality, let $BC = AD = a$ and $AB = AC = BD = CD = 1$. In this case, let the midpoint of BC be $E, \angle AED = \theta$. Then

$$a = AD = 2AE\sin\frac{\theta}{2}$$

$$= 2\sqrt{1 - \frac{a^2}{4}}\sin\frac{\theta}{2}$$

$$= \sqrt{4 - a^2}\sin\frac{\theta}{2}$$

$$< \sqrt{4 - a^2} \quad (0 < 0 < 180°),$$

which is $0 < a < \sqrt{2}$.

Considering the two cases together, we get

$$0 < a < \sqrt{2 + \sqrt{3}}.$$

(3) When $k = 3$, there is always a tetrahedron. If $a \geq 1$, take an equilateral triangle $\triangle ABC$ with side length 1, and let its center be G, and construct a

line l perpendicular to the plane ABC through G. It is easy to know that a point D can be taken on l such that $AD = BD = CD = a$. If $a \leq 1$, take an equilateral triangle $\triangle ABC$ whose side length is a, with the median point and l as before, and take the point D on l such that $AD = BD = CD = 1$. In this case, only $a > 0$ is required.

(4) When $k = 4$, the tetrahedron has four edges of length a and the other two edges of length 1. Consider a similar tetrahedron with four edges of length 1 and the other two edges of lengths $b = \frac{1}{a}$, which returns to (2).

By $0 < b < \sqrt{2 + \sqrt{3}}$, we know

$$a > \frac{1}{b} = \sqrt{2 - \sqrt{3}}.$$

(5) When $k = 5$, similarly it can be translated into (1), and it is easy to know that

$$a > \frac{1}{\sqrt{3}}.$$

【Score Situation】 This particular problem saw the following distribution of scores among contestants: 31 contestants scored 7 points, 13 contestants scored 6 points, 12 contestants scored 5 points, 15 contestants scored 4 points, 5 contestants scored 3 points, 10 contestants scored 2 points, 20 contestants scored 1 point, and 6 contestants scored 0 point. The average score of this problem is 4.196, indicating that it was simple.

Among the top five teams in the team scores, the scores of this problem are as follows: the Hungary team scored 54 points (with a total team score of 247 points), the German Democratic Republic team scored 44 points (with a total team score of 240 points), the Soviet Union team scored 42 points (with a total team score of 231 points), the Romania team scored 42 points (with a total team score of 219 points), and the United Kingdom team scored 50 points (with a total team score of 193 points).

The gold medal cutoff for this IMO was set at 40 points (with 3 contestants earning gold medals), the silver medal cutoff was 30 points (with 20 contestants earning silver medals), and the bronze medal cutoff was 24 points (with 21 contestants earning bronze medals).

In this IMO, only three contestants achieved a perfect score of 40 points, namely Tibor Fiala from Hungary, Vladimir Drinfeld from the Soviet Union and Simon Phillips Norton from the United Kingdom.

Problem 6.2 (IMO 13-2, proposed by the Soviet Union). Consider a convex polyhedron P_1 with nine vertices $A_1, A_2, \ldots, A_9$; let P_i be the

polyhedron obtained from P_1 by a translation that moves vertex A_1 to A_i $(i = 2, 3, \ldots, 9)$. Prove that at least two of the polyhedral $P_1, P_2, \ldots, P_9$ have an interior point in common.

Proof. Create a space Cartesian coordinate system with A_1 as the origin and let

$$\overrightarrow{A_1 B_i} = 2 \overrightarrow{A_1 A_i}, \quad i = 2, 3, \ldots, 9.$$

Denote the polyhedron $A_1 B_2 B_3 \cdots B_9$ as D, that is, D is the polyhedron obtained by doubling P_1; its volume is 8 times that of P_1, and D contains P_1.

Let Q_i be a point in the polyhedron P_i $(i = 2, 3, \ldots, 9)$ and its corresponding point in P_1 be Q_1. Then

$$\overrightarrow{A_1 Q_i} = \overrightarrow{A_1 Q_1} + \overrightarrow{A_1 A_i} = 2 \cdot \frac{1}{2} (\overrightarrow{A_1 Q_1} + \overrightarrow{A_1 A_i}).$$

Since both Q_1 and A_i are in the convex polyhedron P_1, the point $\frac{1}{2}(\overrightarrow{A_1 Q_1} + \overrightarrow{A_1 A_i})$ is also in the convex polyhedron P_1, and thus $\overrightarrow{A_1 Q_i}$ is in the convex polyhedron D, that is, point Q_i is in D.

It can be seen from the above that D contains polyhedra $P_2, P_3, \ldots, P_9$. Since D contains P_1 and $P_2, P_3, \ldots, P_9$ are obtained by translations of P_1, and the volume of D is 8 times that of P_1, there are at least two of these 9 polyhedra $P_1, P_2, \ldots, P_9$, which have an interior point in common.

【Score Situation】This particular problem saw the following distribution of scores among contestants: 2 contestants scored 7 points, 3 contestants scored 6 points, no contestant scored 5 points, 1 contestant scored 4 points, 3 contestants scored 3 points, 3 contestants scored 2 points, 2 contestants scored 1 point, and 14 contestants scored 0 point. The average score of this problem is 1.893, indicating that it was relatively challenging.

Among the top five teams in the team scores, the scores of this problem are as follows: the Hungary team scored 38 points (with a total team score of 255 points), the Soviet Union team scored 21 points (with a total team score of 205 points), the German Democratic Republic team scored 16 points (with a total team score of 142 points), the Poland team scored 20 points (with a total team score of 118 points), the Romania team scored 20 points (with a total team score of 110 points), and the United Kingdom team scored 20 points (with a total team score of 110 points).

The gold medal cutoff for this IMO was set at 35 points (with 7 contestants earning gold medals), the silver medal cutoff was 23 points (with 12 contestants earning silver medals), and the bronze medal cutoff was 11 points (with 29 contestants earning bronze medals).

In this IMO, only one contestant achieved a perfect score of 42 points, namely Imre Ruzsa from Hungary.

Problem 6.3 (IMO 14-6, proposed by the United Kingdom). Given four distinct parallel planes, prove that there exists a regular tetrahedron with a vertex in each plane.

Proof. As shown in Figure 6.4, let the four planes be E_1, E_2, E_3, E_4, and the distance between plane E_i and plane E_{i+1} be d_i $(i = 1, 2, 3)$.

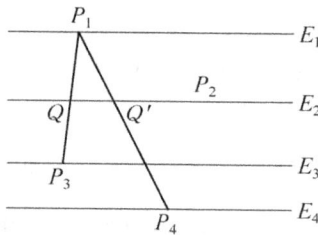

Figure 6.4

Suppose there is a tetrahedron $P_1 P_2 P_3 P_4$, where point P_i falls inside E_i, and let $P_1 P_3$ and $P_1 P_4$ intersect plane E_2 at points Q and Q', respectively.

It is easy to know that
$$\frac{P_1 Q}{P_3 Q} = \frac{d_1}{d_2}, \frac{P_1 Q'}{P_4 Q'} = \frac{d_1}{d_2 + d_3}.$$

Points P_2, Q, Q' are not collinear, and then they can form a plane.

Thus, we can solve the problem in the following way.

First find a regular tetrahedron, let it be $P_1' P_2' P_3' P_4'$, and find points Q_1 and Q_1' on $P_1' P_3'$ and $P_1' P_4'$, respectively, such that
$$\frac{P_1' Q_1}{P_3' Q_1} = \frac{d_1}{d_2}, \frac{P_1' Q_1'}{P_4' Q_1'} = \frac{d_1}{d_2 + d_3}.$$

Construct the plane E_2' crossing points P_2' and Q_1, Q_1', and then construct planes E_1', E_3', E_4' such that they are all parallel to E_2' and crossing points P_1', P_3', P_4', respectively.

Let the distance between E_i' and E_{i+1}' be d_i' $(i = 1, 2, 3)$. It is easy to know that $d_1' : d_2' : d_3' = d_1 : d_2 : d_3$, that is, $E_i'(i = 1, 2, 3, 4)$ is similar to E_i $(i = 1, 2, 3, 4)$.

Then let E_1 and E_1' coincide, P_i' and E_j, E_j' are on the same side of $E_1 (2 \leq i, j \leq 4)$. Let P_1' be P_1, and on the ray $P_1 P_2', P_1 P_3', P_1 P_4'$ respectively find points (may be set as P_2, P_3, P_4) such that
$$P_1 P_2 : P_1 P_2' = P_1 P_3 : P_1 P_3' = P_1 P_4 : P_1 P_4' = d_1 : d_1'.$$

It is easy to know that the tetrahedron $P_1P_2P_3P_4$ is a regular tetrahedron, and the point P_i falls on E_i ($i = 1, 2, 3, 4$).

【Score Situation】This particular problem saw the following distribution of scores among contestants: 11 contestants scored 8 points, no contestant scored 7 points, 2 contestants scored 6 points, 2 contestants scored 5 points, no contestant scored 4 points, 2 contestants scored 3 points, 4 contestants scored 2 points, 4 contestants scored 1 point, and 8 contestants scored 0 point. The average score of this problem is 3.879, indicating that it was relatively straightforward.

Among the top five teams in the team scores, the scores of this problem are as follows: the Soviet Union team scored 55 points (with a total team score of 270 points), the Hungary team scored 52 points (with a total team score of 263 points), the German Democratic Republic team scored 36 points (with a total team score of 239 points), the Romania team scored 44 points (with a total team score of 208 points), and the United Kingdom team scored 30 points (with a total team score of 179 points).

The gold medal cutoff for this IMO was set at 40 points (with 8 contestants earning gold medals), the silver medal cutoff was 30 points (with 16 contestants earning silver medals), and the bronze medal cutoff was 19 points (with 30 contestants earning bronze medals).

In this IMO, a total of eight contestants achieved a perfect score of 40 points.

Problem 6.4 (IMO 15-2, proposed by Poland). Determine whether or not there exists a finite set M of points in the space not lying in the same plane such that, for any two points A and B of M, one can select two other points C and D of M so that lines AB and CD are parallel and not coincident.

Solution. The answer is that there exists such a set of points.

A concrete example of the construction is shown in Figure 6.5.

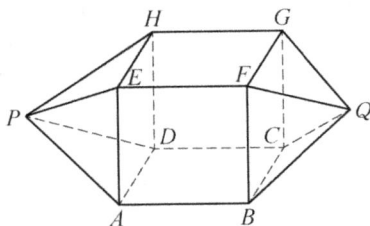

Figure 6.5

Set $M = \{A, B, C, D, E, F, G, H, P, Q\}$, where $ABCD - EFGH$ is a cube, P is the symmetric point of the center of the cube with respect to plane $ADHE$, and Q is the symmetric point of the center of the cube with respect to plane $BCGF$. Then the point set M satisfies the conditions of the problem.

Note. Examples are not unique. For example, let M be composed of the vertex, the center, the center of each face, and the midpoint of each edge of a cube. Then this M also satisfies the problem.

【Score Situation】 This particular problem saw the following distribution of scores among contestants: 48 contestants scored 6 points, 6 contestants scored 5 points, 4 contestants scored 4 points, no contestant scored 3 points, no contestant scored 2 points, 2 contestants scored 1 point, and 65 contestants scored 0 point. The average score of this problem is 2.688, indicating that it had a certain level of difficulty.

Among the top five teams in the team scores, the scores of this problem are as follows: the Soviet Union team scored 41 points (with a total team score of 254 points), the Hungary team scored 31 points (with a total team score of 215 points), the German Democratic Republic team scored 18 points (with a total team score of 188 points), the Poland team scored 36 points (with a total team score of 174 points), and the United Kingdom team scored 30 points (with a total team score of 164 points).

The gold medal cutoff for this IMO was set at 35 points (with 5 contestants earning gold medals), the silver medal cutoff was 27 points (with 15 contestants earning silver medals), and the bronze medal cutoff was 17 points (with 48 contestants earning bronze medals).

In this IMO, only one contestant achieved a perfect score of 40 points, namely Sergei Konyagin from the Soviet Union.

Problem 6.5 (IMO 18-3, proposed by Finland). A rectangular box can be filled completely with unit cubes. If one places as many cubes as possible, each with volume 2, in the box, so that their edges are parallel to the edges of the box, then one can fill exactly 40% of the box. Determine the possible dimensions of such boxes.

Solution. Let the edges of the cuboid box be a_1, a_2, a_3, respectively, where a_1, a_2, a_3 are positive integers satisfying $1 < a_1 \le a_2 \le a_3$. Let $b_i = \left[\frac{a_i}{\sqrt[3]{2}} \right]$, $i = 1, 2, 3$. By the conditions of the problem,

$$40\% \cdot a_1 a_2 a_3 = 2 b_1 b_2 b_3,$$

i.e.,

$$\frac{a_1 a_2 a_3}{b_1 b_2 b_3} = 5. \tag{1}$$

A list the values of $a, b, \frac{a}{b}$ are as follows (where $b = \left[\frac{a}{\sqrt[3]{2}}\right]$):

a	2	3	4	5	6	7	8	9	10
b	1	2	3	3	4	5	6	7	7
$\frac{a}{b}$	2	1.5	1.3	1.6	1.5	1.4	1.3	1.23	1.42

When $a > 10$, we have $b \geq \left[\frac{11}{\sqrt[3]{2}}\right] = 8$, so

$$\frac{a}{b} = \frac{a}{\left[\frac{a}{\sqrt[3]{2}}\right]} < \frac{\sqrt[3]{2}}{\left[\frac{a}{\sqrt[3]{2}}\right]} \left(\left[\frac{a}{\sqrt[3]{2}}\right] + 1\right)$$

$$= \sqrt[3]{2}\left(1 + \frac{1}{b}\right) \leq \sqrt[3]{2}\left(1 + \frac{1}{8}\right) < 1.42.$$

Then

$$\left(\frac{a}{b}\right)^3 < 1.42^3 < 5.$$

If $a_1 > 2$, then

$$\frac{a_1 a_2 a_3}{b_1 b_2 b_3} < \left(\frac{5}{3}\right)^3 < 5,$$

which is not possible. Thus $a_1 = 2$.

If $a_1 = a_2 = 2$, then by (1),

$$\frac{a_3}{b_3} = \frac{5}{4} < \sqrt[3]{2},$$

which is also not possible.

If $a_1 = 2$ and $a_2 \geq 6$, then

$$\frac{a_2 a_3}{b_2 b_3} = \left(\frac{3}{2}\right)^2 < \frac{5}{2},$$

which is in contradiction with (1). Thus $3 \leq a_2 \leq 5$.

When $a_1 = 2, a_2 = 3$, by (1), we get $\frac{a_3}{b_3} = \frac{5}{3}$, and thus $a_3 = 5$.

When $a_1 = 2, a_2 = 4$, by (1), we get $\frac{a_3}{b_3} = \frac{15}{8}$, which is not possible.

When $a_1 = 2, a_2 = 5$, by (1), we get $\frac{a_3}{b_3} = \frac{3}{2}$, and thus $a_3 = 6$.

In summary, the size of the cuboid box can only be $2 \times 3 \times 5$ or $2 \times 5 \times 6$.

【Score Situation】 This particular problem saw the following distribution of scores among contestants: 17 contestants scored 8 points, 8 contestants scored 7 points, 10 contestants scored 6 points, 6 contestants scored 5 points, 7 contestants scored 4 points, 15 contestants scored 3 points, 19 contestants scored 2 points, 40 contestants scored 1 point, and 17 contestants scored 0 point. The average score of this problem is 3.115, indicating that it was relatively straightforward.

Among the top five teams in the team scores, the scores of this problem are as follows: the Soviet Union team scored 46 points (with a total team score of 250 points), the United Kingdom team scored 52 points (with a total team score of 214 points), the United States team scored 32 points (with a total team score of 188 points), the Bulgaria team scored 37 points (with a total team score of 174 points), and the Austria team scored 34 points (with a total team score of 167 points).

The gold medal cutoff for this IMO was set at 34 points (with 9 contestants earning gold medals), the silver medal cutoff was 23 points (with 28 contestants earning silver medals), and the bronze medal cutoff was 15 points (with 45 contestants earning bronze medals).

In this IMO, only one contestant achieved a perfect score of 40 points, namely Laurent Pierre from France.

6.2.2 *Position structure problems*

Problem 6.6 (IMO 2-5, proposed by Czechoslovakia). Consider a cube $ABCDA'B'C'D'$ (with face $ABCD$ directly above face $A'B'C'D'$).

(a) Find the locus of the midpoints of segments XY, where X is any point of AC and Y is any point of $B'D'$.

(b) Find the locus of points Z which lie on the segments XY of part (a) with $ZY = 2XZ$.

Solution. (a) As shown in Figure 6.6, construct plane α that is parallel to plane ABC, and their distance is $\frac{1}{2}AA'$. Then the part of plane α in tetrahedron $ACB'D'$, that is, the square $EFGH$ and its interior, are the desired locus, where E, F, G, H are the centers of the four sides.

Since X, Y are on the surface of tetrahedron $ACB'D'$, the segment XY must pass through its "midsection" $EFGH$, and this intersection point Z is the midpoint of XY, and completeness is proved.

Purity: Let Z be any point in or on the side of the square $EFGH$, construct the plane $D'B'Z$ that intersects AC at X, and then construct the

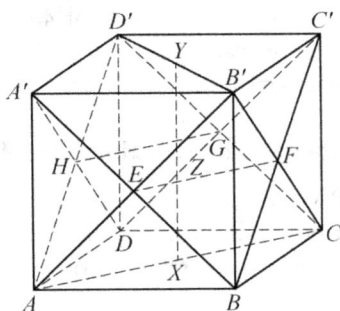

Figure 6.6

line through X, Z that intersects $D'B'$ at Y. From the position of $EFGH$, we know that Z is the midpoint of XY.

(b) Similarly, construct plane β that is parallel to plane ABC, and their distance is $\frac{1}{3}AA'$. Then the part of plane β in tetrahedron $ACB'D'$, that is, the rectangle $E'F'G'H'$ and its interior, is the desired, where E', F', G', H' are the trisection points of AB', CB', CD', AD' respectively.

Note. This problem can also be proved by the coordinate method. We only need to pay attention to the following fact: if $P_1 = (x_1, y_1, z_1)$ and $P_2 = (x_2, y_2, z_2)$ are any two points in the space, then all points P on the segment $P_1 P_2$ can be expressed as

$$P = (1 - \lambda)P_1 + \lambda P_2$$
$$= ((1 - \lambda)x_1 + \lambda x_2, (1 - \lambda)y_1 + \lambda y_2, (1 - \lambda)z_1 + \lambda z_2),$$

where P runs through the whole segment $P_1 P_2$ from P_1 to P_2 when the parameter λ is traversed by $[0, 1]$.

【Score Situation】 This particular problem saw the following distribution of scores among contestants: 1 contestant scored 7 points, no contestant scored 6 points, 1 contestant scored 5 points, 1 contestant scored 4 points, no contestant scored 3 points, no contestant scored 2 points, 2 contestants scored 1 point, and 5 contestants scored 0 point. The average score of this problem is 1.800, indicating that it was relatively challenging.

Among the top five teams in the team scores, the Czechoslovakia team achieved a total score of 257 points, the Hungary team achieved a total score of 248 points, the Romania team achieved a total score of 248 points, the Bulgaria team achieved a total score of 175 points, and the German Democratic Republic team achieved a total score of 38 points.

The gold medal cutoff for this IMO was set at 40 points (with 4 contestants earning gold medals), the silver medal cutoff was 37 points (with 4 contestants earning silver medals), and the bronze medal cutoff was 33 points (with 4 contestants earning bronze medals).

In this IMO, no contestant achieved a perfect score of 44 points.

Problem 6.7 (IMO 3-6, proposed by Romania). Consider a plane ε and three non-collinear points A, B, C on the same side of ε; suppose the plane determined by these three points is not parallel to ε. In the plane take three arbitrary points A', B', C'. Let L, M, N be the midpoints of segments AA', BB', CC'; let G be the centroid of triangle LMN. (We will not consider positions of the points A', B', C' such that the points L, M, N do not form a triangle.) What is the locus of point G as A', B', C' range independently over the plane ε?

Solution. Put all of these points in a space coordinate system, and let ε be $z = 0$. Then A, B, C are in the upper space $z > 0$.

It is easy to know that the midpoint of two points $X = (x_1, x_2, x_3)$ and $Y = (y_1, y_2, y_3)$ is

$$\frac{X+Y}{2} = \left(\frac{x_1 + y_1}{2}, \frac{x_2 + y_2}{2}, \frac{x_3 + y_3}{2}\right),$$

and the median point of $\triangle XYZ$ is $\frac{X+Y+Z}{3}$.

Thus, the midpoints of segments AA', BB', CC' satisfy that $L = \frac{1}{2}(A + A')$, $M = \frac{1}{2}(B+B')$, and $N = \frac{1}{2}(C+C')$, so the median point O of $\triangle LMN$ satisfies

$$O = \frac{1}{3}(L + M + N) = \frac{1}{6}(A + A' + B + B' + C + C')$$

$$= \frac{1}{2}\left[\frac{1}{3}(A + B + C) + \frac{1}{3}(A' + B' + C')\right] = \frac{1}{2}(G + G'),$$

where G, G' are the median points of $\triangle ABC$ and $\triangle A'B'C'$, respectively. It is easy to see that O is in the plane $z = \frac{1}{6}(a_3 + b_3 + c_3)$ (where a_3, b_3, c_3 are the coordinates on the z-axis of A, B, C). This plane is parallel to the plane $z = 0$, and the distance to the plane $z = 0$ is half the distance from G to the plane $z = 0$.

Conversely, choose any point O in the plane $z = \frac{1}{6}(a_3 + b_3 + c_3)$, extend GO to intersect the plane $z = 0$ at point G', find any three points A', B', C' on the plane $z = 0$, such that G' is the median point of $\triangle A'B'C'$, and let the midpoints of AA', BB', CC' be L, M, N respectively. From the previous calculation, we know that: O is the median point of $\triangle LMN$.

Thus, the locus is the plane as described above, which satisfies: it is parallel to plane ε, the distance to ε is half the distance from the median point G of $\triangle ABC$ to ε, and it is on the same side of ε as G.

【Score Situation】 This particular problem saw the following distribution of scores among contestants: 1 contestant scored 7 points, no contestant scored 6 points, 1 contestant scored 5 points, 1 contestant scored 4 points, no contestant scored 3 points, no contestant scored 2 points, 1 contestant scored 1 point, and 5 contestants scored 0 point. The average score of this problem is 1.889, indicating that it was relatively challenging.

Among the top five teams in the team scores, the Hungary team achieved a total score of 270 points, the Poland team achieved a total score of 203 points, the Romania team achieved a total score of 197 points, the Czechoslovakia team achieved a total score of 159 points, and the German Democratic Republic team achieved a total team score of 146 points.

The gold medal cutoff for this IMO was set at 37 points (with 3 contestants earning gold medals), the silver medal cutoff was 34 points (with 4 contestants earning silver medals), and the bronze medal cutoff was 30 points (with 4 contestants earning bronze medals).

In this IMO, only one contestant achieved a perfect score of 40 points, namely Béla Bollobás from Hungary.

Problem 6.8 (IMO 4-3, proposed by Czechoslovakia). Consider a cube $ABCDA'B'C'D'$ ($ABCD$ and $A'B'C'D'$ are the upper and lower bases, respectively, and edges AA', BB', CC', DD' are parallel). A point X moves at constant speed along the perimeter of the square $ABCD$ in the direction $ABCDA$, and a point Y moves at the same rate along the perimeter of the square $B'C'CB$ in the direction $B'C'CBB'$. Points X and Y begin their motion at the same instant from the starting positions A and B', respectively. Determine and draw the locus of the midpoints of the segments XY.

Solution. With A as the origin, AB, AD, and AA' are located in the positive direction of x-axis, y-axis, and z-axis, respectively, and the length of edge is the unit length, and the time of point X passing through one edge is taken as the unit of time.

When $0 \leq t \leq 1$, the coordinates of points X, Y are $(t, 0, 0)$, $(1, t, 1)$, and the midpoint is

$$Z = \frac{X + Y}{2} = \left(\frac{1+t}{2}, \frac{t}{2}, \frac{1}{2} \right).$$

When $1 \leq t \leq 2$, the coordinates of points X, Y are $(1, t-1, 0), (1, 1, 2-t)$, and the midpoint is

$$Z = \frac{X+Y}{2} = \left(1, \frac{t}{2}, \frac{2-t}{2}\right).$$

When $3 \leq t \leq 4$, the coordinates of points X, Y are $(0, 4-t, 0), (1, 0, t-3)$, and the midpoint is

$$Z = \frac{X+Y}{2} = \left(\frac{1}{2}, \frac{4-t}{2}, \frac{t-3}{2}\right).$$

The following proves that if $X = (x, y, z)$ satisfies that x, y, z are all linear functions of the parameter t, this is, $x = x(t) = a_1 t + b_1$, $y = y(t) = a_2 t + b_2$, and $z = z(t) = a_3 t + b_3$ for $t \in [t_0, t_1]$, then the locus of X is a segment with $X_0(x(t_0), y(t_0), z(t_0))$ and $X_1(x(t_1), y(t_1), z(t_1))$ as the endpoints.

It is obviously only necessary to prove that $X_0 X + X_1 X = X_0 X_1$, and from the continuity of the linear function we know that no point on the segment can be excepted.

Just pay attention to the following representation:

$$X_0 X = \sqrt{a_1^2 + a_2^2 + a_3^2}(t - t_0),$$

$$X_1 X = \sqrt{a_1^2 + a_2^2 + a_3^2}(t_1 - t),$$

$$X_0 X_1 = \sqrt{a_1^2 + a_2^2 + a_3^2}(t_1 - t_0).$$

Back to the front, when $t = 0, 1, 2, 3, 4$, the point Z is $Z_1 \left(\frac{1}{2}, 0, \frac{1}{2}\right)$, $Z_2 \left(1, \frac{1}{2}, \frac{1}{2}\right)$, $Z_3(1, 1, 0)$, $Z_4 \left(\frac{1}{2}, \frac{1}{2}, 0\right)$, $Z_1 \left(\frac{1}{2}, 0, \frac{1}{2}\right)$, respectively.

And the coordinates (x, y, z) of point Z are linear functions of t in each time period, so in each time period, Z forms segments $Z_1 Z_2, Z_2 Z_3, Z_3 Z_4, Z_4 Z_1$ respectively.

Z_1 is the midpoint of AB', Z_2 is the midpoint of BC', Z_3 is the point C, and Z_4 is the midpoint of AC. Since $Z_1 Z_2 = Z_3 Z_4$, and $Z_1 Z_2 \parallel Z_3 Z_4$, $Z_1 Z_4 = \frac{1}{2} B'C = \frac{1}{2} AC = Z_3 Z_4$, so the locus of Z is a rhomboid $Z_1 Z_2 Z_3 Z_4$.

【Score Situation】This particular problem saw the following distribution of scores among contestants: 6 contestants scored 8 points, 2 contestants scored 7 points, 1 contestant scored 6 points, no contestant scored 5 points, 3 contestants scored 4 points, 1 contestant scored 3 points, no contestant scored 2 points, no contestant scored 1 point, and 4 contestants scored 0 point. The average score of this problem is 4.882, indicating that it was simple.

Among the top five teams in the team scores, the Hungary team achieved a total team score of 289 points, the Soviet Union team achieved a total score of 263 points, the Romania team achieved a total score of 257 points, the Czechoslovakia team achieved a total score of 212 points, and the Poland team achieved a total score of 212 points.

The gold medal cutoff for this IMO was set at 41 points (with 4 contestants earning gold medals), the silver medal cutoff was 34 points (with 12 contestants earning silver medals), and the bronze medal cutoff was 29 points (with 15 contestants earning bronze medals).

In this IMO, only one contestant achieved a perfect score of 46 points, namely Iosif Bernstein from the Soviet Union.

Problem 6.9 (IMO 4-7, proposed by the Soviet Union). A tetrahedron $SABC$ has the following property: there exist five spheres, each tangent to the edges SA, SB, SC, BC, CA, AB or to their extensions.

(a) Prove that the tetrahedron $SABC$ is regular.
(b) Prove conversely that for every regular tetrahedron five such spheres exist.

Proof. (a) If the sphere Ω is tangent to all the edges or their extensions of the tetrahedron $SABC$, then the sphere intersects each triangular face to form a truncated circle, and this circle is an incircle or excircle of each face, because the circle has a unique common point with each side or extension. Obviously, the circles on two adjacent faces of the same edge have the same tangency point on this edge.

The following two cases are discussed:

(i) The tangency point is inside each edge. In this case, the circles truncated on each face are the incircles of the triangular face. Let the tangency points on the three edges BC, CA, AB of $\triangle ABC$ be P, Q, R, respectively, and the sphere and SA are tangent to K. Since points P, Q, R are not collinear and K is not on the plane ABC, there is a unique sphere that exists and passes through the vertices of the tetrahedron $KPQR$. So, if this first type of sphere exists, then there is only one.

(ii) If the sphere is tangent to the extension line of one edge, might as well let it be the extension line SA, and the tangency point is L, then the truncated circles on the plane SAB, SAC become the excircles of $\triangle SAB, \triangle SAC$. And it is easy to know that the sphere is tangent to the extension lines of SB, SC respectively (also tangent to AB, AC respectively). Let the tangency points be M, N. It is easy to know that M, N are also two tangency points of the excircle of $\triangle SBC$. Thus, this second type of sphere is tangent to the three edges of the plane $\triangle ABC$, and tangent to the extension lines of the edges SA, SB, SC. Similarly, one can prove its existence.

Repeating the above discussion, we know that there are four second type of spheres. In addition, there is no "third type of sphere," so the five spheres required are exactly the first type (one) and the second type (four).

Let AB, BC, CA, SA, SB, SC be c, a, b, a', b', c'. The radius of the first type of sphere is r and the center is O. Then

$$a + a' = BC + SA = BP + CP + SK + AK$$
$$= \sqrt{SO^2 - r^2} + \sqrt{SA^2 - r^2} + \sqrt{SB^2 - r^2} + \sqrt{SC^2 - r^2} = s.$$

Similarly, $b + b', c + c'$ are also the same number s, so let $a' = s - a, b' = s - b$, and $c' = s - c$.

It is easy to know that SL is the half perimeter of $\triangle ASB$ and $\triangle ASC$, and it is not difficult to further deduce that the perimeters of $\triangle ASB, \triangle ASC, \triangle BSC$, and $\triangle ABC$ are equal.

For example, the perimeter of $\triangle SBC$ is $b' + c' + a = 2s + a - b - c$, and the perimeter of $\triangle ASC$ is $a' + c' + b = 2s + b - a - c$, so $a = b$, and similarly, $a = b = c$.

And the perimeter of $\triangle SBC$ is equal to the perimeter of $\triangle ABC$, so $2s - b - c + a = 2s - a = a + b + c = 3a$, and then $s = 2a$. Thus, $SA = s - a = a$. Similarly, $SB = SC = a$, so the six edges of tetrahedron $SABC$ are all equal, and consequently it must be a regular tetrahedron.

(b) Assuming O is the centroid of tetrahedron $SABC$, we can prove that the distance between the centroid and the edges are equal, so the first type of sphere exists.

By symmetry, any sphere whose center is on the line SO must be simultaneously tangent to SA, SB, SC, and must also be simultaneously tangent to AB, BC, CA (as long as there exists the tangent).

Thus we can construct an outer sphere by finding a point X on SO extended which is equidistant from SA and AB. It is easy to know that the locus of points equidistant from SA and AB consists of two planes perpendicular to $\triangle SAB$ through the bisectors of $\angle SAB$. Of these planes, the one containing the internal bisector of $\angle SAB$ passes through O, while the other one intersects SO extended in the desired point X.

The other three spheres are constructed in a completely similar way.

【Score Situation】 This particular problem saw the following distribution of scores among contestants: 1 contestant scored 8 points, no contestant scored 7 points, 1 contestant scored 6 points, 1 contestant scored 5 points, no contestant scored 4 points, no contestant scored 3 points, 2 contestants scored 2 points, 3 contestants scored 1 point, and 9 contestants scored 0 point. The average score of this problem is 1.529, indicating that it was relatively challenging.

Among the top five teams in the team scores, the Hungary team achieved a total score of 289 points, the Soviet Union team achieved a total score of 263 points, the Romania team achieved a total score of 257 points, the Czechoslovakia team achieved a total score of 212 points, and the Poland team achieved a total score of 212 points.

The gold medal cutoff for this IMO was set at 41 points (with 4 contestants earning gold medals), the silver medal cutoff was 34 points (with 12 contestants earning silver medals), and the bronze medal cutoff was 29 points (with 15 contestants earning bronze medals).

In this IMO, only one contestant achieved a perfect score of 46 points, namely Iosif Bernstein from the Soviet Union.

Problem 6.10 (IMO 5-2, proposed by the Soviet Union). Point A and segment BC are given. Determine the locus of points in the space which are vertices of right angles with one side passing through A, and the other side intersecting the segment BC.

Solution. According to the position relationship of point A and segment BC, the discussion is divided into cases.

(i) If point A is on segment BC, as shown in Figure 6.7, then two spheres are constructed with AB and AC as diameters respectively. It is easy to know that there is no point outside the two spheres that satisfies

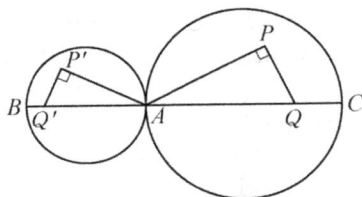

Figure 6.7

the requirements. However, any point inside the two spheres satisfies the requirements (for example, points Q, Q' in the figure). Therefore, the two spheres and their interior form the desired locus (including segment BC), and certainly, when A coincides with B or C, a sphere shrinks to a point.

(ii) If point A is on the extension line of BC, as shown in Figure 6.8, without loss of generality, then let point B lie between A and C. Two spheres are constructed with AB and AC as diameters, respectively. Then the desired locus is the surface of the large sphere and the points inside the large sphere but not inside the small sphere. This is because, if Q is a point in the small sphere, then the plane constructed through Q and perpendicular to AQ intersects AC at AB but not at BC, while for any point P that belongs to the large sphere but not to the small sphere, the plane constructed through P and perpendicular to AP intersects BC.

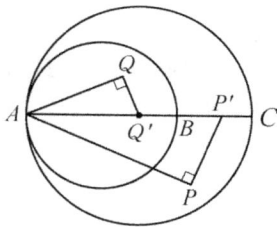

Figure 6.8

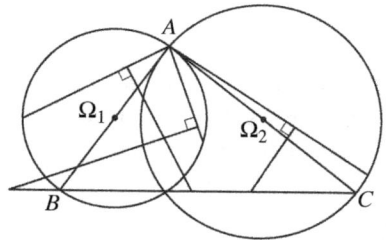

Figure 6.9

(iii) If point A is not on line BC, as shown in Figure 6.9, then two spheres are constructed with AB and AC as diameters, respectively.

It is easy to know that any point on the sphere Ω_1 or Ω_2 belongs to the desired locus, and for any point P, beyond two spheres, the plane through P and perpendicular to AP intersects neither AB nor AC. In this way, this plane is parallel to plane ABC and does not intersect BC. Otherwise, let the intersection line be l, and it is easy to know that since l has no common point with AB, AC, naturally there is no common point with BC. Therefore, any point outside of two spheres does not satisfy the requirements.

And a point in both sphere Ω_1 and sphere Ω_2 also does not satisfy the requirements. This is because, for any point P in this region, the plane constructed through P and perpendicular to AP must intersect AB, AC, and the intersection points are not B, C. So let this plane intersect plane ABC

on line l. Then l "passes through" the interior of AB, AC, and obviously cannot intersect the segment BC. Therefore, such point does not satisfy the requirements.

Any point P that is inside Ω_1 but not inside Ω_2 satisfies the requirements. At this time, the plane through P and perpendicular to AP must intersect AB, not AC. So the intersection line l of this plane and plane ABC "passes through" AB, but does not touch AC, so it has to pass through BC.

Similarly, any point that is inside Ω_2 but not inside Ω_1 also satisfies the requirements.

In summary, if B_1, B_2 represent the set of points inside Ω_1, Ω_2, respectively, and B_1', B_2' represent the set of points on the surface and inside Ω_1, Ω_2, respectively, then the desired locus is $B_1' \cup B_2' \backslash B_1 \cap B_2$.

Note. The solution to this problem repeatedly uses the conclusion that for a line and a triangle on a plane, if they have common points, and the line does not pass through any vertex of the triangle, then the line passes exactly through two sides and the interior, and there is no common point with the third side.

【Score Situation】 This particular problem saw the following distribution of scores among contestants: 2 contestants scored 7 points, no contestant scored 6 points, 3 contestants scored 5 points, 2 contestants scored 4 points, 3 contestants scored 3 points, 1 contestant scored 2 points, 4 contestants scored 1 point, and 1 contestant scored 0 point. The average score of this problem is 3.250, indicating that it was relatively straightforward.

Among the top five teams in the team scores, the Soviet Union team achieved a total score of 271 points, the Hungary team achieved a total score of 234 points, the Romania team achieved a total score of 191 points, the Yugoslavia team achieved a total score of 162 points, and the Czechoslovakia team achieved a total score of 151 points.

The gold medal cutoff for this IMO was set at 35 points (with 7 contestants earning gold medals), the silver medal cutoff was 28 points (with 11 contestants earning silver medals), and the bronze medal cutoff was 21 points (with 17 contestants earning bronze medals).

In this IMO, no contestant achieved a perfect score of 40 points.

Problem 6.11 (IMO 20-2, proposed by the United States). A point P is given inside a sphere. Three mutually perpendicular rays from P intersect the sphere at points A, B, and C; Q denotes the vertex diagonally opposite to P in the parallelepiped determined by PA, PB, and PC. Find the locus of Q for all such triads of rays from P.

Solution. Denote the rectangular cuboid as $APBD - A'CB'Q$. The points A, B, C are on the sphere S. In a Cartesian coordinate system, it is easy to prove by the Pythagorean theorem: for any rectangle $ABCD$ and any point E in the space,

$$AE^2 + CE^2 = BE^2 + DE^2.$$

So let the center of the sphere be O and apply the above conclusion:

$$OQ^2 = AO^2 + B'O^2 - OP^2$$
$$= AO^2 + (OB^2 + OC^2 - OP^2) - OP^2$$
$$= 3r^2 - 2OP^2.$$

Then $OQ = \sqrt{3r^2 - 2OP^2}$ is a constant (where r is the radius of the sphere S). The point Q is on the sphere S' with O as the center and $\sqrt{3r^2 - 2OP^2}$ as the radius.

Next, it is proved that any point Q on the sphere S' has corresponding three points A, B, C. It can be concluded that the locus of Q is exactly the entire sphere S'.

Since $OQ = \sqrt{3r^2 - 2OP^2}$ and $OP < r$, it follows that $OQ > r$, so the sphere with diameter PQ must intersect the sphere S, and its intersection line is a circle. Let C be a point on this circle and consider the rectangle $PCQD$. Since

$$OP^2 + OQ^2 = OC^2 + OD^2,$$

so

$$OD^2 = OP^2 + (3r^2 - 2OP^2) - r^2 = 2r^2 - OP^2 > r^2,$$

and thus, D is outside the sphere S.

Construct plane α, passing through P and perpendicular to PC. Since $\angle CPD = 90°$, the point D is in the plane α.

And let the circle where α intersects S be K, so P is inside K and D is outside K. Then the circle with PD as its diameter in plane α must intersect K. Let the intersection point be B. We see that B is on S and $PB \perp PC$.

And then, construct rectangle $PBDA$, so

$$OP^2 + OD^2 = OA^2 + OB^2;$$

this is

$$OA^2 = OP^2 + 2r^2 - OP^2 - r^2 = r^2.$$

The point A also falls on the sphere S, and edges PA, PB, PC are perpendicular to each other.

【Score Situation】This particular problem saw the following distribution of scores among contestants:4 contestants scored 7 points, 4 contestants scored 6 points, 2 contestants scored 5 points, 10 contestants scored 4 points, 3 contestants scored 3 points, 5 contestants scored 2 points, 4 contestants scored 1 point, and 16 contestants scored 0 point. The average score of this problem is 2.604, indicating that it had a certain level of difficulty.

Among the top five teams in the team scores, the scores of this problem are as follows: the Romania team scored 27 points (with a total team score of 237 points), the United States team scored 36 points (with a total team score of 225 points), the United Kingdom team scored 20 points (with a total team score of 201 points), the Vietnam team scored 20 points (with a total team score of 200 points), and the Czechoslovakia team scored 37 points (with a total team score of 195 points).

The gold medal cutoff for this IMO was set at 35 points (with 5 contestants earning gold medals), the silver medal cutoff was 27 points (with 20 contestants earning silver medals), and the bronze medal cutoff was 22 points (with 38 contestants earning bronze medals).

In this IMO, only one contestant achieved a perfect score of 40 points, namely Mark Kleiman from the United States.

6.2.3 *Quantity relation problems*

Problem 6.12 (IMO 2-6, proposed by Bulgaria). Consider a cone of revolution with an inscribed sphere tangent to the base of the cone. A cylinder is circumscribed about this sphere so that one of its bases lies in the base of the cone. Let V_1 be the volume of the cone and V_2 the volume of the cylinder.

(a) Prove that $V_1 \neq V_2$.
(b) Find the smallest number k for which $V_1 = kV_2$, and for this case, construct the angle subtended by a diameter of the base of the cone at the vertex of the cone.

Solution. (a) The cross section is shown in Figure 6.10, and let the radius of the sphere be R, the radius of the bottom circle of the cone be r, and the height be h.

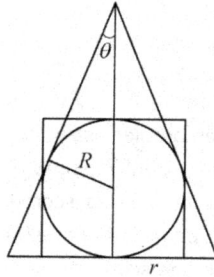

Figure 6.10

It is easy to know that $h > 2R$.

And let half of the apex angle of the cone be θ. Then it is easy to know

$$\frac{R}{h-R} = \frac{r}{\sqrt{r^2+h^2}} = \sin\theta.$$

So,

$$h = \frac{1+\sin\theta}{\sin\theta}R,$$

$$r = h\tan\theta = \frac{1+\sin\theta}{\cos\theta}R.$$

Let $t = \frac{1-\sin\theta}{\sin\theta} > 0, \sin\theta = \frac{1}{1+t}$. Then

$$V_1 = \frac{1}{3}\pi r^2 h = \frac{1}{3}\pi\frac{(1+\sin\theta)^2}{\cos^2\theta}R^2 \cdot \frac{1+\sin\theta}{\sin\theta}R$$

$$= \frac{1}{3}\pi R^3\frac{(1+\sin\theta)^2}{\sin\theta(1-\sin\theta)}$$

$$= \frac{1}{3}\pi R^3\frac{\left(1+\frac{1}{1+t}\right)^2}{\frac{1}{1+t}\cdot\frac{t}{1+t}} = \frac{1}{3}\pi R^3\frac{(2+t)^2}{t}$$

$$= \frac{1}{3}\pi R^3\left(t+\frac{4}{t}+4\right)$$

$$\geq \frac{1}{3}\pi R^3\left(2\sqrt{t\cdot\frac{4}{t}}+4\right) = \frac{8}{3}\pi R^3,$$

while $V_2 = 2\pi R^3$.

Thus, $V_1 \neq V_2$.

(b) By (a), the minimum of $\frac{V_1}{V_2}$ is $\frac{4}{3}$, in which case $t = \frac{4}{t}$, i.e.,

$$\sin\theta = \frac{1}{3}, \quad \theta = \arcsin\frac{1}{3}.$$

Thus, it is easy to construct θ.

Note. In this problem, we can also use half of the base angle of the cone as the parameter (set to α), in which case

$$\frac{V_1}{V_2} = \frac{1}{3\tan^2\alpha(1 - \tan^2\alpha)}.$$

【Score Situation】 This particular problem saw the following distribution of scores among contestants: 1 contestant scored 7 points, 1 contestant scored 6 points, no contestant scored 5 points, no contestant scored 4 points, no contestant scored 3 points, no contestant scored 2 points, 4 contestants scored 1 point, and 4 contestants scored 0 point. The average score of this problem is 1.700, indicating that it was relatively challenging.

Among the top five teams in the team scores, the Czechoslovakia team achieved a total score of 257 points, the Hungary team achieved a total score of 248 points, the Romania team achieved a total score of 248 points, the Bulgaria team achieved a total score of 175 points, and the German Democratic Republic team achieved a total score of 38 points.

The gold medal cutoff for this IMO was set at 40 points (with 4 contestants earning gold medals), the silver medal cutoff was 37 points (with 4 contestants earning silver medals), and the bronze medal cutoff was 33 points (with 4 contestants earning bronze medals).

In this IMO, no contestant achieved a perfect score of 44 points.

Problem 6.13 (IMO 6-6, proposed by Poland). In tetrahedron $ABCD$, vertex D is connected with D_1 the centroid of $\triangle ABC$. Lines parallel to DD_1 are drawn through A, B, and C. These lines intersect the planes BCD, CAD, and ABD at points A_1, B_1, and C_1, respectively. Prove that the volume of $ABCD$ is one third of the volume of $A_1B_1C_1D_1$. Is the result true if point D_1 is selected anywhere within $\triangle ABC$?

Proof. As shown in Figure 6.11, let E, F, G be the midpoints of BC, CA, AB, respectively. Then point A_1 is on ED and $\triangle ED_1D \backsim \triangle EA_1A$, so $AA_1 \parallel DD_1$ and $AA_1 = 3DD_1$.

Similarly, $BB_1 \parallel CC_1 \parallel DD_1$ and $BB_1 = CC_1 = 3DD_1$. Thus, it is not difficult to know that $A_1B_1 = 2EF = AB$. Then similarly, $B_1C_1 = BC$, $C_1A_1 = CA$, $\triangle ABC \cong \triangle A_1B_1C_1$, and the planes of the two triangles are

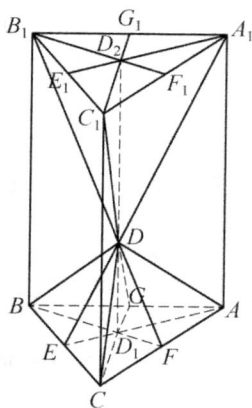

Figure 6.11

parallel. Moreover, the height of tetrahedron $A_1B_1C_1D_1$ is three times the height of tetrahedron $ABCD$. Therefore,

$$V_{A_1B_1C_1D_1} = 3V_{ABCD}.$$

When D_1 is any point inside $\triangle ABC$, the conclusion still holds.

Let DD_1 intersect plane $A_1B_1C_1$ at point D_2, and let $A_1D_2, B_1D_2,$ C_1D_2 intersect B_1C_1, C_1A_1, A_1B_1 at points E_1, F_1, G_1, respectively. The points E, F, G are the intersection points of AD_1, BD_1, CD_1 and $BC,$ CA, AB, respectively. It is easy to see that A_1, B_1, C_1 are on the lines ED, FD, GD, respectively.

Since A is outside the plane B_1BCC_1, we have $AA_1 \parallel BB_1$, so $EE_1 \parallel AA_1$. Then

$$\frac{D_2D}{EE_1} = \frac{A_1D_2}{A_1E_1} = \frac{AD_1}{AE}.$$

And in trapezoid BCC_1B_1, it is easy to know

$$EE_1 = \frac{BE}{BC} \cdot CC_1 + \frac{CE}{BC} \cdot BB_1 = \left(\frac{BE}{BC} \cdot \frac{CG}{D_1G} + \frac{CE}{BC} \cdot \frac{BF}{D_1F}\right) \cdot DD_1.$$

Thus,

$$D_2D = \left(\frac{BE}{BC} \cdot \frac{CG}{D_1G} \cdot \frac{AD_1}{AE} + \frac{CE}{BC} \cdot \frac{BF}{D_1F} \cdot \frac{AD_1}{AE}\right) \cdot DD_1.$$

By Menelaus's theorem, both products in the brackets are 1, so

$$D_1D_2 = 3DD_1. \qquad (1)$$

Construct a plane $\alpha \perp D_1 D_2$, and then arbitrarily construct two parallel planes β_1 and β_2 through the points D_1, D_2, which truncate a cylinder with the lines AA_1, BB_1, and CC_1. Let the cross-section area of this cylinder be S, the height be h, and the dihedral angle between β_1 and α be θ. Then

$$S = \frac{S_0}{\cos\theta}, \quad h = D_1 D_2 \cos\theta,$$

where S_0 is the triangular area formed by the intersection of plane α and lines AA_1, BB_1, CC_1, so the volume V of this cylinder is the constant $S_0 \cdot D_1 D_2$. By the cone volume formula and (1),

$$V_{D_1 A_1 B_1 C_1} = V_{D_2 ABC} = 3V_{DABC}.$$

【Score Situation】This particular problem saw the following distribution of scores among contestants:1 contestant scored 9 points, no contestant scored 8 points, no contestant scored 7 points, 15 contestants scored 6 points, no contestant scored 5 points, no contestant scored 4 points, no contestant scored 3 points, no contestant scored 2 points, no contestant scored 1 point, and 1 contestant scored 0 point. The average score of this problem is 5.824, indicating that it was simple.

Among the top five teams in the team scores, the scores of this problem are as follows: the Soviet Union team scored 52 points (with a total team score of 269 points), the Hungary team scored 47 points (with a total team score of 253 points), the Romania team scored 48 points (with a total team score of 213 points), the Poland team scored 47 points (with a total team score of 209 points), and the Bulgaria team scored 29 points (with a total team score of 198 points).

The gold medal cutoff for this IMO was set at 38 points (with 7 contestants earning gold medals), the silver medal cutoff was 31 points (with 9 contestants earning silver medals), and the bronze medal cutoff was 27 points (with 19 contestants earning bronze medals).

In this IMO, only one contestant achieved a perfect score of 42 points, namely David Bernstein from the Soviet Union.

Problem 6.14 (IMO 7-3, proposed by Czechoslovakia). Given a tetrahedron $ABCD$ whose edges AB and CD have lengths a and b, respectively. The distance between the skew lines AB and CD is d, and the angle between them is ω. The tetrahedron $ABCD$ is divided into two solids by a plane ε, parallel to lines AB and CD. The ratio of the distances of ε from AB and CD is equal to k. Compute the ratio of the volumes of the two solids obtained.

Solution. Since the plane ε is parallel to AB and CD, it intersects the tetrahedron in a parallelogram $PQRS$ with $PQ \parallel SR \parallel CD$ and $QR \parallel PS \parallel AB$, as shown in Figure 6.12.

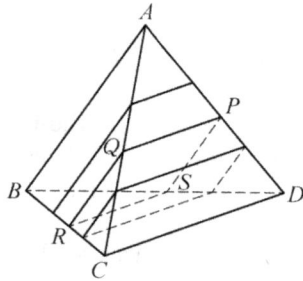

Figure 6.12

In the prismatoid $ABRSPQ$, the midsection that is parallel to $PQRS$, the lengths of its adjacent edges are $\left(1 + \frac{k}{2}\right) RQ$ and $\frac{1}{2}PQ$, respectively. This result is easily obtained by using the proportional property of parallel lines, and hence its area $M_1 = \frac{1}{2}\left(1 + \frac{k}{2}\right) S_{PQRS}$.

Similarly, in the prismatoid $PQRCDS$, there also exists a midsection that is parallel to plane $PQRS$, and its area $M_2 = \frac{1}{2}\left(1 + \frac{1}{2k}\right) S_{PQRS}$.

Using the volume formula of prismatoids and the conditions, it is easy to know that the ratio of the volumes of the above two solids is

$$\frac{k(S_{PQRS} + 4M_1)}{S_{PQRS} + 4M_2} = \frac{k(1 + 2 + k)}{1 + \frac{2k+1}{k}} = \frac{k^2(k+3)}{3k+1}.$$

【Score Situation】 This particular problem saw the following distribution of scores among contestants: 30 contestants scored 8 points, 3 contestants scored 7 points, 2 contestants scored 6 points, 3 contestants scored 5 points, 3 contestants scored 4 points, no contestant scored 3 points, 6 contestants scored 2 points, 10 contestants scored 1 point, and 23 contestants scored 0 point. The average score of this problem is 4.025, indicating that it was simple.

Among the top five teams in the team scores, the scores of this problem are as follows: the Soviet Union team scored 64 points (with a total team score of 281 points), the Hungary team scored 50 points (with a total team score of 244 points), the Romania team scored 47 points (with a total team score of 222 points), the Poland team scored 33 points (with a total team score of 178 points), and the German Democratic Republic team scored 42 points (with a total team score of 175 points).

The gold medal cutoff for this IMO was set at 38 points (with 8 contestants earning gold medals), the silver medal cutoff was 30 points (with 12 contestants earning silver medals), and the bronze medal cutoff was 20 points (with 17 contestants earning bronze medals).

In this IMO, only two contestants achieved a perfect score of 40 points, namely László Lovász from Hungary and Pavel Bleher from the Soviet Union.

Problem 6.15 (IMO 8-3, proposed by Bulgaria). Prove: The sum of the distances of the vertices of a regular tetrahedron from the centre of its circumscribed sphere is less than the sum of the distances of these vertices from any other point in space.

Proof 1. Through four vertices of this regular tetrahedron, make opposite parallel planes, and then enclose them into a large regular tetrahedron. Since the distance from a point outside a plane to any point in the plane is not less than the distance between the point outside the plane and its projection point onto the plane. So, the sum Σ of the distances of any point P to the four vertices of the original tetrahedron is not less than the sum Σ' of the distances of this point P to the four planes of the large tetrahedron, i.e., $\Sigma \geq \Sigma'$. (The equality sign holds only if the connecting lines between the point P and each vertex of the original tetrahedron are perpendicular to their corresponding planes of the large tetrahedron, in which case only the centre of the circumscribed sphere of the original tetrahedron satisfies the requirement.)

When the point P is inside or on the surface of the large tetrahedron, since the product of $\frac{1}{3}\Sigma'$ and the area of one face of the large tetrahedron is the volume of the large tetrahedron, Σ' is the height h of the large tetrahedron, i.e., $\Sigma' = h$. When the point P is outside the large tetrahedron, the product of $\frac{1}{3}\Sigma'$ and the area of one face of the large tetrahedron is greater than the volume of the large tetrahedron, and then $\Sigma' > h$.

Hence $\Sigma \geq h$ (constant); the proposition holds.

Proof 2. The center O of a regular tetrahedron is the point of intersection of six planes, each through an edge and the midpoint of the opposite edge of the tetrahedron. We shall show that if a point P fails to be on one of these planes (i.e., if it is not O), then the sum Σ is not minimal. From this, the proposition holds. This method is called the local adjustment method.

Let the regular tetrahedron be $ABCD$ and E be the midpoint of CD. If the point P is not in the plane ABE, then let l be a line passing through point P and parallel to CD. It is easy to see that l is perpendicular to the plane ABE. And let P' be the intersection of l and $\triangle ABE$. Then

$$P'C + P'D < PC + PD.$$

Indeed triangles CPD and $CP'D$ have the same base and altitude, and since the latter is isosceles, it has the smaller perimeter.

Moreover, PA, PB are the hypotenuses of the right triangles APP' and BPP' respectively, so

$$P'A < PA, P'B < PB.$$

Adding these three inequalities together, we get

$$P'A + P'B + P'C + P'D < PA + PB + PC + PD.$$

Hence the proposition holds.

【Score Situation】 This particular problem saw the following distribution of scores among contestants: 12 contestants scored 7 points, 1 contestant scored 6 points, 2 contestants scored 5 points, 3 contestants scored 4 points, 3 contestants scored 3 points, 2 contestants scored 2 points, 1 contestant scored 1 point, and 3 contestants scored 0 point. The average score of this problem is 4.667, indicating that it was simple.

Among the top five teams in the team scores, the Soviet Union team achieved a total score of 293 points, the Hungary team achieved a total score of 281 points, the German Democratic Republic team achieved a total score of 280 points, the Poland team achieved a total score of 269 points, and the Romania team achieved a total score of 257 points.

The gold medal cutoff for this IMO was set at 39 points (with 13 contestants earning gold medals), the silver medal cutoff was 34 points (with 15 contestants earning silver medals), and the bronze medal cutoff was 31 points (with 11 contestants earning bronze medals).

In this IMO, a total of 11 contestants achieved a perfect score of 40 points.

Problem 6.16 (IMO 9-2, proposed by Czechoslovakia). Prove that if one and only one edge of a tetrahedron is greater than 1, its volume is smaller than or equal to $\frac{1}{8}$.

Proof. Let the tetrahedron be $ABCD$, and the three edges of $\triangle ABC$ be $AB = c, BC = a, CA = b$, respectively, and none greater than 1. Among the six edges, only $AD > 1$.

Let the distance from A to BC be h. As shown in Figure 6.13, construct the perpendicular bisector MN of BC, which meets the parallel line AN of BC at point N. Then $MN = h$ (M is the midpoint of BC).

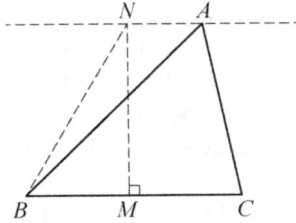

Figure 6.13

Connect MN, and without loss of generality, let point A be to the right of BN. Then it is easy to know that $BN \leq BA \leq 1$.

By the Pythagorean theorem,

$$h^2 + \left(\frac{a}{2}\right)^2 = BN^2,$$

and then

$$h \leq \sqrt{1 - \frac{a^2}{4}}.$$

Let k be the altitude of $\triangle BCD$ from D to BC, and similarly it can be proved that $k \leq \sqrt{1 - \frac{a^2}{4}}$.

Since the height of point D to the bottom $\triangle ABC$ is less than or equal to k,

$$V_{ABCD} \leq \frac{1}{3} S_{\triangle ABC} \cdot k$$

$$= \frac{1}{6} ahk \leq \frac{1}{6} a \left(1 - \frac{a^2}{4}\right)$$

$$= \frac{1}{6} a \left(1 - \frac{a}{2}\right) \left(1 + \frac{a}{2}\right).$$

Since $a \leq 1$,

$$V_{ABCD} \leq \frac{1}{4} a \left(1 - \frac{a}{2}\right)$$

$$\leq \frac{1}{2} \frac{\left(\frac{a}{2} + 1 - \frac{a}{2}\right)^2}{4} = \frac{1}{8}.$$

It is easy to know that if $\triangle ABC$ and $\triangle DBC$ are equilateral triangles with side length 1, which are located in the planes perpendicular to each other, then $V_{ABCD} = \frac{1}{8}$.

【Score Situation】 This particular problem saw the following distribution of scores among contestants: 13 contestants scored 7 points, 1 contestant scored 6 points, 7 contestants scored 5 points, 2 contestants scored 4 points, 2 contestants scored 3 points, 2 contestants scored 2 points, no contestant scored 1 point, and 10 contestants scored 0 point. The average score of this problem is 4.054, indicating that it was simple.

Among the top five teams in the team scores, the Soviet Union team achieved a total score of 275 points, the German Democratic Republic team achieved a total score of 257 points, the Hungary team achieved a total score of 251 points, the United Kingdom team achieved a total score of 231 points, and the Romania team achieved a total score of 214 points.

The gold medal cutoff for this IMO was set at 38 points (with 11 contestants earning gold medals), the silver medal cutoff was 30 points (with 14 contestants earning silver medals), and the bronze medal cutoff was 22 points (with 26 contestants earning bronze medals).

In this IMO, a total of five contestants achieved a perfect score of 42 points.

Problem 6.17 (IMO 12-5, proposed by Bulgaria). In a tetrahedron $ABCD$, angle BDC is a right angle. Suppose that the foot S of the perpendicular from D to the plane ABC is the intersection point of the altitudes of $\triangle ABC$.

Prove that $(AB + BC + CA)^2 \leq 6(AD^2 + BD^2 + CD^2)$.

Proof. As shown in Figure 6.14, $DS\perp$ plane ABC. Thus $DS\perp AC$.

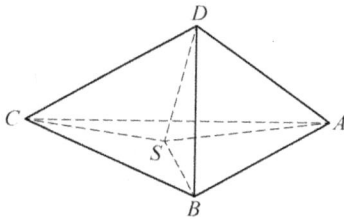

Figure 6.14

And $BS\perp AC$. Then $BD\perp AC$.
Also $BD\perp CD$, so $BD\perp$ plane ADC.
Consequently, $BD\perp AD, \angle ADB = 90°$.
Similarly, $\angle ADC = 90°$.
By the Pythagorean theorem,
$$AB^2 = AD^2 + BD^2, BC^2 = BD^2 + CD^2, CA^2 = CD^2 + AD^2.$$

Using Cauchy's inequality (or the AM-GM inequality), we get

$$(AB + BC + CA)^2 \leq 3(AB^2 + BC^2 + CA^2)$$
$$= 6(AD^2 + BD^2 + CD^2).$$

If the equality holds, then there must be $AB = BC = CA$, which leads to $AD = BD = CD$. In this case, $\triangle ABD, \triangle ACD, \triangle BCD$ are isosceles right triangles, and $\triangle ABC$ is an equilateral triangle.

【Score Situation】 This particular problem saw the following distribution of scores among contestants: 15 contestants scored 6 points, 1 contestant scored 5 points, 1 contestant scored 4 points, 3 contestants scored 3 points, 2 contestants scored 2 points, 3 contestants scored 1 point, and 9 contestants scored 0 point. The average score of this problem is 3.371, indicating that it was relatively straightforward.

Among the top five teams in the team scores, the Hungary team achieved a total score of 233 points, the German Democratic Republic team achieved a total score of 221 points, the Soviet Union team achieved a total score of 221 points, the Yugoslavia team achieved a total score of 209 points, and the Romania team achieved a total score of 208 points.

The gold medal cutoff for this IMO was set at 37 points (with 7 contestants earning gold medals), the silver medal cutoff was 30 points (with 11 contestants earning silver medals), and the bronze medal cutoff was 19 points (with 40 contestants earning bronze medals).

In this IMO, only three contestants achieved a perfect score of 40 points, namely Wolfgang Burmeister from the German Democratic Republic, Imre Ruzsa from Hungary and Andrei Hodulev from the Soviet Union.

6.3 Summary

In the first 64th IMOs, there are 17 problems on solid geometry as depicted in Figure 6.15, which can be broadly categorized into three types.

Problems 6.1–6.5 focus on "existence problems;" among these five problems, the one with the lowest average score is Problem 6.2 (IMO 13-2), proposed by the Soviet Union. Problems 6.6–6.11 deal with "position structure problems;" among these six problems, the one with the lowest average score is Problem 6.9 (IMO 4-7), proposed by the Soviet Union. Problems 6.12–6.17 are about "quantity relation problems;" among these six problems, the one with the lowest average score is Problem 6.12 (IMO 2-6), proposed by Bulgaria.

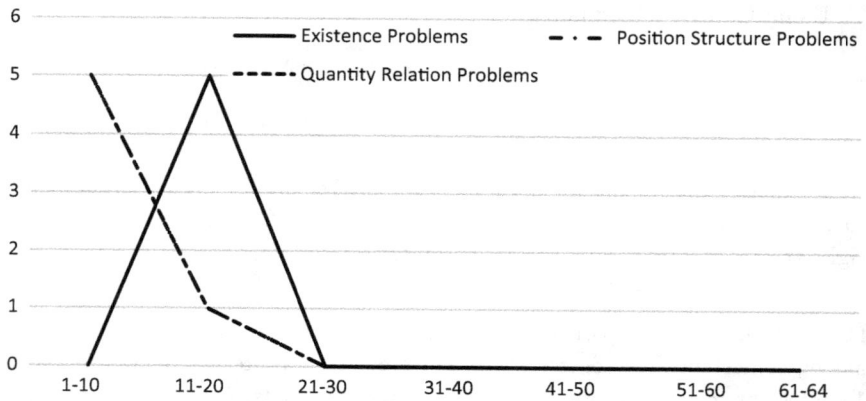

Figure 6.15   Numbers of Solid Geometry Problems in the First 64 IMOs.

These problems were proposed by nine countries and regions, with Czechoslovakia contributing the most, totaling four problems. Bulgaria and the Soviet Union each proposed three problems, while Poland contributed two problems.

From Table 6.2, it can be observed that in the first 64 IMOs, there were five problems with average scores of 1–2 points; two problems with average scores of 2–3 points; four problems with average scores of 3–4 points; six problems with average scores above 4 points. Overall, solid geometry problems had a certain difficulty.

In the 24th to 64th IMOs, there were no solid geometry problems. This indicates that solid geometry problems were primarily featured in the first 20 IMOs, as shown in Table 6.3. Moreover, the score information of the contestants was missing at that time, so there were many gaps in the table.

From Table 6.2, it can be observed that, in most cases, the average score of the top five teams is typically about 1–2 points higher than the average score of the problem. There are exceptions where the average score of the top five teams is lower than the average score of the problem, such as Problem 6.13 (IMO 6-6). Moreover, the average score of the 6th–15th teams is generally lower than the average score of the problem, such as Problem 6.2 (IMO 13-2), Problem 6.4 (IMO 15-2), Problem 6.5 (IMO 18-3), and Problem 6.11 (IMO 20-2). This phenomenon is due to the smaller number of participating teams in early IMOs. It was not until the 30th IMO in 1989 that the number of participating teams exceeded 50. However,

Table 6.2 Score Details of Solid Geometry Problems in the First 64 IMOs.

Problem	6.1	6.2	6.3	6.4	6.5	6.6
Full points	7.000	7.000	8.000	6.000	8.000	7.000
Average score	4.196	1.893	3.879	2.688	3.115	1.800
Top five mean	5.800	2.813	5.425	3.900	5.025	
6th–15th mean		1.299		2.175	2.463	
16th–25th mean						
Problem number in the IMO	11-3	13-2	14-6	15-2	18-3	2-5
Proposing country	Poland	Soviet Union	United Kingdom	Poland	Finland	Czechoslovakia

Problem	6.7	6.8	6.9	6.10	6.11	6.12
Full points	7.000	8.000	8.000	7.000	7.000	7.000
Average score	1.889	4.882	1.529	3.250	2.604	1.700
Top five mean		5.900			3.500	
6th–15th mean					2.355	
16th–25th mean						
Problem number in the IMO	3-6	4-3	4-7	5-2	20-2	2-6
Proposing country	Romania	Czechoslovakia	Soviet Union	Soviet Union	United States	Bulgaria

Problem	6.13	6.14	6.15	6.16	6.17
Full points	9.000	8.000	7.000	7.000	6.000
Average score	5.824	4.025	4.667	4.054	3.371
Top five mean	5.575	5.900			
6th–15th mean					
16th–25th mean					
Problem number in the IMO	6-6	7-3	8-3	9-2	12-5
Proposing country	Poland	Czechoslovakia	Bulgaria	Czechoslovakia	Bulgaria

Note. Top five mean = Total score of the top five teams–Total number of contestants from the top five teams,
6th–15th mean = Total score of the 6th–15th teams/Total number of contestants from the 6th–15th teams,
16th–25th mean = Total score of the 16th–25th teams/Total number of contestants from the 16th–25th teams.

Table 6.3 Numbers of Solid Geometry Problems in the 24th–64th IMOs.

Solid Geometry Problem	Problem Number			Number of Problems in the First 64 IMOs
	1, 4	2, 5	3, 6	
Existence problems	0	0	0	5
Position structure problems	0	0	0	6
Quantity relation problems	0	0	0	6
Total	0	0	0	17

after the 20th IMO, the type of problems in this chapter has not appeared again.

Chapter 7

Geometric Inequalities

In geometry, the expression of an inequality relation of quantities is called a geometric inequality, which can be divided into segments inequalities, angles inequalities, and area inequalities. When solving problems of geometric inequalities, it is often necessary to use some basic theorems that have been learned and conclusions that have been proved. Many problems can be solved and proved by geometric, trigonometric, algebraic, and other methods. At the same time, it is necessary to consider characteristics and properties of geometric figures.

In the first 64th IMOs, there had been a total of 24 problems related to geometric inequalities, accounting for 19.5% of all geometry problems. These problems are mainly categorized into three types: (1) existence problems, totaling five problems; (2) equality or inequality proof problems, totaling 16 problems; (3) conditional evaluation problems, totaling three problems. The statistical distribution of these three types of problems in the previous IMOs is presented in Table 7.1.

It can be seen that geometric inequality problems constitute an important content in the field of geometry, and the frequency of occurrence in the 21st to the 50th IMOs is more than one quarter. However, after the 50th IMO, geometric inequality problems did not appear again, partly because the knowledge of inequalities was already examined in the field of algebra.

Secondly, geometric inequality problems focus on proving an equality or inequality relation, and this type of problems is mainly concentrated in the 21st to 50th IMOs. In particular, among equality or inequality proof problems, there are 10 problems related to lengths, three problems related to angles, and three problems related to areas.

Table 7.1   Numbers of Geometric Inequality Problems in the First 64 IMOs

Content	Session							Total
	1–10	11–20	21–30	31–40	41–50	51–60	61–64	
Existence problems	3	1	0	1	0	0	0	5
Equality or inequality proof problems	2	0	4	5	5	0	0	16
Conditional evaluation problems	0	1	2	0	0	0	0	3
Geometry problems	29	18	18	15	20	17	6	123
Percentage of geometric inequality problems among geometry problems	17.2%	11.1 %	33.3%	40.0%	25.0%	0.0%	0.0%	19.5%

Therefore, this chapter will be divided into three parts. The first part introduces common geometric inequalities and relevant important theorems, and explains them by some examples to help understand and master their conditions of use.

The second part focuses on three kinds of problems: "existence problems," "equality or inequality proof problems," and "conditional evaluation problems." These problems are presented in chronological order, and some problems include various solutions and generalizations.

It is important to note that for each problem, the solutions are followed by information on the scores, including the number of contestants in each score range, the average score, the scores of the top five teams. However, early IMOs often lacked information on contestant scores, so the number of contestants in each score range only represents the counted number of contestants, and some problems lack scores of the top five teams.

The third part provides a brief summary of this chapter.

7.1   Related Properties, Theorems, and Methods

7.1.1   *Basic properties of geometric inequalities*

Property 7.1. In a triangle, the sum of the lengths of any two sides is greater than the length of the third side, and the difference of the lengths of any two sides is less than the length of the third side.

Property 7.2. In a triangle, the longest side is opposite the largest angle, and the shortest side is opposite the smallest angle, and vice versa.

Property 7.3. In two triangles where two sets of corresponding sides are equal, the third side with the larger angle is also longer, and vice versa.

Property 7.4. The sum of the distances from any point in a triangle to two vertices is less than the sum of the distances from the other vertex to these two vertices.

Property 7.5. The median on one side of a triangle is less than half the sum of the other sides.

Property 7.6. In $\triangle ABC$, if P is any point on the side BC, then

$$PA \leq \max\{AB, AC\},$$

and the equality holds when point P coincides with point B or point C.

Property 7.7. If a convex polygon is inside another polygon, then the perimeter of the outer polygon is greater than the perimeter of the inner convex polygon.

Property 7.8. The length of a segment inside a convex polygon does not exceed either the length of the longest side of the convex polygon or the length of the longest diagonal of the convex polygon.

Example 7.1. Let P be a point in a triangle ABC. Prove:

$$\sqrt{PA \cdot PB} + \sqrt{PB \cdot PC} + \sqrt{PC \cdot PA} \leq \frac{\sqrt{5}+1}{4}(AB + BC + CA).$$

Proof. Let us prove a lemma first.

Lemma. *Let P be a point in a triangle ABC. If $PA \leq PB \leq PC$, then*

$$PA + PC \leq \frac{1}{2}(AB + BC + CA).$$

Proof of the Lemma. Extend CP to intersect AB at point D. Then

$$PA + PC \leq DA + DC,$$

$$PA + PC \leq PB + PC \leq DB + DC,$$

so

$$PA + PC \leq \min\{DA, DB\} + DC.$$

Without loss of generality, let $DB \leq DA$, and the midpoints of AB, AC be E, F, respectively (see Figure 7.1).

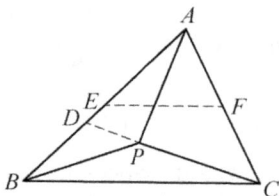

Figure 7.1

Then

$$DB + DC \leq BE + EF + FC = \frac{1}{2}(AB + BC + CA),$$

so

$$PA + PC \leq \frac{1}{2}(AB + BC + CA).$$

The lemma is proved.

Without loss of generality, let $PA \leq PB \leq PC$. Then

$$\sqrt{PA \cdot PB} + \sqrt{PB \cdot PC} + \sqrt{PC \cdot PA} \leq \sqrt{PA \cdot PC} + PC + \sqrt{PC \cdot PA}$$
$$= PC + 2\sqrt{PA \cdot PC}.$$

By the AM–GM inequality

$$2\sqrt{PA \cdot PC} \leq \frac{\sqrt{5}+1}{2}PA + \frac{\sqrt{5}-1}{2}PC.$$

Hence, combined with the lemma we get

$$PC + 2\sqrt{PA \cdot PC} \leq \frac{\sqrt{5}+1}{2}(PA + PC) \leq \frac{\sqrt{5}+1}{4}(AB + BC + CA).$$

7.1.2 *Important theorems of geometric inequalities*

(1) *Ptolemy inequality*

In a convex quadrilateral $ABCD$,

$$AB \cdot CD + AD \cdot BC \geq AC \cdot BD,$$

where the equality holds if and only if the quadrilateral $ABCD$ can be inscribed in a circle.

Proof. As shown in Figure 7.2, take a point M on AB or its extension and a point N on AD or its extension such that $AB \cdot AM = AC^2 = AD \cdot AN$, and connect to form MC, NC, MN, and thus

$$\triangle ABC \backsim \triangle ACM,$$

so

$$MC = BC \cdot \frac{AC}{AB}.$$

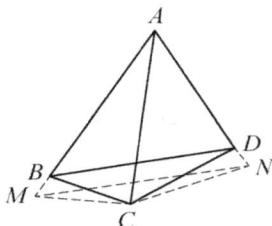

Figure 7.2

Similarly,

$$NC = BC \cdot \frac{AC}{AD}.$$

Also $\triangle ABD \backsim \triangle ANM$, from which

$$MN = BD \cdot \frac{AM}{AD} = \frac{BD}{AD} \cdot \frac{AC^2}{AB}.$$

Since $MN \leq CM + CN$, combining the above equalities, we get

$$\frac{BD}{AD} \cdot \frac{AC^2}{AB} \leq BC \cdot \frac{AC}{AB} + CD \cdot \frac{AC}{AD}.$$

By simplifying the above, we obtain

$$AB \cdot CD + AD \cdot BC \geq AC \cdot BD.$$

A condition for the equality is that the three points M, C, N are collinear. In this case,

$$\angle ABC + \angle ADC = \angle ACM + \angle ACN = 180°,$$

which means that the four points of A, B, C, D are concyclic.

Example 7.2. As shown in Figure 7.3, construct $\triangle DBC, \triangle EAC, \triangle FAB$ outward from $\triangle ABC$ such that

$$\angle DBC = \angle ECA = \angle FAB, \quad \text{and} \quad \angle DCB = \angle EAC = \angle FBA.$$

Prove: $FA + FB + BD + DC + CE + EA \geq AD + BE + CF$.

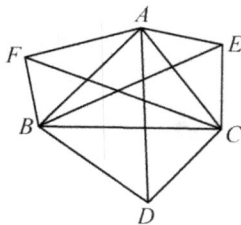

Figure 7.3

Proof. For quadrilateral $ABCD$, using Ptolemy's inequality, we get

$$AB \cdot CD + AC \cdot BD \geq AD \cdot BC,$$

and then

$$\frac{AB \cdot CD}{BC} + \frac{AC \cdot BD}{BC} \geq AD.$$

By the condition $\triangle BCD \backsim \triangle CAE \backsim \triangle ABF$,

$$\frac{AB \cdot CD}{BC} = BF,$$

$$\frac{AC \cdot BD}{BC} = CE.$$

Thus,

$$BF + CE \geq AD.$$

Similarly,

$$AF + CD \geq BE, \quad AE + BD \geq CF.$$

Adding the three inequalities, we find that the original inequality holds.

From the condition for the equality of Ptolemy's inequality to hold, we know that the equality holds if and only if A, B, C, D; B, C, E, A; and C, A, F, B are concyclic.

Then

$$\angle BAC = 180° - \angle BDC = 180° - \angle CEA = \angle ABC.$$

Similarly,

$$\angle BCA = \angle ABC = \angle CAB.$$

Thus, $\triangle ABC$ is an equilateral triangle. In this case,

$$\angle BDC = \angle CEA = \angle AFB = 180° - \angle ACB = 120°.$$

Then the equality holds if and only if $\triangle ABC$ is an equilateral triangle and $\angle BDC = 120°$.

(2) Fermat point

Find a point P in the plane that minimizes the sum of the distances to the three vertices of $\triangle ABC$. This point is called the Fermat point.

(i) When all the internal angles of $\triangle ABC$ are less than 120°, the Fermat point is the point inside $\triangle ABC$ that spreads at the angle of 120° facing each of the three sides.

(ii) When there is an internal angle in $\triangle ABC$ greater than or equal to 120°, the vertex of this internal angle is the Fermat point.

Proof. First of all, we prove that the point P is not outside of $\triangle ABC$. Divide the outside of $\triangle ABC$ into six regions, as shown in Figure 7.4.

If point P is in region I, as shown in Figure 7.5, then

$$AB + AC < PB + PC < PA + PB + PC.$$

That is, the sum of distances from point A to the three vertices is less than the sum of distances from point P to the three vertices. Similarly, for the cases that point P is in regions III and V, we get the same result.

If point P is in region VI, as shown in Figure 7.6. Let BP intersect AC at point Q. Then

$$QA + QB + QC = QB + AC < BP + AC < PA + PB + PC.$$

That is, the sum of distances from point Q to the three vertices is less than the sum of distances from point P to the three vertices. Similarly, for the cases that point P is in regions II and IV, we also get the same result.

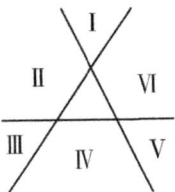

Figure 7.4

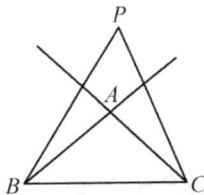

Figure 7.5

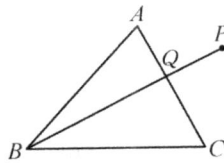

Figure 7.6

Thus, the point P must be in the interior or on the sides of $\triangle ABC$.

(i) When the three internal angles of $\triangle ABC$ are all less than $120°$, construct equilateral $\triangle BCD$, equilateral $\triangle CAE$, and equilateral $\triangle ABF$ by taking BC, CA, AB as sides, respectively, towards the exterior of $\triangle ABC$, and then construct the circumcircles of the three equilateral triangles, respectively. The intersection point of the three circumcircles in $\triangle ABC$, that is, the point that spreads at the angle of $120°$ facing the three sides of $\triangle ABC$, is denoted as point P, as shown in Figure 7.7. The following is to prove: for any point Q inside $\triangle ABC$,

$$PA + PB + PC \leq QA + QB + QC.$$

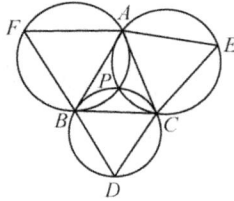

Figure 7.7

Through the three points A, B, C, construct three perpendicular lines of PA, PB, PC, respectively, and the triangle formed by the intersection of the three perpendicular lines is denoted as $\triangle A_1B_1C_1$. Since point P spreads at the angle of $120°$ facing the three sides of $\triangle ABC$, then $\angle B_1A_1C_1 = \angle C_1B_1A_1 = \angle A_1C_1B_1 = 60°$, that is, $\triangle A_1B_1C_1$ is an equilateral triangle, and let its side length be a.

Take any point Q different from P and make perpendicular lines to the three sides of $\triangle A_1B_1C_1$ to obtain three distances h_1, h_2, h_3. By the fact "the sum of the distances from any point inside the equilateral triangle to the three sides is equal to the altitude of the equilateral triangle," we then get

$$2S_{\triangle A_1B_1C_1} = a(PA + PB + PC) = a(h_1 + h_2 + h_3)$$
$$\leq a(QA + QB + QC),$$

so

$$PA + PB + PC \leq QA + QB + QC,$$

and the equality holds only if Q, P coincide.

(ii) When there is an internal angle of $\triangle ABC$ greater than or equal to $120°$, without loss of generality, let $\angle A \geq 120°$ and Q be a point inside or on the side of $\triangle ABC$. Rotate BA around point A such that B becomes point B' on the CA extension line, and Q becomes Q', as shown in Figure 7.8. Since the rotation angle is less than or equal to $60°$, so $QQ' \le AQ$. Thus

$$CA + AB = CB' \le QQ' + Q'B' + QC \le QA + QB + QC,$$

and the equality holds if and only if the point Q coincides with A.

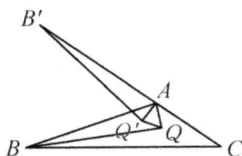

Figure 7.8

In summary, when all the internal angles of $\triangle ABC$ are less than $120°$, the Fermat point is the point inside $\triangle ABC$ that spreads at the angle of $120°$ facing each of the three sides. When there is an internal angle in $\triangle ABC$ greater than or equal to $120°$, the vertex of this internal angle is the Fermat point.

(3) Erdös–Mordell inequality

As shown in Figure 7.9, let P be any point inside triangle ABC, and the distances from point P to the three sides BC, CA, AB are $PD = p$, $PE = q, PF = r$ respectively; denote $PA = x, PB = y, PC = z$. Then

$$x + y + z \ge 2(p + q + r),$$

where the equality holds if and only if the triangle ABC is equilateral and P is the center of this triangle.

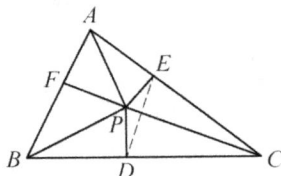

Figure 7.9

Proof. Note that $\angle DPE = 180° - \angle C$, and by the law of cosines,

$$DE = \sqrt{p^2 + q^2 + 2pq \cos C}$$
$$= \sqrt{p^2 + q^2 + 2pq \sin A \sin B - 2pq \cos A \cos B}$$
$$= \sqrt{(p \sin B + q \sin A)^2 + (p \cos B - q\cos A)^2}$$
$$\geq \sqrt{(p \sin B + q \sin A)^2}$$
$$= p \sin B + q \sin A.$$

And since the four points P, D, C, E are concyclic, the segment CP is the diameter of this circle, so

$$z = \frac{DE}{\sin C} \geq \left(\frac{\sin B}{\sin C}\right) p + \left(\frac{\sin A}{\sin C}\right) q.$$

Similarly,

$$x \geq \left(\frac{\sin B}{\sin A}\right) r + \left(\frac{\sin C}{\sin A}\right) q,$$

$$y \geq \left(\frac{\sin A}{\sin B}\right) r + \left(\frac{\sin C}{\sin B}\right) p.$$

Adding the three inequalities, we get

$$x + y + z \geq \left(\frac{\sin B}{\sin C} + \frac{\sin C}{\sin B}\right) p + \left(\frac{\sin A}{\sin C} + \frac{\sin C}{\sin A}\right) q + \left(\frac{\sin B}{\sin A} + \frac{\sin A}{\sin B}\right) r$$
$$\geq 2(p + q + r).$$

The theorem is proved.

Example 7.3. Three circles intersect at one point P.
Prove: The sum of three common chords is not greater than the sum of the radii of three circles.

Proof. Let the three circles be $\odot O_1, \odot O_2, \odot O_3$ with radii r_1, r_2, r_3, respectively. The three common chords are PA, PB, PC, as shown in Figure 7.10. The intersection points of PA, PB, PC and O_2O_3, O_3O_1, O_1O_2 are D, E, F, respectively. Then O_2O_3, O_3O_1, O_1O_2 are perpendicular bisectors of PA, PB, PC, respectively.

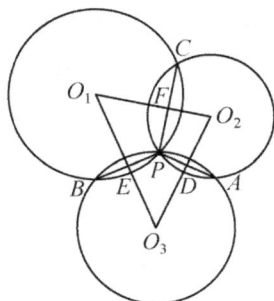

Figure 7.10

In the triangle $O_1 O_2 O_3$, by the Erdös–Mordell inequality,

$$PO_1 + PO_2 + PO_3 \geq 2(PD + PE + PF)$$
$$= PA + PB + PC,$$

and consequently, the conclusion holds.

The Erdös–Mordell inequality is a very classic inequality, and its proof method contains many good ideas. Many proofs of geometric inequalities can be inspired by the proof of the Erdös–Mordell inequality.

Example 7.4. Let the radius of the circumcircle of $\triangle ABC$ be R. A point P is inside $\triangle ABC$. Prove:

$$\frac{PA}{BC^2} + \frac{PB}{CA^2} + \frac{PC}{AB^2} \geq \frac{1}{R}.$$

Proof. As shown in Figure 7.11, through the point P, construct $PD \perp BC, PE \perp CA, PF \perp AB$, and the perpendicular feet are D, E, F, respectively. Let $x = PD, y = PE, z = PF$. Since the points P, E, A, F

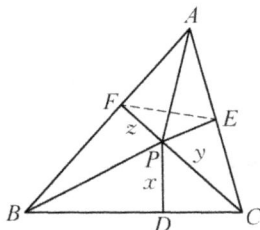

Figure 7.11

are on the circle with diameter PA, so by the law of sines and the law of cosines,

$$PA^2 \sin^2 A = EF^2 = y^2 + z^2 - 2yz \cos \angle EPF$$
$$= y^2 + z^2 - 2yz \cos(B + C)$$
$$= (y \sin C + z \sin B)^2 + (y \cos C - z \cos B)^2$$
$$\geq (y \sin C + z \sin B)^2,$$

thus

$$PA \geq \frac{\sin C}{\sin A} y + \frac{\sin B}{\sin A} z = \frac{c}{a} y + \frac{b}{a} z,$$
$$\frac{PA}{BC^2} \geq \frac{c}{a^3} y + \frac{b}{a^3} z.$$

Similarly,

$$\frac{PB}{CA^2} \geq \frac{c}{b^3} x + \frac{a}{b^3} z, \quad \frac{PC}{AB^2} \geq \frac{a}{c^3} y + \frac{b}{c^3} x.$$

Hence

$$\frac{PA}{BC^2} + \frac{PB}{CA^2} + \frac{PC}{AB^2} \geq \left(\frac{c}{b^3} + \frac{b}{c^3} \right) x + \left(\frac{c}{a^3} + \frac{a}{c^3} \right) y + \left(\frac{b}{a^3} + \frac{a}{b^3} \right) z$$
$$\geq \frac{2x}{bc} + \frac{2y}{ca} + \frac{2z}{ab}$$
$$= \frac{2ax + 2by + 2cz}{abc}$$
$$= \frac{4S}{abc} = \frac{1}{R}.$$

(4) *Isoperimetric problems*

Theorem 7.1. *Among "n-sided" polygons ($n \geq 3$) with a fixed perimeter, the area of the regular "n-sided" polygon is the largest.*

Theorem 7.2. *Among "n-sided" polygons ($n \geq 3$) with a fixed area, the perimeter of the regular "n-sided" polygon is the smallest.*

Theorem 7.3. *Among planar closed curves with a fixed perimeter, the area of the circle is the largest.*

Theorem 7.4. *Among planar closed curves with fixed area, the perimeter of the circle is the smallest.*

Example 7.5. Let QOP be a right angle. Try to find points A, B, C on the sides of OP, OQ and inside this angle, such that $BC + CA = 1$ (fixed length) and the area of quadrilateral $ACBO$ is maximized.

Solution. As shown in Figure 7.12, taking OQ as the axis of symmetry, we make the symmetric image of quadrilateral $ACBO$ as quadrilateral $A'OBC'$, then taking $A'A$ as the axis of symmetry, we make the symmetric image of pentagon $A'ACBC'$ as pentagon $A'C'''B'C''A$, and then get the octagon $ACBC'A'C'''B'C''$, whose perimeter is 4 and area is four times that of the quadrilateral $ACBO$.

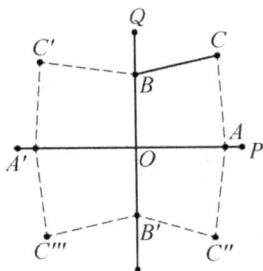

Figure 7.12

By the isoperimetric theorem, the area is a maximum when the octagon is a regular octagon, and thus the area of the quadrilateral $ACBO$ is a maximum when A, B, C are three continuous vertices of the regular octagon with exactly O as the center and $\frac{1}{2}$ as the side length.

7.2 Problems and Solutions

7.2.1 *Existence problems*

Problem 7.1 (IMO 3-4, proposed by the German Democratic Republic). Consider triangle $P_1 P_2 P_3$ and a point P within the triangle. Lines $P_1 P, P_2 P, P_3 P$ intersect the opposite sides at points Q_1, Q_2, Q_3, respectively. Prove that, of the numbers $\frac{P_1 P}{P Q_1}, \frac{P_2 P}{P Q_2}, \frac{P_3 P}{P Q_3}$, at least one is less than or equal to 2 and at least one is greater than or equal to 2.

Proof. As shown in Figure 7.13,

$$\frac{PQ_1}{P_1 Q_1} = \frac{S_{\triangle PP_2 P_3}}{S_{\triangle P_1 P_2 P_3}}, \quad \frac{PQ_2}{P_2 Q_2} = \frac{S_{\triangle PP_1 P_3}}{S_{\triangle P_1 P_2 P_3}}, \quad \frac{PQ_3}{P_3 Q_3} = \frac{S_{\triangle PP_1 P_2}}{S_{\triangle P_1 P_2 P_3}}.$$

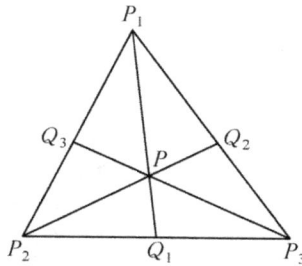

Figure 7.13

Thus

$$\frac{PQ_1}{P_1Q_1} + \frac{PQ_2}{P_2Q_2} + \frac{PQ_3}{P_3Q_3} = 1.$$

Therefore, one of $\frac{P_1P}{PQ_1}, \frac{P_2P}{PQ_2}, \frac{P_3P}{PQ_3}$ must have a value $\geq \frac{1}{3}$, and one of them must have a value $\leq \frac{1}{3}$. Without loss of generality, let

$$\frac{PQ_1}{P_1Q_1} \geq \frac{1}{3}, \quad \frac{PQ_2}{P_2Q_2} \leq \frac{1}{3},$$

and so

$$\frac{P_1P}{PQ_1} \leq 2, \quad \frac{P_2P}{PQ_2} \geq 2.$$

Note. One can also introduce the median point G to solve the problem, and then discuss different positions of point P in different cases.

【Score Situation】 This particular problem saw the following distribution of scores among contestants: 3 contestants scored 6 points, no contestant scored 5 points, no contestant scored 4 points, no contestant scored 3 points, 3 contestants scored 2 points, no contestant scored 1 point, and 3 contestants scored 0 point. The average score of this problem is 2.667, indicating that it had a certain level of difficulty.

Among the top five teams in the team scores, the Hungary team achieved a total score of 270 points, the Poland team achieved a total score of 203 points, the Romania team achieved a total score of 197 points, the Czechoslovakia team achieved a total score of 159 points, and the German Democratic Republic team achieved a total score of 146 points.

The gold medal cutoff for this IMO was set at 37 points (with 3 contestants earning gold medals), the silver medal cutoff was 34 points (with 4 contestants earning silver medals), and the bronze medal cutoff was 30 points (with 4 contestants earning bronze medals).

In this IMO, only one contestant achieved a perfect score of 40 points, namely Béla Bollobás from Hungary.

Problem 7.2 (IMO 8-6, proposed by Poland). In the interior of sides BC, CA, AB of triangle ABC, points K, L, M, respectively, are selected. Prove that the area of at least one of the triangles AML, BKM, CLK is less than or equal to one quarter of the area of triangle ABC.

Proof 1. As shown in Figure 7.14, without loss of generality, let $AB = c, BC = a, CA = b, LA = l, MB = m, KC = k$.

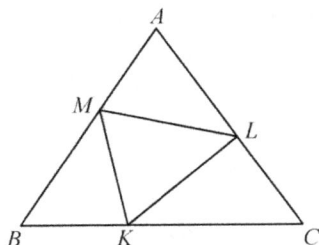

Figure 7.14

It is easy to know

$$S_{\triangle ALM} \cdot S_{\triangle MBK} \cdot S_{\triangle KCL}$$

$$= \frac{1}{2}l(c-m)\sin A \cdot \frac{1}{2}m(a-k)\sin B \cdot \frac{1}{2}k(b-l)\sin C$$

$$= \frac{1}{8}l(b-l) \cdot m(c-m) \cdot k(a-k)\sin A \sin B \sin C.$$

By the AM-GM inequality, $x(s-x) \le \frac{(x+s-x)^2}{4} = \frac{s^2}{4}$, so

$$l(b-l) \le \frac{b^2}{4}, \quad m(c-m) \le \frac{c^2}{4}, \quad k(a-k) \le \frac{a^2}{4}.$$

Thus

$$S_{\triangle ALM} \cdot S_{\triangle MBK} \cdot S_{\triangle KCL} \le \frac{1}{512}a^2b^2c^2 \sin A \sin B \sin C$$

$$= \frac{1}{64} \cdot \frac{1}{2}ab\sin C \cdot \frac{1}{2}bc\sin A \cdot \frac{1}{2}ac\sin B$$

$$= \left(\frac{1}{4}S_{\triangle ABC}\right)^3,$$

from which

$$\min(S_{\triangle ALM}, S_{\triangle MBK}, S_{\triangle KCL}) \le \frac{1}{4}S_{\triangle ABC}.$$

Proof 2. We use the geometric local adjustment method. A very simple geometric conclusion is applied here: If points A, B are on the same side of line CD, and extension AB intersects line CD, then $S_{\triangle ACD} > S_{\triangle BCD}$.

The following is a proof. Without loss of generality, let point M be in AC', where A', B', C' are the midpoints of BC, AC, AB, respectively.

If point L is inside segment AB', then.

$$S_{\triangle AML} \leq S_{\triangle AB'C'} = \frac{1}{4} S_{\triangle ABC},$$

so, the problem has been solved.

If point L is inside $B'C$, likewise, point K needs to be inside $A'B$, as shown in Figure 7.15.

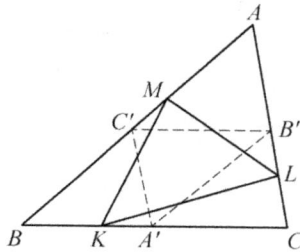

Figure 7.15

It is shown below that in this case, $S_{\triangle MKL} > \frac{1}{4} S_{\triangle ABC}$. By the properties of the inequality,

$$\min(S_{\triangle ALM}, S_{\triangle MBK}, S_{\triangle KCL}) \leq \frac{1}{4} S_{\triangle ABC}.$$

Since $A'C' \parallel AC$, the segment KM intersects $A'C'$ and the extension KM intersects the line AC, and thus

$$S_{\triangle MKL} > S_{\triangle MKB'}.$$

Since $A'B' \parallel AB$, the extension $B'K$ intersects the line AB, and so

$$S_{\triangle MKB'} > S_{\triangle C'KB'} = S_{\triangle A'B'C'} = \frac{1}{4} S_{\triangle ABC}.$$

According to the above two inequalities,

$$S_{\triangle MKL} > \frac{1}{4} S_{\triangle ABC}.$$

【Score Situation】 This particular problem saw the following distribution of scores among contestants: 21 contestants scored 8 points, 1 contestant scored 7 points,

1 contestant scored 6 points, 1 contestant scored 5 points, 3 contestants scored 4 points, no contestant scored 3 points, no contestants scored 2 points, no contestants scored 1 point, and no contestant scored 0 point. The average score of this problem is 7.333, indicating that it was simple.

Among the top five teams in the team scores, the Soviet Union team achieved a total score of 293 points, the Hungary team achieved a total score of 281 points, the German Democratic Republic team achieved a total score of 280 points, the Poland team achieved a total score of 269 points, and the Romania team achieved a total score of 257 points.

The gold medal cutoff for this IMO was set at 39 points (with 13 contestants earning gold medals), the silver medal cutoff was 34 points (with 15 contestants earning silver medals), and the bronze medal cutoff was 31 points (with 11 contestants earning bronze medals).

In this IMO, a total of 11 contestants achieved a perfect score of 40 points.

Problem 7.3 (IMO 10-4, proposed by Poland). Prove that in every tetrahedron there is a vertex such that the three edges meeting there have lengths which are the sides of a triangle.

Proof. Let the tetrahedron be $ABCD$ and the longest edge be AB. Then

$$AB < AC + BC, \quad AB < AD + BD,$$

so

$$2AB < AC + BC + AD + BD$$
$$= (AC + AD) + (BC + BD)$$
$$\leq 2\max\{AC + AD, BC + BD\}. \tag{1}$$

Without loss of generality, let $AC + AD \geq BC + BD$, from which (1) is

$$AB < AC + AD.$$

The definition of AB implies that AB, AC, AD can be the three sides of a triangle.

【Score Situation】 This particular problem saw the following distribution of scores among contestants: 58 contestants scored 5 points, 1 contestant scored 4 points, 3 contestants scored 3 points, 10 contestants scored 2 points, 2 contestants scored 1 point, and 17 contestants scored 0 point. The average score of this problem is 3.571, indicating that it was relatively straightforward.

Among the top five teams in the team scores, the German Democratic Republic team achieved a total score of 304 points, the Soviet Union team achieved a total score of 298 points, the Hungary team achieved a total score of 291 points, the United Kingdom team achieved a total score of 263 points, and the Poland team achieved a total score of 262 points.

The gold medal cutoff for this IMO was set at 39 points (with 22 contestants earning gold medals), the silver medal cutoff was 33 points (with 22 contestants earning silver medals), and the bronze medal cutoff was 26 points (with 20 contestants earning bronze medals).

In this IMO, a total of 16 contestants achieved a perfect score of 40 points.

Problem 7.4 (IMO 13-4, proposed by the Netherlands). All the faces of tetrahedron $ABCD$ are acute-angled triangles. We consider all closed polygonal paths of the form $XYZTX$ defined as follows: X is a point on edge AB distinct from A and B; similarly, Y, Z, T are interior points of edges BC, CD, DA, respectively. Prove:

(a) If $\angle DAB + \angle BCD \neq \angle CDA + \angle ABC$, then among the polygonal paths, there is none of minimal length.
(b) If $\angle DAB + \angle BCD = \angle CDA + \angle ABC$, then there are infinitely many shortest polygonal paths, their perimeter being $2AC \sin \frac{\alpha}{2}$, where $\alpha = \angle BAC + \angle CAD + \angle DAB$.

Proof.   (a) As shown in Figure 7.16, unfold the four faces of the tetrahedron $ABCD$ in the plane (where A and A_1, D and D_1 are coincident). Obviously $CA_1 = CA, BD_1 = BD, A_1D_1 = AD$.

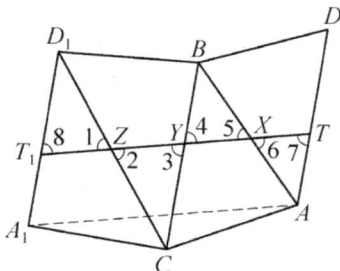

Figure 7.16

Consider the polyline XYZ. If point Y is not on the line XZ, then $XY + YZ > XZ$.

Since $\triangle ABC, \triangle BCD_1$ are acute triangles, XZ and BC must intersect. Let the intersection point be Y', and replace Y by Y' to get a smaller polygonal path $XY'ZTX$. Consequently, if $XYZTX$ has a shortest polygonal path, then its expansion must be straight, that is, segment TT_1.

As shown in the figure, since $\angle 1 = \angle 2, \angle 3 = \angle 4, \angle 5 = \angle 6$, if we place $\triangle ADB$ on the left side of A_1D_1, then $\angle 7 = \angle 8$. Thus,

$$\angle DAB + \angle BCD_1 = 360° - (\angle 2 + \angle 3 + \angle 6 + \angle 7)$$
$$= 360° - (\angle 4 + \angle 5 + \angle 1 + \angle 8)$$
$$= \angle ABC + \angle A_1D_1C. \tag{1}$$

Back to the original tetrahedron, formula (1) is

$$\angle DAB + \angle BCD = \angle ABC + \angle ADC.$$

Therefore, the problem (a) is proved by the local adjustment method.

(b) We know that $A_1T_1 = AT, AD \parallel A_1D_1$. Thus, the quadrilateral ATT_1A_1 is a parallelogram, so

$$TT_1 = AA_1 = 2AC \sin \frac{\angle ACA_1}{2}.$$

While,

$$\angle ACA_1 = \angle ACB + \angle BCD_1 + \angle A_1CD_1$$
$$= \angle 4 + \angle 5 - \angle BAC + 180° - \angle 2 - \angle 3 + \angle 1 + \angle 8 - \angle CA_1D_1$$
$$= \angle 5 + \angle 8 - \angle BAC - \angle CA_1D_1 + 180°$$
$$= \angle 6 + \angle 7 - \angle BAC - \angle CA_1D_1 + 180°$$
$$= 360° - (\angle BAD + \angle BAC + \angle CA_1D_1)$$
$$= 360° - \alpha,$$

the minimum length of $XYZTX$ is $2AC \sin \left(180° - \frac{\alpha}{2}\right) = 2AC \sin \frac{\alpha}{2}$. By the scalability of AT, it follows that there are infinitely many such quadrilaterals.

【Score Situation】This particular problem saw the following distribution of scores among contestants: 3 contestants scored 6 points, 5 contestants scored 5 points, 3 contestants scored 4 points, 3 contestants scored 3 points, 2 contestants scored 2 points, 2 contestants scored 1 point, and 10 contestants scored 0 point. The average score of this problem is 2.500, indicating that it had a certain level of difficulty.

Among the top five teams in the team scores, the scores of this problem are as follows: the Hungary team scored 38 points (with a total team score of 255 points), the Soviet Union team scored 34 points (with a total team score of 205 points), the German Democratic Republic team scored 30 points (with a total team score of 142 points), the Poland team scored 13 points (with a total team score of 118 points), the United Kingdom team scored 27 points (with a total team score of 110 points), and the Romania team scored 22 points (with a total team score of 110 points).

The gold medal cutoff for this IMO was set at 35 points (with 7 contestants earning gold medals), the silver medal cutoff was 23 points (with 12 contestants earning silver medals), and the bronze medal cutoff was 11 points (with 29 contestants earning bronze medals).

In this IMO, only one contestant achieved a perfect score of 42 points, namely Imre Ruzsa from Hungary.

Problem 7.5 (IMO 32-5, proposed by France). Let ABC be a triangle and P an interior point in ABC. Show that at least one of the angles $\angle PAB, \angle PBC, \angle PCA$ is less than or equal to $30°$.

Proof 1.   As shown in Figure 7.17, let PD, PE, PF be perpendicular to BC, CA, AB, respectively, and the feet of the altitudes are D, E, F.

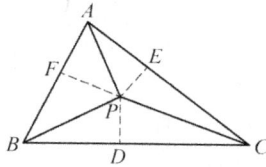

Figure 7.17

By the Erdős–Mordell inequality,

$$PA + PB + PC \geq 2(PD + PE + PF).$$

Thus, at least one of the three inequalities, $PA \geq 2PF, PB \geq 2PD, PC \geq 2PE$, is valid. Without loss of generality, let $PB \geq 2PD$. Then

$$\sin \angle PBC = \frac{PD}{PB} \leq \frac{1}{2},$$

so

$$\angle PBC \leq 30° \quad \text{or} \quad \angle PBC \geq 150°.$$

The first inequality above shows that the conclusion holds, and the second one shows that $\angle PAB < \angle CAB < 30°$ in the case. (Similarly, $\angle PCA < 30°$.)

To sum up, the conclusion is true.

Proof 2. Let$\angle PAB = \alpha, \angle PBC = \beta, \angle PCA = \gamma$. By the law of sines,

$$\frac{\sin \alpha}{\sin(B - \beta)} = \frac{BP}{AP},$$

$$\frac{\sin \beta}{\sin(C - \gamma)} = \frac{CP}{BP},$$

$$\frac{\sin \gamma}{\sin(A - \alpha)} = \frac{AP}{CP}.$$

Multiplying them, we get

$$\sin \alpha \sin \beta \sin \gamma = \sin(A - \alpha) \sin(B - \beta) \sin(C - \gamma).$$

Clearly

$$\sin \alpha \sin(A - \alpha) = \frac{1}{2}[\cos(A - 2\alpha) - \cos A]$$

$$\leq \frac{1}{2}(1 - \cos A) = \sin^2 \frac{A}{2}.$$

Then,

$$\sin^2 \alpha \sin^2 \beta \sin^2 \gamma$$

$$= \sin \alpha \sin(A - \alpha) \sin \beta \sin(B - \beta) \sin \gamma \sin(C - \gamma)$$

$$\leq \sin^2 \frac{A}{2} \sin^2 \frac{B}{2} \sin^2 \frac{C}{2}.$$

Hence,

$$\sin \alpha \sin \beta \sin \gamma \leq \sin \frac{A}{2} \sin \frac{B}{2} \sin \frac{C}{2}$$

$$\leq \sin^2 \frac{A+B}{4} \sin \frac{C}{2}$$

$$= \sin^2 \left(45° - \frac{C}{4}\right) \sin \frac{C}{2}$$

$$= \frac{1}{2}\left(1 - \sin \frac{C}{2}\right) \sin \frac{C}{2}$$

$$\leq \frac{1}{8},$$

so at least one of the values of $\sin \alpha, \sin \beta, \sin \gamma$ is not greater than $\frac{1}{2}$, and the remaining discussion is the same as that of Proof I.

Proof 3. As shown in Figure 7.18, let $PD \perp BC, PE \perp AC, PF \perp AB$, and D, E, F are the feet of the altitudes.

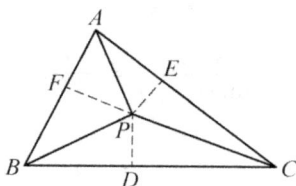

Figure 7.18

(i) If the conclusion does not hold, that is,

$$30° < \angle PAB, \quad \angle PBC, \quad \angle PCA < 120°,$$

then

$$\frac{PF}{PA} = \sin \angle PAB > \sin 30° = \frac{1}{2},$$

that is, $2PF > PA$.

Similarly, $2PE > PC, 2PD > PB$.

Thus, $PA + PB + PC < 2(PD + PE + PF)$, which contradicts the Erdős–Mordell inequality.

(ii) Suppose one of the angles $\angle PAB, \angle PBC, \angle PCA$ has a degree at least $120°$. Without loss of generality, let $\angle PAB \geq 120°$. Then

$$\angle PBC + \angle PCA \leq 60°.$$

So, there must be an angle not greater than $30°$.

To sum up, the conclusion holds.

【Score Situation】This particular problem saw the following distribution of scores among contestants: 134 contestants scored 7 points, 7 contestants scored 6 points, 6 contestants scored 5 points, 14 contestants scored 4 points, 14 contestants scored 3 points, 18 contestants scored 2 points, 42 contestants scored 1 point, and 77 contestants scored 0 point. The average score of this problem is 3.801, indicating that it was relatively straightforward.

Among the top five teams in the team scores, the scores of this problem are as follows: the Soviet Union team scored 41 points (with a total team score of 241 points), the China team scored 42 points (with a total team score of 231 points), the Romania team scored 42 points (with a total team score of 225 points), the Germany team scored 42 points (with a total team score of 222 points), and the United States team scored 39 points (with a total team score of 212 points).

The gold medal cutoff for this IMO was set at 39 points (with 20 contestants earning gold medals), the silver medal cutoff was 31 points (with 51 contestants

earning silver medals), and the bronze medal cutoff was 19 points (with 84 contestants earning bronze medals).

In this IMO, a total of nine contestants achieved a perfect score of 42 points.

7.2.2 *Equality or inequality proof problems*

(1) *Problems related to length*

Problem 7.6 (IMO 5-3, proposed by Hungary). In an n-gon all of whose interior angles are equal, the lengths of consecutive sides satisfy the relation

$$a_1 \geq a_2 \geq \cdots \geq a_n.$$

Prove that $a_1 = a_2 = \cdots = a_n$.

Proof. It is discussed in two cases.

(i) When $n = 2k + 1$, make the angle bisector l of the interior angle with a_1 and a_{2k+1} as the sides. Then $l \perp a_{k+1}$. Let the projections of a_j $(1 \leq j \leq 2k + 1)$ onto l be a_j^* $(1 \leq j \leq 2k + 1)$, respectively. Clearly $a_{k+1}^* = 0$.

By the conditions, the angle between a_i and l is equal to the angle between a_j and l $(i + j = 2k + 2, 1 \leq i \leq j \leq 2k + 1)$. Thus, the length relation between a_i and a_j is the same as the length relation between a_i^* and a_j^*.

Since $a_i^* \geq a_j^*$ $(1 \leq i < j \leq 2k + 1, i + j = 2k + 2)$,

$$\sum_{i=1}^{k} a_i^* \geq \sum_{i=k+2}^{2k+1} a_i^*.$$

By the convexity of n-gon,

$$\sum_{i=1}^{k} a_i^* = \text{the distance from the vertex of the angle with the sides}$$

$$a_1, a_{2k+1} \text{ to } a_{k+1} = \sum_{i=k+2}^{2k+1} a_i^*.$$

So, there can only be $a_1 = a_{2k+1}$. Thus $a_1 = \cdots = a_{2k+1}$, and the proposition is proved.

(ii) When $n = 2k$, make the angle bisector l of the interior angle with a_1 and a_{2k} as the sides. The definition of a_i^* $(1 \leq i \leq a_{2k})$ is the same as before.

By the conditions, the angle between a_i and l is equal to the angle between a_j and l $(i + j = 2k + 1)$. Thus $a_i \geq a_j$ $(1 \leq i < j \leq 2k, i + j = 2k + 1)$. We know $a_i^* \geq a_j^*$, from which

$$\sum_{i=1}^{k} a_i^* = \sum_{i=k+1}^{2k} a_i^*.$$

So $a_1 = a_{2k}$, and consequently $a_1 = a_2 = \cdots = a_{2k}$. The proposition is proved.

【Score Situation】 This particular problem saw the following distribution of scores among contestants: 2 contestants scored 7 points, no contestant scored 6 points, 5 contestants scored 5 points, 1 contestant scored 4 points, no contestant scored 3 points, 1 contestant scored 2 points, 1 contestant scored 1 point, and 9 contestants scored 0 point. The average score of this problem is 1.938, indicating that it was relatively challenging.

Among the top five teams in the team scores, the Soviet Union team achieved a total score of 271 points, the Hungary team achieved a total score of 234 points, the Romania team achieved a total score of 191 points, the Yugoslavia team achieved a total score of 162 points, and the Czechoslovakia team achieved a total score of 151 points.

The gold medal cutoff for this IMO was set at 35 points (with 7 contestants earning gold medals), the silver medal cutoff was 28 points (with 11 contestants earning silver medals), and the bronze medal cutoff was 21 points (with 17 contestants earning bronze medals).

In this IMO, no contestant achieved a perfect score of 40 points.

Problem 7.7 (IMO 9-1, proposed by Poland). Let $ABCD$ be a parallelogram with side lengths $AB = a$ and $AD = 1$, and with $\angle BAD = \alpha$. If $\triangle ABD$ is acute, prove that the four circles of radius 1 with centres A, B, C, D cover the parallelogram if and only if

$$a \leq \cos \alpha + \sqrt{3} \sin \alpha.$$

Proof. Let O be the circumcenter of $\triangle ABD$, which must be inside $\triangle ABD$, and let the radius of the circumcircle be r.

If $r > 1$, then circles $A, B,$ and D cannot cover circle O, and circle C alone cannot cover circle O.

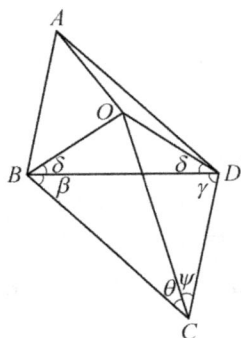

Figure 7.19

This is because (see Figure 7.19),

$$2\delta + \beta + \gamma = 180° - 2(\theta + \psi) + \beta + \gamma$$
$$> 180° - 2(\theta + \psi) + \theta + \psi$$
$$= 180° - (\theta + \psi)$$
$$> \theta + \psi,$$

and here we have used the facts that $\beta, \gamma, \theta + \psi < 90°$.

The above is just $\angle OBC + \angle ODC > \angle BCD$, so at least one of $\angle OBC > \angle OCB, \angle ODC > \angle OCD$ holds. Hence $OC > r$.

Conversely, if $r \leq 1$, then the point O can be covered. Let P be any point inside $\triangle ABD$, which must be inside one of $\triangle ABO, \triangle BDO, \triangle ADO$. Without loss of generality, let P be inside $\triangle ABO$. Then

$$\min\{AP, BP\} \leq \frac{1}{2}(AP + BP) \leq \frac{1}{2}(AO + BO) = r \leq 1,$$

and thus, $\triangle ABD$ is covered.

By symmetry, $\triangle BCD$ is also covered. So $r \leq 1$ is a necessary and sufficient condition.

In the following we prove that $r \leq 1$ is equivalent to $a \leq \cos\alpha + \sqrt{3}\sin\alpha$.

By the laws of sines and cosines,

$$BD^2 = 1 + a^2 - 2a\cos\alpha,$$
$$BD = 2r\sin\alpha.$$

So $r \leq 1$ if and only if

$$4 \sin^2 \alpha \geq 1 + a^2 - 2a \cos \alpha,$$

which is,

$$3 \sin^2 \alpha \geq (a - \cos \alpha)^2. \tag{1}$$

Let $DQ \perp AB$ at point Q. Since $\triangle ABD$ is an acute triangle, the point Q is inside AB. Then $\cos \alpha = AQ < AB = a$, and (1) is equivalent to

$$a - \cos \alpha \leq \sqrt{3} \sin \alpha.$$

【Score Situation】This particular problem saw the following distribution of scores among contestants: 10 contestants scored 6 points, 6 contestants scored 5 points, 4 contestants scored 4 points, 7 contestants scored 3 points, 3 contestants scored 2 points, 5 contestants scored 1 point, and 2 contestants scored 0 point. The average score of this problem is 3.730, indicating that it was relatively straightforward.

Among the top five teams in the team scores, the Soviet Union team achieved a total score of 275 points, the German Democratic Republic team achieved a total score of 257 points, the Hungary team achieved a total score of 251 points, the United Kingdom team achieved a total score of 231 points, and the Romania team achieved a total score of 214 points.

The gold medal cutoff for this IMO was set at 38 points (with 11 contestants earning gold medals), the silver medal cutoff was 30 points (with 14 contestants earning silver medals), and the bronze medal cutoff was 22 points (with 26 contestants earning bronze medals).

In this IMO, a total of five contestants achieved a perfect score of 42 points.

Problem 7.8 (IMO 25-5, proposed by Mongolia). Let d be the sum of the lengths of all the diagonals of a plane convex polygon with n vertices ($n > 3$), and let p be its perimeter.

Prove that $n - 3 < \frac{2d}{p} < \left[\frac{n}{2}\right] \cdot \left[\frac{n+1}{2}\right] - 2$, where $[x]$ denotes the greatest integer not exceeding x.

Proof. Let the convex polygon be $A_1 A_2 \ldots A_n$ and the diagonal be $A_i A_j, j > i + 1$. Since this is a convex polygon, diagonals $A_i A_j$ and $A_{i+1} A_{j+1}$ must intersect. Let the intersection point be O. Then

$$A_i A_j + A_{i+1} A_{j+1} = A_i O + A_{i+1} O + A_j O + A_{j+1} O > A_i A_{i+1} + A_j A_{j+1}.$$

Since in a convex n-gon, there exists

$$C_n^2 - n = \frac{n(n-3)}{2}$$

diagonals. For each diagonal A_iA_j, the above inequality holds, and by adding these inequalities, each diagonal appears twice on the left side of the inequalities. So, the sum of the left sides of these inequalities is $2d$. The right side of the inequality has $n(n-3)$ terms, which means that each side occurs exactly $n-3$ times, so the sum on the right sides is $(n-3)p$. Thus

$$2d > (n-3)p,$$

so

$$n - 3 < \frac{2d}{p}.$$

On the other hand, for

$$A_iA_j < A_iA_{i+1} + \cdots + A_{j-1}A_j \quad \text{and} \quad A_iA_j < A_jA_{j+1} + \cdots + A_{i-1}A_i.$$

choose the inequality where the number of terms on the right side does not exceed $\frac{n}{2}$.

When $n = 2k - 1$ is an odd number, the number of terms on the right side of the above inequality is at least 2 and at most $k-1$. Adding all such inequalities, the sum of the left side of the resulting inequality is d, and the number of repetitions of each side of the polygon on the right side of the inequality is

$$2 + 3 + \cdots + (k-1) = \frac{k(k-1)}{2} - 1 = \frac{1}{2}\left[\frac{n}{2}\right]\left[\frac{n+1}{2}\right] - 1.$$

Thus

$$d < \left(\frac{1}{2}\left[\frac{n}{2}\right]\left[\frac{n+1}{2}\right] - 1\right)p,$$

so

$$\frac{2d}{p} < \left[\frac{n}{2}\right]\left[\frac{n+1}{2}\right] - 2.$$

When $n = 2k$ is an even number, there are k diagonals with $A_iA_{i+k} \leq \frac{p}{2}$, and the other diagonals still conform to the above inequality. Adding the two parts, we get

$$d < k \cdot \frac{p}{2} + \left(\frac{k(k-1)}{2} - 1\right)p$$

$$= \frac{1}{2}(k^2 - 2)p,$$

and hence

$$\frac{2d}{p} < \left[\frac{n}{2}\right]\left[\frac{n+1}{2}\right] - 2.$$

【Score Situation】This particular problem saw the following distribution of scores among contestants: 41 contestants scored 7 points, 7 contestants scored 6 points, 4 contestants scored 5 points, 19 contestants scored 4 points, 12 contestants scored 3 points, 4 contestants scored 2 points, 17 contestants scored 1 point, and 88 contestants scored 0 point. The average score of this problem is 2.531, indicating that it had a certain level of difficulty.

Among the top five teams in the team scores, the scores of this problem are as follows: the Soviet Union team scored 40 points (with a total team score of 235 points), the Bulgaria team scored 36 points (with a total team score of 203 points), the Romania team scored 30 points (with a total team score of 199 points), the Hungary team scored 42 points (with a total team score of 195 points), and the United States team scored 27 points (with a total team score of 195 points).

The gold medal cutoff for this IMO was set at 40 points (with 14 contestants earning gold medals), the silver medal cutoff was 26 points (with 35 contestants earning silver medals), and the bronze medal cutoff was 17 points (with 49 contestants earning bronze medals).

In this IMO, a total of eight contestants achieved a perfect score of 42 points.

Problem 7.9 (IMO 30-4, proposed by Iceland).

Let $ABCD$ be a convex quadrilateral such that the sides AB, AD, BC satisfy $AB = AD + BC$. There exists a point P inside the quadrilateral at a distance h from the line CD such that $AP = AD + h$ and $BP = BC + h$.

Show that $\frac{1}{\sqrt{h}} \geq \frac{1}{\sqrt{AD}} + \frac{1}{\sqrt{BC}}$.

Proof.   Let $AD = a, BC = b$. As shown in Figure 7.20, let AM, PQ, BN be perpendicular to CD, and the feet of the altitudes be M, Q, N, respectively. We have $AM = a \sin D, BN = b \sin C$.

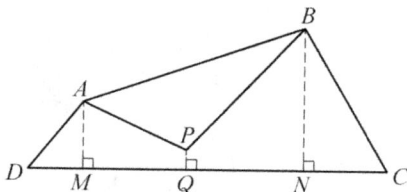

Figure 7.20

Even if $\angle D < 90°$, the point Q cannot be on segment DM; otherwise $AP < AD$. Similarly, the point Q cannot be on segment CN. So, the point Q is on segment MN.

Since $MQ + QN = MN$, it is equivalent to

$$\sqrt{(a+h)^2 - (a\sin D - h)^2} + \sqrt{(b+h)^2 - (b\sin C - h)^2}$$
$$= \sqrt{(a+b)^2 - (a\sin D - b\sin C)^2}.$$

Squaring the both sides of the equality, we get

$$\sqrt{(a+h)^2 - (a\sin D - h)^2} \cdot \sqrt{(b+h)^2 - (b\sin C - h)^2}$$
$$= (a\sin D - h)(b\sin C - h) + ab - ah - bh - h^2. \tag{1}$$

Without loss of generality, let $AM \le BN$. Since the quadrangle $ABCD$ is convex, there must be $PQ < BN$. In the following we discuss separately according to the cases.

(i) $PQ \le AM, BN$, which is

$$h < a\sin D, \quad b\sin C.$$

Thus

$$\sqrt{(a+h)^2 - (a\sin D - h)^2} \ge \sqrt{(a+h)^2 - (a-h)^2} = 2\sqrt{ah}.$$

Similarly,

$$2\sqrt{bh} \le \sqrt{(b+h)^2 - (b\sin C - h)^2}.$$

And

$$(a\sin D - h)(b\sin C - h) \le (a-h)(b-h).$$

By (1),

$$4h\sqrt{ab} \le 2ab - 2ah - 2bh.$$

Then

$$\sqrt{h} \le \frac{\sqrt{ab}}{\sqrt{a} + \sqrt{b}},$$
$$\frac{1}{\sqrt{h}} \ge \frac{1}{\sqrt{a}} + \frac{1}{\sqrt{b}}.$$

(ii) $AM < PQ < BN$.

Now, the right side of equality (1) $< ab - ah - bh - h^2 < ab - ah - bh$.

The left side of equality (1) is as follows: the first part still holds, which is

$$2\sqrt{bh} \le \sqrt{(b+h)^2 - (b\sin C - h)^2}.$$

While, the second part is as follows.

$$\sqrt{(a+h)^2 - (a\sin D - h)^2} > \sqrt{(a+h)^2 - (h^2 + a^2)}$$
$$= \sqrt{2ah} > \sqrt{ah}.$$

The equality (1) can still lead to

$$2h\sqrt{ab} < ab - ah - bh,$$

and then the inequality strictly holds, that is

$$\frac{1}{\sqrt{h}} > \frac{1}{\sqrt{a}} + \frac{1}{\sqrt{b}}.$$

Combining the above two cases, we know that the equality holds if $\angle C = \angle D = 90°$. In this case, the circles with A, B, P as the centers and AD, BC, h as the radii respectively are externally tangent to each other and tangent to CD.

【Score Situation】 This particular problem saw the following distribution of scores among contestants: 132 contestants scored 7 points, 5 contestants scored 6 points, 9 contestants scored 5 points, 8 contestants scored 4 points, 23 contestants scored 3 points, 12 contestants scored 2 points, 37 contestants scored 1 point, and 65 contestants scored 0 point. The average score of this problem is 3.990, indicating that it was relatively straightforward.

Among the top five teams in the team scores, the scores of this problem are as follows: the China team scored 41 points (with a total team score of 237 points), the Romania team scored 41 points (with a total team score of 223 points), the Soviet Union team scored 41 points (with a total team score of 217 points), the German Democratic Republic team scored 36 points (with a total team score of 216 points), and the United States team scored 42 points (with a total team score of 207 points).

The gold medal cutoff for this IMO was set at 38 points (with 20 contestants earning gold medals), the silver medal cutoff was 30 points (with 55 contestants earning silver medals), and the bronze medal cutoff was 18 points (with 72 contestants earning bronze medals).

In this IMO, a total of 10 contestants achieved a perfect score of 42 points.

Problem 7.10 (IMO 32-1, proposed by the Soviet Union). Given a triangle ABC, let I be the centre of its inscribed circle. The internal bisectors of the angles A, B, C meet the opposite sides at A', B', C', respectively.
Prove that $\frac{1}{4} < \frac{AI \cdot BI \cdot CI}{AA' \cdot BB' \cdot CC'} \le \frac{8}{27}$.

Proof. Let $BC = a, CA = b, AB = c$. Then

$$\frac{AI}{A'I} = \frac{b+c}{a},$$

so

$$\frac{AI}{AA'} = \frac{b+c}{a+b+c}.$$

Similarly,

$$\frac{BI}{BB'} = \frac{c+a}{a+b+c}, \frac{CI}{CC'} = \frac{a+b}{a+b+c}.$$

Hence, this problem is equivalent to proving that

$$\frac{1}{4} < \frac{(a+b)(b+c)(c+a)}{(a+b+c)^3} \le \frac{8}{27}. \tag{1}$$

By the AM-GM inequality,

$$(a+b)(b+c)(c+a) \le \left(\frac{a+b+b+c+c+a}{3}\right)^3$$

$$= \frac{8}{27}(a+b+c)^3.$$

Therefore, the right side inequality of (1) holds (the equality holds only if $a = b = c$).

And because the sum of the lengths of any two sides of a triangle must be greater than the length of the third side, $\frac{a+b}{a+b+c}, \frac{b+c}{a+b+c}, \frac{c+a}{a+b+c}$ are all greater than $\frac{1}{2}$.

Let $\frac{a+b}{a+b+c} = \frac{1+\varepsilon_1}{2}, \frac{b+c}{a+b+c} = \frac{1+\varepsilon_2}{2}, \frac{c+a}{a+b+c} = \frac{1+\varepsilon_3}{2}$, where $\varepsilon_1, \varepsilon_2, \varepsilon_3$ are positive numbers. It is easy to know that $\varepsilon_1 + \varepsilon_2 + \varepsilon_3 = 1$. Thus

$$\frac{(a+b)(b+c)(c+a)}{(a+b+c)^3} = \frac{(1+\varepsilon_1)(1+\varepsilon_2)(1+\varepsilon_3)}{8}$$

$$> \frac{1+\varepsilon_1+\varepsilon_2+\varepsilon_3}{8} = \frac{1}{4}.$$

Therefore, the left side of inequality of (1) holds.

Note. If I is any point inside $\triangle ABC$, then the right side inequality of (1) still holds.

【Score Situation】 This particular problem saw the following distribution of scores among contestants: 131 contestants scored 7 points, 6 contestants scored 6 points, 18 contestants scored 5 points, 34 contestants scored 4 points, 3 contestants scored 3 points, 10 contestants scored 2 points, 27 contestants scored 1 point, and 43 contestants scored 0 point. The average score of this problem is 3.798, indicating that it was relatively straightforward.

Among the top five teams in the team scores, the scores of this problem are as follows: the Soviet Union team scored 40 points (with a total team score of 241 points), the China team scored 42 points (with a total team score of 231 points), the Romania team scored 42 points (with a total team score of 225 points), the Germany team scored 28 points (with a total team score of 222 points), and the United States team scored 42 points (with a total team score of 212 points).

The gold medal cutoff for this IMO was set at 39 points (with 20 contestants earning gold medals), the silver medal cutoff was 31 points (with 51 contestants earning silver medals), and the bronze medal cutoff was 19 points (with 84 contestants earning bronze medals).

In this IMO, a total of nine contestants achieved a perfect score of 42 points.

Problem 7.11 (IMO 34-4, proposed by Macedonia). For three points P, Q, R in a plane, we define $m(PQR)$ to be the minimum of the lengths of the altitudes of the triangle PQR (where $m(PQR) = 0$ when P, Q, R are collinear.) Let A, B, C be given points in the plane. Prove that, for any point X in the plane,

$$m(ABC) \leq m(ABX) + m(AXC) + m(XBC).$$

Proof. Let $a = BC, b = CA, c = AB, p = AX, q = BX, r = CX$. Obviously, the shortest altitude of a triangle is on the longest side.

Here are two cases.

(i) If $\max\{a, b, c, p, q, r\} \in \{a, b, c\}$, without loss of generality, let $\max\{a, b, c, p, q, r\} = a$. Then

$$m(ABC) = \frac{2}{a} S_{\triangle ABC}$$

$$\leq \frac{2}{a} (S_{\triangle ABX} + S_{\triangle AXC} + S_{\triangle XBC})$$

$$\leq \frac{2S_{\triangle ABX}}{\max\{c, p, q\}} + \frac{2S_{\triangle AXC}}{\max\{b, r, p\}} + \frac{2S_{\triangle XBC}}{\max\{a, q, r\}}$$

$$= m(ABX) + m(AXC) + m(XBC).$$

(ii) If $\max(a, b, c, p, q, r) \in \{p, q, r\}$, without loss of generality, let $\max(a, b, c, p, q, r) = p$, and let $b \leq c$. Then

$$m(ABC) \leq b \sin \angle BAC$$

$$= b|\sin(\angle BAX \pm \angle CAX)|$$

$$= b|\sin \angle BAX \cos \angle CAX \pm \sin \angle CAX \cos \angle BAX|$$

$$\leq b(|\sin \angle BAX| + |\sin \angle CAX|)$$

$$\leq c|\sin \angle BAX| + b|\sin \angle CAX|$$

$$= m(ABX) + m(ACX)$$

$$\leq m(ABX) + m(AXC) + m(XBC).$$

【Score Situation】 This particular problem saw the following distribution of scores among contestants: 39 contestants scored 7 points, 15 contestants scored 6 points, 16 contestants scored 5 points, 35 contestants scored 4 points, 55 contestants scored 3 points, 55 contestants scored 2 points, 43 contestants scored 1 point, and 145 contestants scored 0 point. The average score of this problem is 2.303, indicating that it had a certain level of difficulty.

Among the top five teams in the team scores, the scores of this problem are as follows: the China team scored 39 points (with a total team score of 215 points), the Germany team scored 41 points (with a total team score of 189 points), the Bulgaria team scored 32 points (with a total team score of 178 points), the Russia team scored 30 points (with a total team score of 177 points), and the Chinese Taiwan team scored 22 points (with a total team score of 162 points).

The gold medal cutoff for this IMO was set at 30 points (with 35 contestants earning gold medals), the silver medal cutoff was 20 points (with 66 contestants earning silver medals), and the bronze medal cutoff was 11 points (with 97 contestants earning bronze medals).

In this IMO, only two contestants achieved a perfect score of 42 points, namely Hong Zhou from China and Hung-Wu Wu from Chinese Taiwan.

Problem 7.12 (IMO 36-5, proposed by New Zealand). Let $ABCDEF$ be a convex hexagon with $AB = BC = CD$, $DE = EF = FA$, and $\angle BCD = \angle EFA = 60°$. Let G, H be two points in the interior of this hexagon such that $\angle AGB = \angle DHE = 120°$.

Prove that $AG + GB + GH + DH + HE \geq CF$.

Proof. As shown in Figure 7.21, connect to form BD, AE. Then the quadrilateral $ABDE$ is a kite.

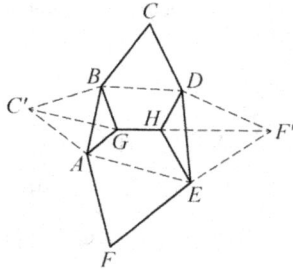

Figure 7.21

Make equilateral triangles ABC' and $DF'E$ on the sides of AB and DE outwards, and join $C'G, F'H$.

It is easy to see that C and C' are symmetric with respect to BE, so are F and F'. Then $CF = C'F'$.

By Ptolemy's theorem,

$$C'G = AG + GB, HF' = DH + HE.$$

Thus,

$$AG + GB + GH + DH + HE = C'G + GH + HF'$$
$$\geq C'F' = CF.$$

【Score Situation】 This particular problem saw the following distribution of scores among contestants: 172 contestants scored 7 points, 5 contestants scored 6 points, 7 contestants scored 5 points, 1 contestant scored 4 points, 8 contestants scored 3 points, 30 contestants scored 2 points, 48 contestants scored 1 point, and 141 contestants scored 0 point. The average score of this problem is 3.410, indicating that it was relatively straightforward.

Among the top five teams in the team scores, the scores of this problem are as follows: the China team scored 42 points (with a total team score of 236 points), the Romania team scored 42 points (with a total team score of 230 points), the Russia team scored 42 points (with a total team score of 227 points), the Vietnam team scored 42 points (with a total team score of 220 points), and the Hungary team scored 42 points (with a total team score of 210 points).

The gold medal cutoff for this IMO was set at 37 points (with 30 contestants earning gold medals), the silver medal cutoff was 29 points (with 71 contestants earning silver medals), and the bronze medal cutoff was 19 points (with 100 contestants earning bronze medals).

In this IMO, a total of 14 contestants achieved a perfect score of 42 points.

Problem 7.13 (IMO 37-5, proposed by Armenia). Let $ABCDEF$ be a convex hexagon such that AB is parallel to ED, also BC is parallel to EF, and CD is parallel to AF. Let R_A, R_C, R_E denote the circumradii of triangles FAB, BCD, DEF respectively, and let p denote the perimeter of the hexagon.

Prove that $R_A + R_C + R_E \geq \frac{p}{2}$.

Proof. Denote by a, b, c, d, e, f the lengths of the sides $AB, BC, \ldots, FA$, and by $\angle A, \angle B, \ldots, \angle F$ the interior angles of the hexagon.

Since the opposite sides are parallel, $\angle A = \angle D, \angle B = \angle E, \angle C = \angle F$.

Let the distance between BC and EF be h, the distances from A to BC, EF be $a \sin B$, $f \sin F$, respectively. Then $h = a \sin B + f \sin F$. Similarly, $h = c \sin C + d \sin E$.

Thus,

$$2BF \geq 2h = a \sin B + f \sin F + c \sin C + d \sin E.$$

Similarly,

$$2DF \geq a \sin A + b \sin C + d \sin D + e \sin F,$$

$$2BD \geq f \sin A + e \sin E + b \sin B + c \sin D.$$

Hence

$$\begin{aligned} R_A + R_C + R_E &= \frac{1}{4}\left(\frac{2BF}{\sin A} + \frac{2BD}{\sin C} + \frac{2DF}{\sin E}\right) \\ &\geq \frac{1}{4}\left[a\left(\frac{\sin B}{\sin A} + \frac{\sin A}{\sin E}\right) + b\left(\frac{\sin B}{\sin C} + \frac{\sin C}{\sin E}\right)\right. \\ &\quad + c\left(\frac{\sin C}{\sin A} + \frac{\sin D}{\sin C}\right) + d\left(\frac{\sin E}{\sin A} + \frac{\sin D}{\sin E}\right) \\ &\quad \left. + e\left(\frac{\sin E}{\sin C} + \frac{\sin F}{\sin E}\right) + f\left(\frac{\sin F}{\sin A} + \frac{\sin A}{\sin C}\right)\right]. \end{aligned}$$

Considering that $\angle A = \angle D, \angle B = \angle E, \angle C = \angle F$, we see that the sum in each pair in the brackets of the above is ≥ 2.

Therefore,

$$R_A + R_C + R_E \geq \frac{1}{4}(2a + 2b + 2c + 2d + 2e + 2f) = \frac{p}{2}.$$

【Score Situation】 This particular problem saw the following distribution of scores among contestants: 6 contestants scored 7 points, no contestant scored 6 points, 4 contestants scored 5 points, 4 contestants scored 4 points, 7 contestants scored

3 points, 18 contestants scored 2 points, 74 contestants scored 1 point, and 311 contestants scored 0 point. The average score of this problem is 0.493, indicating that it was extremely difficult.

Among the top five teams in the team scores, the scores of this problem are as follows: the Romania team scored 16 points (with a total team score of 187 points), the United States team scored 11 points (with a total team score of 185 points), the Hungary team scored 6 points (with a total team score of 167 points), the Russia team scored 5 points (with a total team score of 162 points), and the United Kingdom team scored 4 points (with a total team score of 161 points).

The gold medal cutoff for this IMO was set at 28 points (with 35 contestants earning gold medals), the silver medal cutoff was 20 points (with 66 contestants earning silver medals), and the bronze medal cutoff was 12 points (with 99 contestants earning bronze medals).

In this IMO, only one contestant achieved a perfect score of 42 points, namely Ciprian Manolescu from Romania.

Problem 7.14 (IMO 43-6, proposed by Ukraine). There are drawn $n \geq 3$ circles of radius 1 in a plane so that no line meets more than two of the circles. Their centers are $O_1, O_2, \ldots, O_n$.

Show that $\sum_{1 \leq i < j \leq n} \frac{1}{O_i O_j} \leq \frac{(n-1)\pi}{4}$.

Proof.   Denote the set of points formed by $O_1, O_2, \ldots, O_n$ as A. Let us prove a lemma first.

Lemma. *Suppose O_i, O_j, O_k are any three points in the set A. Let $\angle O_j O_i O_k = \theta$. Then*

$$\frac{1}{O_i O_j} < \frac{\theta}{2}.$$

Proof of the Lemma. As shown in Figure 7.22, make the median line l. The distances from O_i, O_j, O_k to l are greater than 1. Then the distance from O_j to $O_i O_k$ is greater than 2, or $O_i O_j \sin \theta > 2$. Thus

$$\frac{1}{O_i O_j} < \frac{\sin \theta}{2} < \frac{\theta}{2} \text{ (Radian)}.$$

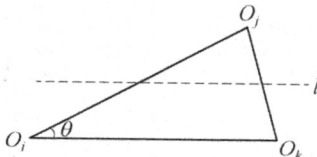

Figure 7.22

For the convenience of the argument, we can change the lemma to a more symmetric form, i.e.,

$$\frac{1}{O_iO_j} + \frac{1}{O_iO_k} < \theta.$$

In the following we prove it.

Let the convex hull of A be $P_1P_2\ldots P_m$, and the inner points be $Q_1, Q_2, \ldots, Q_l$. Then $m + l = n$.

Let us start with $\sum_i \frac{1}{P_iO_j}$, where P_i is a fixed point, and O_j runs through all the elements of A, except P_i.

It is easy to know that all rays P_iO_j are "squeezed" into an angle (we call it α) of less than π, and the angle is divided into $n-2$ disjoint angles. So by the lemma,

$$2\sum_j \frac{1}{P_iO_j} < \alpha + \frac{\beta + \gamma}{2}, \tag{1}$$

where β, γ are the "most borderline" two angles.

Let us think about any inner point Q_i, where we need a little bit of skills. For any point O_j ($\neq Q_i$) in A, we introduce its symmetric point with respect to Q_i, so that we get a total of $2n - 2$ points, half of which belong to A and half do not belong to A.

So, this is divided into $2n - 2$ angles that add up to 2π, and the corresponding $2n - 2$ points are $O_1', O_2', \ldots, O_{2n-2}'$.

Now consider the "adjacent" two points O_k', O_{k+1}' and let $\angle O_k'O_i$ $O_{k+1}' = \theta$.

If $O_k', O_{k+1}' \in A$ or $O_k', O_{k+1}' \notin A$, then obviously there is

$$\frac{1}{Q_iO_k'} + \frac{1}{Q_iO_{k+1}'} < \theta.$$

If either O_k' or O_{k+1}' belongs to A, then the other does not belong to A. For example, when $O_{k+1}' \notin A$, then its symmetric point $O_{k+1}'' \in A$ with respect to Q_i, as shown in Figure 7.23. Considering the distance from O_k'

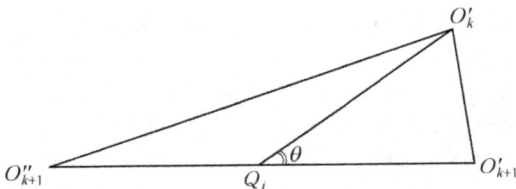

Figure 7.23

to $Q_iO'_{k+1}$ and the distance from O'_{k+1} to $Q_iO'_k$, which are both altitudes in $\triangle O'_k Q_i O''_{k+1}$, so they are greater than 2. Then by the lemma,

$$\frac{1}{O'_k Q_i} + \frac{1}{O'_{k+1} Q_i} < \theta.$$

Sum up the above. Considering that half of the sides are "introduced", we obtain

$$\sum_j \frac{1}{Q_i O'_j} < \frac{2\pi}{4} = \frac{\pi}{2}. \tag{2}$$

Combining (1) with (2), we get

$$\sum_{i<j} \frac{1}{O_i O_j} = \sum_i \sum_j \frac{1}{P_i O_j} + \sum_i \sum_j \frac{1}{Q_i O'_j}$$

$$< \frac{1}{2}\left(\frac{1}{2}\sum \alpha + \frac{1}{4}\sum(\beta + \gamma) + \frac{\pi}{2}l \right)$$

$$\leq \frac{1}{2}\left[\frac{1}{2}(m-2)\pi + \frac{\pi}{2} + \frac{\pi}{2}l \right] = \frac{n-1}{4}\pi.$$

Here the coefficient on the outside is $\frac{1}{2}$, since each side $O_i O_j$ is counted twice. When there is only a convex m-gon, $\sum(\beta + \gamma) = 2\pi$ $(m \geq 4)$ or $\pi(m = 3)$, and once the inner point is added, it may be smaller.

【Score Situation】 This particular problem saw the following distribution of scores among contestants: 12 contestants scored 7 points, no contestant scored 6 points, 1 contestant scored 5 points, no contestant scored 4 points, 12 contestants scored 3 points, 25 contestants scored 2 points, 21 contestants scored 1 point, and 408 contestants scored 0 point. The average score of this problem is 0.409, indicating that it was extremely difficult.

Among the top five teams in the team scores, the scores of this problem are as follows: the China team scored 22 points (with a total team score of 212 points), the Russia team scored 20 points (with a total team score of 204 points), the United States team scored 31 points (with a total team score of 171 points), the Bulgaria team scored 14 points (with a total team score of 167 points), and the Vietnam team scored 0 points (with a total team score of 166 points).

The gold medal cutoff for this IMO was set at 29 points (with 39 contestants earning gold medals), the silver medal cutoff was 23 points (with 73 contestants earning silver medals), and the bronze medal cutoff was 14 points (with 120 contestants earning bronze medals).

In this IMO, only three contestants achieved a perfect score of 42 points, namely Yunhao Fu and Botong Wang from China, and Andrei Khaliavine from Russia.

Problem 7.15 (IMO 47-1, proposed by South Korea). Let ABC be a triangle with incentre I. A point P in the interior of the triangle satisfies $\angle PBA + \angle PCA = \angle PBC + \angle PCB$. Show that $AP \geq AI$, and the equality holds if and only if $P = I$.

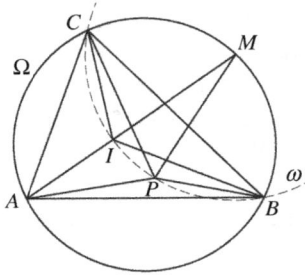

Figure 7.24

Proof. As shown in Figure 7.24, let $\angle A = \alpha$, $\angle B = \beta$, and $\angle C = \gamma$. Since

$$\angle PBA + \angle PCA + \angle PBC + \angle PCB = \beta + \gamma,$$

by the assumption, $\angle PBC + \angle PCB = \frac{\beta + \gamma}{2}$.

Since points P, I are on the same side of BC, the points B, C, I, P are concyclic, that is, the point P is on the circumcircle ω of $\triangle BCI$.

Denote Ω the circumcircle of $\triangle ABC$. Then the midpoint M of the arc BC is the center of circumcircle ω, that is, M is also the intersection point where the bisector AI of $\angle A$ intersects the circle Ω.

From triangle $\triangle APM$,

$$AP + PM \geq AM = AI + IM = AI + PM.$$

Therefore $AP \geq AI$. The equality holds if and only if P lies on the segment AI, that is $P = I$.

【Score Situation】 This particular problem saw the following distribution of scores among contestants: 358 contestants scored 7 points, 7 contestants scored 6 points, 27 contestants scored 5 points, 10 contestants scored 4 points, 12 contestants scored 3 points, 14 contestants scored 2 points, 9 contestants scored 1 point, and 61 contestants scored 0 point. The average score of this problem is 5.614, indicating that it was simple.

Among the top five teams in the team scores, the scores of this problem are as follows: the China team scored 42 points (with a total team score of 214 points), the Russia team scored 42 points (with a total team score of 174 points), the South Korea team scored 42 points (with a total team score of 170 points), the Germany team scored 42 points (with a total team score of 157 points), and the United States team scored 42 points (with a total team score of 154 points).

The gold medal cutoff for this IMO was set at 28 points (with 42 contestants earning gold medals), the silver medal cutoff was 19 points (with 89 contestants earning silver medals), and the bronze medal cutoff was 15 points (with 122 contestants earning bronze medals).

In this IMO, only three contestants achieved a perfect score of 42 points, namely Zhiyu Liu from China, Iurie Boreico from Moldova and Alexander Magazinov from Russia.

2. Problems related to angles

Problem 7.16 (IMO 39-5, proposed by Ukraine). Let I be the incentre of triangle ABC. Let the incircle of ABC touch the sides BC, CA, and AB at K, L, and M, respectively. The line through B parallel to MK meets the lines LM and LK at R and S, respectively. Prove that $\angle RIS$ is acute.

Proof.   As shown in Figure 7.25, connect to form MI, BI. We have $MI \perp AB$, and then $BM < BI$.

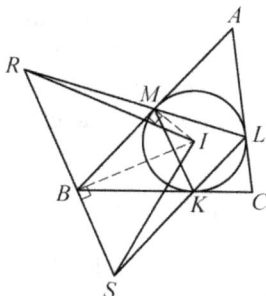

Figure 7.25

Also $BI \perp MK$, while $RS \parallel MK$, so $BI \perp RS$.
And by $RS \parallel MK$, we know that

$$\angle BSK = \angle MKL,$$

$$\angle KBS = \angle MKB = \angle MLK.$$

Thus, $\triangle KBS \backsim \triangle MLK$. Consequently

$$\frac{BS}{BK} = \frac{LK}{LM}.$$

Similarly,

$$\frac{BR}{BM} = \frac{LM}{LK}.$$

Then we know that $BR \cdot BS = BM \cdot BK = BM^2 < BI^2$.

And by $BI \perp RS$, there exists a point P inside BI such that $BR \cdot BS = BP^2$. In the triangle $\triangle RPS$, by the geometric mean theorem or similarity, $\angle RPS = 90°$. Then the point I is outside the circle with diameter RS, and thus $\angle RIS < 90°$.

【Score Situation】 This particular problem saw the following distribution of scores among contestants: 139 contestants scored 7 points, 6 contestants scored 6 points, 5 contestants scored 5 points, 9 contestants scored 4 points, 22 contestants scored 3 points, 38 contestants scored 2 points, 16 contestants scored 1 point, and 184 contestants scored 0 point. The average score of this problem is 2.931, indicating that it had a certain level of difficulty.

Among the top five teams in the team scores, the scores of this problem are as follows: the Iran team scored 38 points (with a total team score of 211 points), the Bulgaria team scored 37 points (with a total team score of 195 points), the Hungary team scored 38 points (with a total team score of 186 points), the United States team scored 17 points (with a total team score of 186 points), and the Chinese Taiwan team scored 41 points (with a total team score of 184 points).

The gold medal cutoff for this IMO was set at 31 points (with 37 contestants earning gold medals), the silver medal cutoff was 24 points (with 66 contestants earning silver medals), and the bronze medal cutoff was 14 points (with 102 contestants earning bronze medals).

In this IMO, only one contestant achieved a perfect score of 42 points, namely Omid Amini from Iran.

Problem 7.17 (IMO 42-1, proposed by South Korea). Let ABC be an acute-angled triangle with circumcenter O. Let P on BC be the foot of the altitude from A. Suppose that $\angle BCA \geq \angle ABC + 30°$. Prove that $\angle CAB + \angle COP < 90°$.

Proof. Let the radius of the circumcircle of $\triangle ABC$ be R. As shown in Figure 7.26, connect to form AO, and make $OQ \perp BC$, where Q is the midpoint of BC.

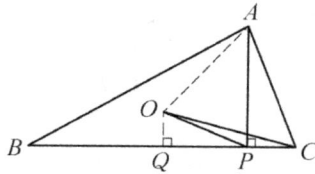

Figure 7.26

Since

$$90° > \angle OAP = \angle OAC - \angle PAC$$

$$= (90° - \angle ABC) - (90° - \angle ACB)$$

$$= \angle ACB - \angle ABC \geq 30°,$$

so

$$QP = OA \sin \angle OAP \geq \frac{1}{2}R.$$

While $QC < OC = R$, we have $CP < \frac{1}{2}R$. Therefore $CP < QP < OP$, and then

$$\angle COP < \angle OCB = 90° - \angle BAC.$$

【Score Situation】This particular problem saw the following distribution of scores among contestants: 177 contestants scored 7 points, 20 contestants scored 6 points, 12 contestants scored 5 points, 12 contestants scored 4 points, 45 contestants scored 3 points, 47 contestants scored 2 points, 28 contestants scored 1 point, and 132 contestants scored 0 point. The average score of this problem is 3.645, indicating that it was relatively straightforward.

Among the top five teams in the team scores, the scores of this problem are as follows: the China team scored 42 points (with a total team score of 225 points), the United States team scored 41 points (with a total team score of 196 points), the Russia team scored 41 points (with a total team score of 196 points), the South Korea team scored 42 points (with a total team score of 185 points), and the Bulgaria team scored 41 points (with a total team score of 185 points).

The gold medal cutoff for this IMO was set at 30 points (with 39 contestants earning gold medals), the silver medal cutoff was 20 points (with 81 contestants earning silver medals), and the bronze medal cutoff was 11 points (with 122 contestants earning bronze medals).

In this IMO, a total of four contestants achieved a perfect score of 42 points.

Problem 7.18 (IMO 44-3, proposed by Poland). A convex hexagon has the property that for any pair of opposite sides the distance between their midpoints is $\sqrt{3}/2$ times the sum of their lengths. Show that all the hexagon's angles are equal.

Proof. Let us prove a lemma first.

Lemma. *Consider a triangle $\triangle PQR$ with $\angle QPR \geq 60°$. Let L be the midpoint of QR. Then $PL \leq \frac{\sqrt{3}}{2}QR$. The equality holds if and only if the triangle PQR is equilateral.*

Proof of the Lemma. Let S be a point such that the $\triangle SQR$ is equilateral, where the points S and P lie in the same half-plane bounded by the line QR. Then the point P lies inside the circumcircle (including the circumference of the circle) of the $\triangle SQR$. Thus,

$$PL \leq OL + OP \leq OL + OS = \frac{\sqrt{3}}{2}QR,$$

where O is the center of the circumcircle of the $\triangle SQR$. This completes the proof of the lemma.

For a convex hexagon, the main diagonals form a triangle (or a common point). Thus, we may choose two of these three diagonals that form an angle greater than or equal to 60°. Without loss of generality, we may assume that the diagonals AD and BE of the given hexagon $ABCDEF$ satisfy $\angle APB \geq 60°$, where P is the intersection of these two diagonals (see Figure 7.27). By the lemma and the conditions,

$$MN = \frac{\sqrt{3}}{2}(AB + DE) \geq PM + PN \geq MN,$$

where M, N are the midpoints of AB, DE, respectively.

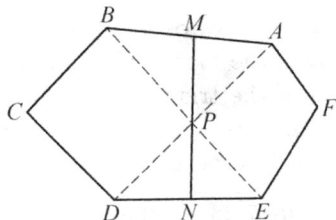

Figure 7.27

Thus, it follows from the lemma that the triangles ABP and DEP are equilateral triangles.

At this moment, the diagonal CF forms an angle greater than or equal to $60°$ with one of the diagonals AD and BE. Without loss of generality, let $\angle AQF \geq 60°$, where Q is the intersection point of AD and CF. Similarly, the triangles AQF and CQD are equilateral triangles. This implies that $\angle BRC = 60°$, where R is the intersection point of BE and CF. Using the lemma again, we get that the triangles BRC and EFR are equilateral triangles. This completes the proof.

【Score Situation】This particular problem saw the following distribution of scores among contestants: 23 contestants scored 7 points, 1 contestant scored 6 points, no contestant scored 5 points, 1 contestant scored 4 points, 3 contestants scored 3 points, no contestant scored 2 points, 5 contestants scored 1 point, and 424 contestants scored 0 point. The average score of this problem is 0.405, indicating that it was extremely difficult.

Among the top five teams in the team scores, the scores of this problem are as follows: the Bulgaria team scored 41 points (with a total team score of 227 points), the China team scored 28 points (with a total team score of 211 points), the United States team scored 18 points (with a total team score of 188 points), the Vietnam team scored 14 points (with a total team score of 172 points), and the Russia team scored 14 points (with a total team score of 167 points).

The gold medal cutoff for this IMO was set at 29 points (with 37 contestants earning gold medals), the silver medal cutoff was 19 points (with 69 contestants earning silver medals), and the bronze medal cutoff was 13 points (with 104 contestants earning bronze medals).

In this IMO, only three contestants achieved a perfect score of 42 points, namely Yunhao Fu from China, and Bảo Lê Hùng Việt and Trọng Cảnh Nguyễn from Vietnam.

3. Problems related to areas

Problem 7.19 (IMO 29-5, proposed by Greece). A triangle ABC is right-angled at A, and D is the foot of the altitude from A. The straight line joining the incenters of the triangles ABD, ACD intersects the sides AB, AC at the points K, L respectively. Let S and T denote the areas of the triangles ABC and AKL respectively. Show that $S \geq 2T$.

Proof.   As shown in Figure 7.28, let the incenters of $\triangle ABD$ and $\triangle ACD$ be M, N respectively. It is easy to see that $\triangle ABC \backsim \triangle DMN$. Then $\angle B = \angle NMD$, and hence B, K, M, D are concyclic.

So $\angle AKL = \angle BDM = 45° = \angle ADM$.

And $\angle KAM = \angle DAM, AM = AM$. Thus $\triangle AMK \cong \triangle AMD$, from which $AK = AD$.

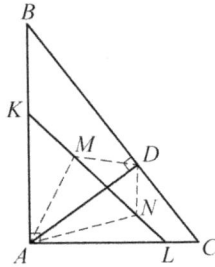

Figure 7.28

Similarly, $AL = AD$. Then,

$$S = \frac{1}{2}AB \cdot AC,$$

$$T = \frac{1}{2}AD^2 = \frac{1}{2}\frac{AB^2 \cdot AC^2}{BC^2} = \frac{AB^2 \cdot AC^2}{2(AB^2 + AC^2)}.$$

Hence

$$\frac{S}{2T} = \frac{AB^2 + AC^2}{2AB \cdot AC} \geq 1,$$

i.e., $S \geq 2T$.

The equality holds only if $AB = AC$.

【Score Situation】 This particular problem saw the following distribution of scores among contestants: 86 contestants scored 7 points, 7 contestants scored 6 points, 11 contestants scored 5 points, 20 contestants scored 4 points, 10 contestants scored 3 points, 23 contestants scored 2 points, 35 contestants scored 1 point, and 76 contestants scored 0 point. The average score of this problem is 3.321, indicating that it was relatively straightforward.

Among the top five teams in the team scores, the scores of this problem are as follows: the Soviet Union team scored 42 points (with a total team score of 217 points), the China team scored 42 points (with a total team score of 201 points), the Romania team scored 31 points (with a total team score of 201 points), the Germany team scored 32 points (with a total team score of 174 points), and the Vietnam team scored 36 points (with a total team score of 166 points).

The gold medal cutoff for this IMO was set at 32 points (with 17 contestants earning gold medals), the silver medal cutoff was 23 points (with 48 contestants earning silver medals), and the bronze medal cutoff was 14 points (with 65 contestants earning bronze medals).

In this IMO, a total of five contestants achieved a perfect score of 42 points.

Problem 7.20 (IMO 30-2, proposed by Australia). In an acute-angled triangle ABC, the internal bisector of angle A meets the circumcircle of the triangle again at A_1. Points B_1 and C_1 are defined similarly. Let A_0 be the point of intersection of the line AA_1 with the external bisectors of angles B and C. Points B_0 and C_0 are defined similarly. Prove that

    (a) the area of the triangle $A_0B_0C_0$ is twice the area of the hexagon $AC_1BA_1CB_1$;

    (b) the area of the triangle $A_0B_0C_0$ is at least four times the area of the triangle ABC.

Proof.   (a) As shown in Figure 7.29, it is easy to know

$$\angle A_1IC = \frac{1}{2}(\angle A + \angle C) = \angle A_1CI.$$

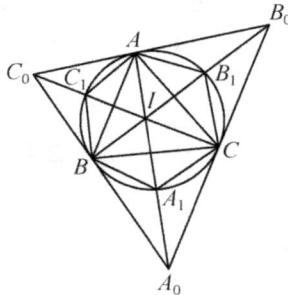

Figure 7.29

Thus $IA_1 = CA_1$. Here I is the incenter of $\triangle ABC$.

And $\angle ICA_0 = 90°$, so A_1 is the midpoint of A_0I, and then

$$S_{IBA_0C} = 2S_{IBA_1C}.$$

Similarly

$$S_{AIBC_0} = 2S_{AIBC_1},$$

$$S_{AICB_0} = 2S_{AICB_1}.$$

Adding the above three equalities gives

$$S_{\triangle A_0B_0C_0} = 2S_{\text{hexagon } AC_1BA_1CB_1}.$$

    (b) Obviously we just have to prove that $S_{\text{hexagon } AC_1BA_1CB_1} \geq 2S_{\triangle ABC}$.

Let the circumcenter of $\triangle ABC$ be O. It is easy to know that $OA_1 \perp BC$. Let the lengths of the three sides of $\triangle ABC$ be a, b, c, respectively and the radius of the circumcircle be R. Then

$$S_{OBA_1C} = \frac{1}{2}Ra.$$

Similarly

$$S_{OAC_1B} = \frac{1}{2}Rc,$$

$$S_{AOCB_1} = \frac{1}{2}Rb.$$

Thus

$$S_{\text{hexagon } AC_1BA_1CB_1} = \frac{1}{2}R(a+b+c).$$

The problem becomes proving that

$$R(a+b+c) \geq 4S_{\triangle ABC}.$$

By the law of sines and the area formula, the above inequality is equivalent to proving that

$$\sin A + \sin B + \sin C \geq 4\sin A \sin B \sin C.$$

From the AM–GM inequality,

$$\text{The left side of inequality} \geq 3\sqrt[3]{\sin A \sin B \sin C}.$$

The proof is complete if we can show that

$$3\sqrt[3]{\sin A \sin B \sin C} \geq 4\sin A \sin B \sin C.$$

The above inequality is equivalent to

$$\sin A \sin B \sin C \leq \frac{3\sqrt{3}}{8}.$$

The inequality can be solved by trigonometric formulas, and note that the equality holds only if $\angle A = \angle B = \angle C = 60°$.

【Score Situation】This particular problem saw the following distribution of scores among contestants: 108 contestants scored 7 points, 15 contestants scored 6 points, 28 contestants scored 5 points, 14 contestants scored 4 points, 49 contestants scored 3 points, 9 contestants scored 2 points, 23 contestants scored 1 point, and 45 contestants scored 0 point. The average score of this problem is 4.227, indicating that it was simple.

Among the top five teams in the team scores, the scores of this problem are as follows: the China team scored 42 points (with a total team score of 237 points), the Romania team scored 42 points (with a total team score of 223 points), the Soviet Union team scored 41 points (with a total team score of 217 points), the German Democratic Republic team scored 40 points (with a total team score of 216 points), and the United States team scored 36 points (with a total team score of 207 points).

The gold medal cutoff for this IMO was set at 38 points (with 20 contestants earning gold medals), the silver medal cutoff was 30 points (with 55 contestants earning silver medals), and the bronze medal cutoff was 18 points (with 72 contestants earning bronze medals).

In this IMO, a total of 10 contestants achieved a perfect score of 42 points.

Problem 7.21 (IMO 47-6, proposed by Serbia–Montenegro). Assign to each side b of a convex polygon P the maximum area of a triangle that has b as a side and is contained in P. Show that the sum of the areas assigned to the sides of P is at least twice the area of P.

Proof.　First, we prove a lemma.

Lemma. *Every convex $2n$-gon, of area S has a side and a vertex that jointly span a triangle of area not less than $\frac{S}{2}$.*

Proof of the Lemma. By the main diagonals of the $2n$-gon we shall mean those which partition the $2n$-gon into two polygons with equally many sides. For any side b of the $2n$-gon, denote by $\triangle_b$ the triangle ABP, where A, B are the endpoints of b and P is the intersection point of the main diagonals AA', BB'. We claim that the union of triangles $\triangle_b$, taken over all sides, covers the whole polygon.

To show this, choose any side AB and consider the main diagonal AA' as a directed line segment. Let X be any point in the polygon, not on any main diagonal. For definiteness, let X lie on the left side of the ray AA'. Consider the sequence of the main diagonals $AA', BB', CC', \ldots$, where $A, B, C, \ldots$ are consecutive vertices, situated to the right of AA'.

The nth item in this sequence is the diagonal $A'A$, having X on its right side. So there are two successive vertices K, L in the sequence $A, B, C, \ldots$ before A' such that X still lies to the left of KK' but to the right of LL'. This means that X is in the triangle $\triangle_{l'}$, where $l' = K'L'$.

We can apply the analogous reasoning to points X on the right of AA' (the points lying on the main diagonals can be safely ignored). Thus, the union of triangles $\triangle_b$ covers the whole polygon. The sum of their areas is no less than S. Therefore, we can find two opposite sides, say $b = AB$ and $b' = A'B'$ (with AA', BB' being main diagonals), such that

$$[\triangle_b] + [\triangle_{b'}] \geq \frac{S}{n},$$

where $[\cdots]$ stands for the area of a region. Let AA' and BB' intersect at P. Without loss of generality, assume that $PB \geq PB'$.

Then

$$[ABA'] = [ABP] + [PBA'] \geq [ABP] + [PA'B']$$

$$= [\triangle_b] + [\triangle_{b'}] \geq \frac{S}{n},$$

proving the lemma.

Now, let P be any convex polygon of area S, with m sides $a_1, a_2, \ldots, a_m$. Let S_i be the area of the greatest triangle in P with side a_i. Suppose, contrary to the assertion, that

$$\sum_{i=1}^{m} \frac{S_i}{S} < 2.$$

Then there exist rational numbers $q_1, q_2, \ldots, q_m$ such that $\sum_{i=1}^{m} q_i = 2$ and $q_i > \frac{S_i}{S}$ for each i.

Let n be a common denominator of the m fractions $q_1, q_2, \ldots, q_m$, and let $q_i = \frac{k_i}{n}$. Then $\sum k_i = 2n$.

Partition each side a_i of P into k_i equal segments, creating a convex $2n$-gon of area S (with some angles of size $180°$), to which we apply the lemma. Accordingly, this induced polygon has a side b and a vertex H spanning a triangle T of area $[T] \geq \frac{S}{n}$. If b is a portion of a side a_i of P, then the triangle W with base a_i and summit H has area

$$[W] = k_i \cdot [T] \geq k_i \cdot \frac{S}{n} = q_i \cdot S > S_i,$$

in contradiction with the definition of S_i. This completes the proof.

【Score Situation】 This particular problem saw the following distribution of scores among contestants: 8 contestants scored 7 points, 1 contestant scored 6 points, 1 contestant scored 5 points, no contestant scored 4 points, 3 contestants scored 3 points, 3 contestants scored 2 points, 11 contestants scored 1 point, and 471 contestants scored 0 point. The average score of this problem is 0.187, indicating that it was extremely difficult.

Among the top five teams in the team scores, the scores of this problem are as follows: the China team scored 16 points (with a total team score of 214 points), the Russia team scored 15 points (with a total team score of 174 points), the South Korea team scored 7 points (with a total team score of 170 points), the Germany team scored 8 points (with a total team score of 157 points), and the United States team scored 3 points (with a total team score of 154 points).

The gold medal cutoff for this IMO was set at 28 points (with 42 contestants earning gold medals), the silver medal cutoff was 19 points (with 89 contestants earning silver medals), and the bronze medal cutoff was 15 points (with 122 contestants earning bronze medals).

In this IMO, only three contestants achieved a perfect score of 42 points, namely Zhiyu Liu from China, Iurie Boreico from Moldova and Alexander Magazinov from Russia.

7.2.3 *Conditional evaluation problems*

Problem 7.22 (IMO 18-1, proposed by Czechoslovakia). In a plane convex quadrilateral of area 32, the sum of the lengths of two opposite sides and one diagonal is 16. Determine all possible lengths of the other diagonal.

Solution. Without loss of generality, let in a convex quadrilateral $ABCD$, $AB + AC + CD = 16$. We need to figure out the length of BD.

Since

$$32 = S_{ABCD} = S_{\triangle ABC} + S_{\triangle ACD}$$

$$= \frac{1}{2}AB \cdot AC \cdot \sin\angle BAC + \frac{1}{2}AC \cdot CD \cdot \sin\angle ACD$$

$$\leq \frac{1}{2}AC\,(AB + CD) \leq \frac{1}{8}\,(AC + AB + CD)^2$$

$$= \frac{256}{8} = 32,$$

the equality holds. The former requires $\angle BAC = \angle ACD = 90°$ and the latter requires $AC = AB + CD = 8$.

Make a rectangle $ABA'C$ with AC, AB as sides, it is easy to know that

$$A'B = AC = 8, \ DA' = DC + AB = 8,$$

$$\angle DA'B = 90°,$$

thus in the isosceles right triangle $BA'D$, we have $BD = 8\sqrt{2}$ cm.

【Score Situation】 This particular problem saw the following distribution of scores among contestants: 60 contestants scored 5 points, 5 contestants scored 4 points, 3 contestants scored 3 points, 7 contestants scored 2 points, 20 contestants scored 1 point, and 44 contestants scored 0 point. The average score of this problem is 2.612, indicating that it had a certain level of difficulty.

Among the top five teams in the team scores, the scores of this problem are as follows: the Soviet Union team scored 36 points (with a total team score of 250 points), the United Kingdom team scored 23 points (with a total team score of 214 points), the United States team scored 25 points (with a total team score of 188 points), the Bulgaria team scored 22 points (with a total team score of 174 points), and the Austria team scored 26 points (with a total team score of 167 points).

The gold medal cutoff for this IMO was set at 34 points (with 9 contestants earning gold medals), the silver medal cutoff was 23 points (with 28 contestants earning silver medals), and the bronze medal cutoff was 15 points (with 45 contestants earning bronze medals).

In this IMO, only one contestant achieved a perfect score of 40 points, namely Laurent Pierre from France.

Problem 7.23 (IMO 21-4, proposed by the United States). Given a plane π, a point P in this plane, and a point Q not in π, find all points R in π such that the ratio $\frac{QP+PR}{QR}$ is a maximum.

Solution. Let the projection of the point Q onto the plane π be T, $TP = a$, $QT = h$, and a, h are known quantities. And let $PR = x$. Then

$$QP = \sqrt{h^2 + a^2},$$

$$QR = \sqrt{QT^2 + TR^2} \geq \sqrt{h^2 + (a-x)^2}.$$

Thus

$$\frac{QP + PR}{QR} \leq \frac{\sqrt{h^2 + a^2} + x}{\sqrt{h^2 + (a-x)^2}}.$$

Using local adjustments, with fixed $PR = x$, we show that point R must be on ray PT (discussed separately when P and T coincide).

As shown in Figure 7.30, extend TP to S such that $QP = PS$, and then S is also a fixed point. Let $\angle S = \frac{\theta}{2}$ be a fixed angle. By the law of sines, we know that

$$\frac{QP + PR}{QR} = \frac{RS}{RQ} = \frac{\sin \angle RQS}{\sin \frac{\theta}{2}} \leq \frac{1}{\sin \frac{\theta}{2}},$$

and the equality holds only if $\angle RQS = 90°$.

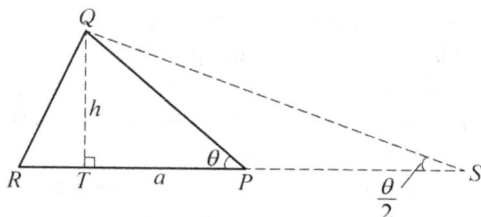

Figure 7.30

So, the point R is found like this:

(i) When points P and T do not coincide, let QP and π form an angle θ, the projection of Q on the plane π is T, and R is on the ray PT, satisfying $\angle QRP = 90° - \frac{\theta}{2}$. And then $\frac{QP+PR}{QR}$ reaches a maximum of $\frac{1}{\sin\frac{\theta}{2}}$.

(ii) When points P and T coincide, R is any point on a circle in the plane π with P as the center and QP as the radius, where the maximum is $\frac{1}{\sin 45°} = \sqrt{2}$.

【Score Situation】This particular problem saw the following distribution of scores among contestants: 69 contestants scored 6 points, 38 contestants scored 5 points, 20 contestants scored 4 points, 4 contestants scored 3 points, 5 contestants scored 2 points, 16 contestants scored 1 point, and 14 contestants scored 0 point. The average score of this problem is 4.349, indicating that it was simple.

Among the top five teams in the team scores, the scores of this problem are as follows: the Soviet Union team scored 48 points (with a total team score of 267 points), the Romania team scored 42 points (with a total team score of 240 points), the Germany team scored 37 points (with a total team score of 235 points), the United Kingdom team scored 40 points (with a total team score of 218 points), and the United States team scored 28 points (with a total team score of 199 points).

The gold medal cutoff for this IMO was set at 37 points (with 8 contestants earning gold medals), the silver medal cutoff was 29 points (with 32 contestants earning silver medals), and the bronze medal cutoff was 20 points (with 42 contestants earning bronze medals).

In this IMO, a total of four contestants achieved a perfect score of 40 points.

Problem 7.24 (IMO 22-1, proposed by the United Kingdom). A point P is inside a given triangle ABC. Points D, E, F are the feet of the perpendiculars from P to the lines BC, CA, AB respectively. Find all P for which $\frac{BC}{PD} + \frac{CA}{PE} + \frac{AB}{PF}$ is the least.

Solution. As shown in Figure 7.31, let the lengths of BC, CA, AB be a, b, c respectively; the lengths of PD, PE, PF be p, q, r respectively.

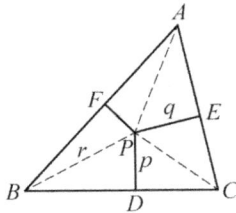

Figure 7.31

We need to find a point P in the interior of $\triangle ABC$ such that $\frac{a}{p} + \frac{b}{q} + \frac{c}{r}$ achieves the minimum.

Since

$$ap + bq + cr = 2S_{\triangle PBC} + 2S_{\triangle PCA} + 2S_{\triangle PAB} = 2S_{\triangle ABC}$$

is a constant independent of point P, so "$(ap + bq + cr)(\frac{a}{p} + \frac{b}{q} + \frac{c}{r})$ achieves the minimum" can be used instead of "$\frac{a}{p} + \frac{b}{q} + \frac{c}{r}$ achieves the minimum".

By Cauchy's inequality,

$$(ap + bq + cr)\left(\frac{a}{p} + \frac{b}{q} + \frac{c}{r}\right) \geq (a + b + c)^2.$$

And when $ap \cdot \frac{p}{a} = bq \cdot \frac{q}{b} = cr \cdot \frac{r}{c}$, i.e., $p = q = r$, the equality holds. Consequently when $p = q = r$, the quantity $(ap + bq + cr)\left(\frac{a}{p} + \frac{b}{q} + \frac{c}{r}\right)$ achieves the minimum.

In other words, the position of the desired point P is the incenter of $\triangle ABC$.

【Score Situation】 This particular problem saw the following distribution of scores among contestants: 36 contestants scored 7 points, 3 contestants scored 6 points, no contestant scored 5 points, 1 contestant scored 4 points, no contestant scored 3 points, no contestant scored 2 points, 6 contestants scored 1 point, and 5 contestants scored 0 point. The average score of this problem is 5.490, indicating that it was simple.

Among the top five teams in the team scores, the scores of this problem are as follows: the United States team scored 55 points (with a total team score of 314 points), the Germany team scored 49 points (with a total team score of 312 points), the United Kingdom team scored 38 points (with a total team score of 301 points), the Austria team scored 49 points (with a total team score of 290 points), and the Bulgaria team scored 44 points (with a total team score of 287 points).

The gold medal cutoff for this IMO was set at 41 points (with 36 contestants earning gold medals), the silver medal cutoff was 34 points (with 37 contestants

earning silver medals), and the bronze medal cutoff was 26 points (with 30 contestants earning bronze medals).

In this IMO, a total of 26 contestants achieved a perfect score of 42 points.

7.3 Summary

In the first 64 IMOs, there are 24 problems on geometric inequalities, as depicted in Figure 7.32, which can be broadly categorized into three types.

Problems 7.1–7.5 focus on "existence problems;" among these five problems, the one with the lowest average score is Problem 7.4 (IMO 13-4), proposed by the Netherlands. Problems 7.6–7.21 deal with "equality or inequality proof problems;" among these 16 problems, the one with the lowest average score is Problem 7.21 (IMO 47-6), proposed by Serbia and Montenegro. Problems 7.22–7.24 are about "conditional evaluation problems;" among these three problems, the one with the lowest average score is Problem 7.22 (IMO 18-1), proposed by Czechoslovakia. In addition, Problem 7.13 (IMO 37-5), Problem 7.14 (IMO 43-6), and Problem 7.18 (IMO 44-3) all have extremely low average scores.

These problems were proposed by 19 countries and regions. Poland proposed four problems, while South Korea and Ukraine each contributed two problems.

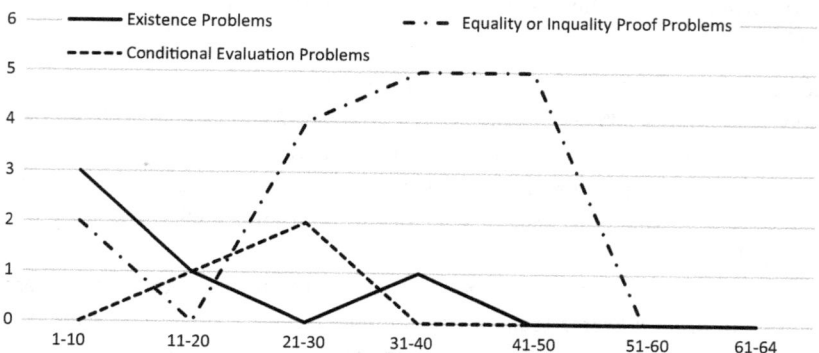

Figure 7.32 Numbers of Geometric Inequality Problems in the First 64 IMOs

From Table 7.2, it can be observed that in the first 64 IMOs, there were four problems with an average score of 0–1 points; one problem with average

Table 7.2 Score Details of Geometric Inequality Problems in the First 64 IMOs

Problem	7.1	7.2	7.3	7.4	7.5	7.6
Full points	6.000	8.000	5.000	6.000	7.000	7.000
Average score	2.667	7.333	3.571	2.500	3.801	1.938
Top five mean				3.417	6.867	
6th–15th mean				1.567	6.183	
16th–25th mean					4.383	
Problem number in the IMO	3-4	8-6	10-4	13-4	32-5	5-3
Proposing country	The German Democratic Republic	Poland	Poland	Netherlands	France	Hungary

Problem	7.7	7.8	7.9	7.10	7.11	7.12
Full points	6.000	7.000	7.000	7.000	7.000	7.000
Average score	3.730	2.531	3.990	3.798	2.303	3.410
Top five mean		5.833	6.700	6.467	5.467	7.000
6th–15th mean		3.467	5.7	6.25	3.364	5.933
16th–25th mean		1.467	4.35	4.483	2.926	4.467
Problem number in the IMO	9-1	25-5	30-4	32-1	34-4	36-5
Proposing country	Poland	Mongolia	Iceland	Soviet Union	Macedonia	New Zealand

Problem	7.13	7.14	7.15	7.16	7.17	7.18
Full points	7.000	7.000	7.000	7.000	7.000	7.000
Average score	0.493	0.409	5.614	2.931	3.645	0.405
Top five mean	1.400	2.900	7.000	5.700	6.900	3.833
6th–15th mean	1.15	0.833	6.848	4.967	6.6	0.583
16th-25th mean	0.636	0.533	6.833	3.733	5.533	0.271
Problem number in the IMO	37-5	43-6	47-1	39-5	42-1	44-3
Proposing country	Armenia	Ukraine	South Korea	Ukraine	South Korea	Poland

Problem	7.19	7.20	7.21	7.22	7.23	7.24
Full points	7.000	7.000	7.000	5.000	6.000	7.000
Average score	3.321	4.227	0.187	2.612	4.349	5.490
Top five mean	6.100	6.700	1.633	3.300	4.875	5.875
6th–15th mean	4.288	5.75	0.333	2.75	5.013	4.041
16th–25th mean	3.379	4.783	0.093			2.415
Problem number in IMO	29-5	30-2	47-6	18-1	21-4	22-1
Proposing country	Greece	Australia	Serbia and Montenegro	Czechoslovakia	United States	United Kingdom

Note. Top five mean = Total score of the top five teams/Total number of contestants from the top five teams,

6th–15th mean = Total score of the 6th–15th teams/Total number of contestants from the 6th–15th teams,

16th–25th mean = Total score of the 16th–25th teams/Total number of contestants from the 16th–25th teams.

scores of 1–2 points; six problems with average scores of 2–3 points; eight problems with average scores of 3–4 points; five problems with average scores above 4 points. Overall, geometric inequality problems were relatively difficult.

In the 24th to 64th IMOs, there were a total of 15 geometric inequality problems. Among these, four had an average score of 0–1 point; three had an average score of 2–3 points; six had an average score of 3–4 points; two had an average score above 4 points. Further analysis of the problem numbers of these 15 geometric inequality problems, as shown in Table 7.3, reveals that these problems frequently appeared as the 1st/4th or 2nd/5th problem. Most of these problems were the type of equality or inequality proof problems, with fourteen in total. For the other two types of geometric inequality problems, only one existence problem appeared.

Table 7.3 Numbers of Geometric Inequality Problems in the 24th-64th IMOs

Geometric Inequality Problems	Problem Number			Number of Problems in the First 64 IMOs
	1, 4	2, 5	3, 6	
Existence problem	0	1	0	5
Equality or inequality proof problems	5	6	3	16
Conditional evaluation problems	0	0	0	3
Total	5	7	3	24

From Table 7.2, it can be observed that, in most cases, the average score of the top five teams is typically about 1–3 points higher than the average score of the problem, except for individual problems. However, in simpler problems, the difference in average scores between the top five teams and 6th–15th teams is generally not significant, as seen in problems such as Problem 7.10 (IMO 32-1), Problem 7.15 (IMO 47-1), Problem 7.17 (IMO 42-1), and Problem 7.23 (IMO 21-4).

From Table 7.2, it can also be observed that in most cases, the average scores of the 6th–15th teams and 16th–25th teams also tend to be higher than the average score of the problem. This phenomenon is due to the smaller number of participating teams in early IMOs. It was not until the 30th IMO in 1989 that the number of participating teams exceeded 50. In addition, the appearance of extremely difficult geometry problem can have a depressing effect on the scores of many contestants, thus affecting the average score of the problem.

Appendix A

IMO General Information

Session	Year	Host	Number of Participating Teams	Number of Contestants	Gold Silver Bronze (Cutoffs/Numbers of Medalists)		
IMO 1	1959	Romania	7	52	37/3	36/3	33/5
IMO 2	1960	Romania	5	39	40/4	37/4	33/4
IMO 3	1961	Hungary	6	48	37/3	34/4	30/4
IMO 4	1962	Czechoslovakia	7	56	41/4	34/12	29/15
IMO 5	1963	Poland	8	64	35/7	28/11	21/17
IMO 6	1964	The Union of Soviet Socialist Republics	9	72	38/7	31/9	27/19
IMO 7	1965	The German Democratic Republic	10	80	38/8	30/12	20/17
IMO 8	1966	Bulgaria	9	72	39/13	34/15	31/11
IMO 9	1967	Yugoslavia	13	99	38/11	30/14	22/26
IMO 10	1968	The Union of Soviet Socialist Republics	12	96	39/22	33/22	26/20
IMO 11	1969	Romania	14	112	40/3	30/20	24/21
IMO 12	1970	Hungary	14	112	37/7	30/11	19/40
IMO 13	1971	Czechoslovakia	15	115	35/7	23/12	11/29
IMO 14	1972	Poland	14	107	40/8	30/16	19/30
IMO 15	1973	The Union of Soviet Socialist Republics	16	125	35/5	27/15	17/48

(*Continued*)

(*Continued*)

Session	Year	Host	Number of Partici-pating Teams	Number of Contes-tants	Gold (Cutoffs/Numbers of Medalists)	Silver	Bronze
IMO 16	1974	The German Democratic Republic	18	140	38/10	30/24	23/37
IMO 17	1975	Bulgaria	17	135	38/8	32/25	23/36
IMO 18	1976	Austria	18	139	34/9	23/28	15/45
IMO 19	1977	Yugoslavia	21	155	34/13	24/29	17/35
IMO 20	1978	Romania	17	132	35/5	27/20	22/38
IMO 21	1979	The United Kingdom	23	166	37/8	29/32	20/42
IMO 22	1981	The United States of America	27	185	41/36	34/37	26/30
IMO 23	1982	Hungary	30	119	37/10	30/20	21/31
IMO 24	1983	France	32	186	38/9	26/27	15/57
IMO 25	1984	Czechoslovakia	34	192	40/14	26/35	17/49
IMO 26	1985	Finland	38	209	34/14	22/35	15/52
IMO 27	1986	Poland	37	210	34/18	26/41	17/48
IMO 28	1987	Cuba	42	237	42/22	32/42	18/56
IMO 29	1988	Australia	49	268	32/17	23/48	14/65
IMO 30	1989	Germany	50	291	38/20	30/55	18/72
IMO 31	1990	The People's Republic of China	54	308	34/23	23/56	16/76
IMO 32	1991	Sweden	56	318	39/20	31/51	19/84
IMO 33	1992	The Russian Federation	56	322	32/26	24/55	14/74
IMO 34	1993	Turkey	73	413	30/35	20/66	11/97
IMO 35	1994	Chinese Hong Kong	69	385	40/30	30/64	19/98
IMO 36	1995	Canada	73	412	37/30	29/71	19/100
IMO 37	1996	India	75	424	28/35	20/66	12/99
IMO 38	1997	Argentina	82	460	35/39	25/70	15/122
IMO 39	1998	Chinese Taiwan	76	419	31/37	24/66	14/102
IMO 40	1999	Romania	81	450	28/38	19/70	12/118
IMO 41	2000	Republic of Korea	82	461	30/39	21/71	11/119
IMO 42	2001	The United States of America	83	473	30/39	20/81	11/122
IMO 43	2002	The United Kingdom	84	479	29/39	23/73	14/120
IMO 44	2003	Japan	82	457	29/37	19/69	13/104
IMO 45	2004	Greece	85	486	32/45	24/78	16/120

(*Continued*)

(*Continued*)

Session	Year	Host	Number of Participating Teams	Number of Contestants	Gold	Silver	Bronze
					(Cutoffs/Numbers of Medalists)		
IMO 46	2005	Mexico	91	513	35/42	23/79	12/128
IMO 47	2006	Slovenia	90	498	28/42	19/89	15/122
IMO 48	2007	Vietnam	93	520	29/39	21/83	14/131
IMO 49	2008	Spain	97	535	31/47	22/100	15/120
IMO 50	2009	Germany	104	565	32/49	24/98	14/135
IMO 51	2010	Kazakhstan	95	522	27/47	21/103	15/115
IMO 52	2011	The Netherlands	101	563	28/54	22/90	16/137
IMO 53	2012	Argentina	100	547	28/51	21/88	14/137
IMO 54	2013	Colombia	97	527	31/45	24/92	15/141
IMO 55	2014	South Africa	101	560	29/49	22/113	16/133
IMO 56	2015	Thailand	104	577	26/39	19/100	14/143
IMO 57	2016	Chinese Hong Kong	109	602	29/44	22/101	16/135
IMO 58	2017	Brazil	111	615	25/48	19/90	16/153
IMO 59	2018	Romania	107	594	31/48	25/98	16/143
IMO 60	2019	The United Kingdom	112	621	31/52	24/94	17/156
IMO 61	2020	The Russian Federation	105	616	31/49	24/112	15/155
IMO 62	2021	The Russian Federation	107	619	24/52	19/103	12/148
IMO 63	2022	Norway	104	589	34/44	29/101	23/140
IMO 64	2023	Japan	112	618	32/54	25/90	18/170

Appendix B

IMO Geometry Problem Index

Problem Number in the IMO	Proposing Country	Category	Problem Number in the Book	Page Number
IMO 1-4	Hungary	Trigonometry, areas, and analytic geometry, existence problems	Problem 5.1	261
IMO 1-5	Romania	Trigonometry, areas, and analytic geometry, position structure problems	Problem 5.4	270
IMO 1-6	Czechoslovakia	Basic properties of circles and four points on a circle, existence problems	Problem 2.1	89
IMO 2-3	Romania	Trigonometry, areas, and analytic geometry, quantity relation problems	Problem 5.10	286
IMO 2-4	Hungary	Similarity and congruence, existence problems	Problem 1.1	42
IMO 2-5	Czechoslovakia	Solid geometry, position structure problems	Problem 6.6	326
IMO 2-6	Bulgaria	Solid geometry, quantity relation problems	Problem 6.12	337
IMO 2-7	The German Democratic Republic	Basic properties of circles and four points on a circle, existence problems	Problem 2.2	90
IMO 3-4	The German Democratic Republic	Geometric inequalities, existence problems	Problem 7.1	363

(Continued)

(Continued)

Problem Number in the IMO	Proposing Country	Category	Problem Number in the Book	Page Number
IMO 3-5	Czechoslovakia	Basic properties of circles and four points on a circle, existence problems	Problem 2.3	92
IMO 3-6	Romania	Solid geometry, position structure problems	Problem 6.7	328
IMO 4-3	Czechoslovakia	Solid geometry, position structure problems	Problem 6.8	329
IMO 4-5	Bulgaria	Basic properties of circles and four points on a circle, existence problems	Problem 2.4	94
IMO 4-6	The German Democratic Republic	Special points and special lines in a triangle, quantity relation problems	Problem 4.6	237
IMO 4-7	The Soviet Union	Solid geometry, position structure problems	Problem 6.9	331
IMO 5-2	The Soviet Union	Solid geometry, position structure problems	Problem 6.10	333
IMO 5-3	Hungary	Geometric inequalities, equality or inequality proof problems	Problem 7.6	373
IMO 6-3	Yugoslavia	Similarity and congruence, quantity relation problems	Problem 1.10	61
IMO 6-6	Poland	Solid geometry, quantity relation problems	Problem 6.13	339
IMO 7-3	Czechoslovakia	Solid geometry, quantity relation problems	Problem 6.14	341
IMO 7-5	Romania	Special points and special lines in a triangle, position structure problems	Problem 4.1	227
IMO 8-2	Hungary	Trigonometry, areas, and analytic geometry, position structure problems	Problem 5.5	275
IMO 8-3	Bulgaria	Solid geometry, quantity relation problems	Problem 6.15	343
IMO 8-6	Poland	Geometric inequalities, existence problems	Problem 7.2	365

(Continued)

(*Continued*)

Problem Number in the IMO	Proposing Country	Category	Problem Number in the Book	Page Number
IMO 9-1	Poland	Geometric inequalities, equality or inequality proof problems	Problem 7.7	374
IMO 9-2	Czechoslovakia	Solid geometry, quantity relation problems	Problem 6.16	344
IMO 9-4	Italy	Similarity and congruence, existence problems	Problem 1.2	44
IMO 10-1	Romania	Similarity and congruence, existence problems	Problem 1.3	46
IMO 10-4	Poland	Geometric inequalities, existence problems	Problem 7.3	367
IMO 11-3	Poland	Solid geometry, existence problems	Problem 6.1	318
IMO 11-4	The Netherlands	Trigonometry, areas, and analytic geometry, position structure problems	Problem 5.6	276
IMO 12-1	Poland	Special points and special lines in a triangle, quantity relation problems	Problem 4.7	239
IMO 12-5	Bulgaria	Solid geometry, quantity relation problems	Problem 6.17	346
IMO 13-2	The Soviet Union	Solid geometry, existence problems	Problem 6.2	320
IMO 13-4	The Netherlands	Geometric inequalities, existence problems	Problem 7.4	368
IMO 14-2	The Netherlands	Basic properties of circles and four points on a circle, existence problems	Problem 2.5	97
IMO 14-6	The United Kingdom	Solid geometry, existence problems	Problem 6.3	322
IMO 15-1	Czechoslovakia	Trigonometry, areas, and analytic geometry, quantity relation problems	Problem 5.11	288
IMO 15-2	Poland	Solid geometry, existence problems	Problem 6.4	323
IMO 15-4	Yugoslavia	Trigonometry, areas, and analytic geometry, existence problems	Problem 5.2	263

(*Continued*)

IMO Problems, Theorems, and Methods: Geometry

(Continued)

Problem Number in the IMO	Proposing Country	Category	Problem Number in the Book	Page Number
IMO 16-2	Finland	Trigonometry, areas, and analytic geometry, existence problems	Problem 5.3	266
IMO 17-3	The Netherlands	Trigonometry, areas, and analytic geometry, quantity relation problems	Problem 5.12	290
IMO 18-1	Czechoslovakia	Geometric inequalities, conditional evaluation problems	Problem 7.22	400
IMO 18-3	The Netherlands	Solid geometry, existence problems	Problem 6.5	324
IMO 19-1	The Netherlands	Trigonometry, areas, and analytic geometry, position structure problems	Problem 5.7	280
IMO 20-2	The United States	Solid geometry, position structure problems	Problem 6.11	335
IMO 20-4	The United States	Basic properties of circles and four points on a circle, position structure problems	Problem 2.7	100
IMO 21-3	The Soviet Union	Basic properties of circles and four points on a circle, existence problems	Problem 2.6	98
IMO 21-4	The United States	Geometric inequalities, conditional evaluation problems	Problem 7.23	401
IMO 22-1	The United Kingdom	Geometric inequalities, conditional evaluation problems	Problem 7.24	402
IMO 22-5	The Soviet Union	Special points and special lines in a triangle, position structure problems	Problem 4.2	229
IMO 23-2	The Netherlands	Similarity and congruence, position structure problems	Problem 1.5	50
IMO 23-5	The Netherlands	Basic properties of circles and four points on a circle, quantity relation problems	Problem 2.25	145

(Continued)

(Continued)

Problem Number in the IMO	Proposing Country	Category	Problem Number in the Book	Page Number
IMO 24-2	The Soviet Union	Similarity and congruence, quantity relation problems	Problem 1.11	63
IMO 25-4	Romania	Basic properties of circles and four points on a circle, position structure problems	Problem 2.8	102
IMO 25-5	Mongolia	Geometric inequalities, equality or inequality proof problems	Problem 7.8	376
IMO 26-1	The United Kingdom	Trigonometry, areas, and analytic geometry, quantity relation problems	Problem 5.13	294
IMO 26-5	The Soviet Union	Power of a point, radical axis, and radical center, position structure problems	Problem 3.1	177
IMO 27-2	China	Trigonometry, areas, and analytic geometry, position structure problems	Problem 5.8	282
IMO 27-4	Ireland	Basic properties of circles and four points on a circle, position structure problems	Problem 2.9	104
IMO 28-2	The Soviet Union	Trigonometry, areas, and analytic geometry, quantity relation problems	Problem 5.14	297
IMO 29-1	Luxembourg	Trigonometry, areas, and analytic geometry, position structure problems	Problem 5.9	283
IMO 29-5	Greece	Geometric inequalities, equality or inequality proof problems	Problem 7.19	394
IMO 30-2	Australia	Geometric inequalities, equality or inequality proof problems	Problem 7.20	396
IMO 30-4	Ireland	Geometric inequalities, equality or inequality proof problems	Problem 7.9	378

(Continued)

(Continued)

Problem Number in the IMO	Proposing Country	Category	Problem Number in the Book	Page Number
IMO 31-1	India	Similarity and congruence, quantity relation problems	Problem 1.12	65
IMO 32-1	The Soviet Union	Geometric inequalities, equality or inequality proof problems	Problem 7.10	381
IMO 32-5	France	Geometric inequalities, existence problems	Problem 7.5	370
IMO 33-4	France	Similarity and congruence, existence problems	Problem 1.4	48
IMO 34-2	The United Kingdom	Similarity and congruence, quantity relation problems	Problem 1.13	67
IMO 34-4	Macedonia	Geometric inequalities, equality or inequality proof problems	Problem 7.11	382
IMO 35-2	Armenia-Australia	Basic properties of circles and four points on a circle, quantity relation problems	Problem 2.26	147
IMO 36-1	Bulgaria	Power of a point, radical axis, and radical center, position structure problems	Problem 3.2	180
IMO 36-5	New Zealand	Geometric inequalities, equality or inequality proof problems	Problem 7.12	383
IMO 37-2	Canada	Basic properties of circles and four points on a circle, position structure problems	Problem 2.10	106
IMO 37-5	Armenia	Geometric inequalities, equality or inequality proof problems	Problem 7.13	385
IMO 38-2	The United Kingdom	Basic properties of circles and four points on a circle, quantity relation problems	Problem 2.27	148
IMO 39-1	Luxembourg	Basic properties of circles and four points on a circle, quantity relation problems	Problem 2.28	150

(Continued)

(*Continued*)

Problem Number in the IMO	Proposing Country	Category	Problem Number in the Book	Page Number
IMO 39-5	Ukraine	Geometric inequalities, equality or inequality proof problems	Problem 7.16	390
IMO 40-5	Russia	Power of a point, radical axis, and radical center, position structure problems	Problem 3.3	181
IMO 41-1	Russia	Power of a point, radical axis, and radical center, quantity relation problems	Problem 3.10	203
IMO 41-6	Russia	Special points and special lines in a triangle, position structure problems	Problem 4.3	231
IMO 42-1	South Korea	Geometric inequalities, equality or inequality proof problems	Problem 7.17	391
IMO 42-5	Israel	Basic properties of circles and four points on a circle, quantity relation problems	Problem 2.29	151
IMO 43-2	South Korea	Special points and special lines in a triangle, position structure problems	Problem 4.4	233
IMO 43-6	Ukraine	Geometric inequalities, equality or inequality proof problems	Problem 7.14	386
IMO 44-3	Poland	Geometric inequalities, equality or inequality proof problems	Problem 7.18	393
IMO 44-4	Finland	Special points and special lines in a triangle, quantity relation problems	Problem 4.8	241
IMO 45-1	Romania	Basic properties of circles and four points on a circle, position structure problems	Problem 2.11	108
IMO 45-5	Poland	Similarity and congruence, quantity relation problems	Problem 1.14	68

(*Continued*)

(Continued)

Problem Number in the IMO	Proposing Country	Category	Problem Number in the Book	Page Number
IMO 46-1	Romania	Special points and special lines in a triangle, position structure problems	Problem 4.5	235
IMO 46-5	Poland	Basic properties of circles and four points on a circle, position structure problems	Problem 2.12	109
IMO 47-1	South Korea	Geometric inequalities, equality or inequality proof problems	Problem 7.15	389
IMO 47-6	Serbia-Montenegro	Geometric inequalities, equality or inequality proof problems	Problem 7.21	398
IMO 48-2	Luxembourg	Similarity and congruence, position structure problems	Problem 1.6	53
IMO 48-4	The Czech Republic	Trigonometry, areas, and analytic geometry, quantity relation problems	Problem 5.15	300
IMO 49-1	Russia	Power of a point, radical axis, and radical center, position structure problems	Problem 3.4	186
IMO 49-6	Russia	Similarity and congruence, position structure problems	Problem 1.7	55
IMO 50-2	Russia	Power of a point, radical axis, and radical center, quantity relation problems	Problem 3.11	204
IMO 50-4	Belgium	Basic properties of circles and four points on a circle, quantity relation problems	Problem 2.30	154
IMO 51-2	Chinese Hong Kong	Similarity and congruence, position structure problems	Problem 1.8	58
IMO 51-4	Poland	Basic properties of circles and four points on a circle, quantity relation problems	Problem 2.31	157

(Continued)

(*Continued*)

(Continued)

(Continued)

Problem Number in the IMO	Proposing Country	Category	Problem Number in the Book	Page Number
IMO 59-1	Greece	Basic properties of circles and four points on a circle, position structure problems	Problem 2.20	130
IMO 59-6	Poland	Trigonometry, areas, and analytic geometry, quantity relation problems	Problem 5.16	301
IMO 60-2	Ukraine	Basic properties of circles and four points on a circle, position structure problems	Problem 2.21	132
IMO 60-6	India	Basic properties of circles and four points on a circle, position structure problems	Problem 2.22	133
IMO 61-1	Poland	Basic properties of circles and four points on a circle, position structure problems	Problem 2.23	141
IMO 62-3	Ukraine	Power of a point, radical axis, and radical center, position structure problems	Problem 3.7	191
IMO 62-4	Poland	Basic properties of circles and four points on a circle, quantity relation problems	Problem 2.32	159
IMO 63-4	Slovakia	Basic properties of circles and four points on a circle, position structure problems	Problem 2.24	142
IMO 64-2	Portugal	Power of a point, radical axis, and radical center, position structure problems	Problem 3.8	196
IMO 64-6	The United States	Power of a point, radical axis, and radical center, position structure problems	Problem 3.9	198